Lecture Notes in Computer Science

Lecture Notes in Artificial Intelligence 14845

Founding Editor

Jörg Siekmann

The series Lecture Notes in Artificial Intelligence (LNAI) was established in 1988 as a topical subseries of LNCS devoted to artificial intelligence.

The series publishes state-of-the-art research results at a high level. As with the LNCS mother series, the mission of the series is to serve the international R & D community by providing an invaluable service, mainly focused on the publication of conference and workshop proceedings and postproceedings.

Joseph Finkelstein · Robert Moskovitch ·
Enea Parimbelli
Editors

Artificial Intelligence in Medicine

22nd International Conference, AIME 2024
Salt Lake City, UT, USA, July 9–12, 2024
Proceedings, Part II

Springer

Editors
Joseph Finkelstein
University of Utah
Salt Lake City, UT, USA

Robert Moskovitch
Ben-Gurion University of the Negev
Beer Sheva, Israel

Enea Parimbelli
University of Pavia
Pavia, Italy

ISSN 0302-9743 ISSN 1611-3349 (electronic)
Lecture Notes in Artificial Intelligence
ISBN 978-3-031-66534-9 ISBN 978-3-031-66535-6 (eBook)
https://doi.org/10.1007/978-3-031-66535-6

LNCS Sublibrary: SL7 – Artificial Intelligence

This Springer imprint is published by the registered company Springer Nature Switzerland AG
The registered company address is: Gewerbestrasse 11, 6330 Cham, Switzerland

Preface

The Society for Artificial Intelligence in Medicine (AIME) was established in 1986 following a very successful workshop held in Pavia, Italy, the year before. The principal aims of AIME are to foster fundamental and applied research in the application of Artificial Intelligence (AI) techniques to medical care and medical research, and to provide a forum at conferences for discussing any progress made. The main activity of the society thus far has been the organization of a series of international conferences, held in Marseille, France (1987), London, UK (1989), Maastricht, Netherlands (1991), Munich, Germany (1993), Pavia, Italy (1995), Grenoble, France (1997), Aalborg, Denmark (1999), Cascais, Portugal (2001), Protaras, Cyprus (2003), Aberdeen, UK (2005), Amsterdam, Netherlands (2007), Verona, Italy (2009), Bled, Slovenia (2011), Murcia, Spain (2013), Pavia, Italy (2015), Vienna, Austria (2017), Poznan, Poland (2019), Minneapolis, USA (2020), Porto, Portugal (2021), Halifax, Canada (2022), and Portoroz, Slovenia (2023).

The AIME 2024 goals were to present and consolidate the international state of the art of AI in biomedical research from the perspectives of theory, methodology, systems, and applications. The conference included three invited keynotes, presentations of full, short, and demonstration papers, tutorials, workshops, a doctoral consortium, and posters. Overall, AIME 2024 received 335 submissions from which 54 full papers and 22 short papers were chosen for oral presentation at the main conference and for publication in two volumes of conference proceedings. Submissions came from authors in 46 countries, including submissions from Europe, North and South America, Africa, Australia, and Asia.

All papers were carefully peer-reviewed by experts from the Program Committee, and subsequently by members of the Senior Program Committee. Each submission was single-blindly reviewed in most cases by three reviewers, and all papers by at least two reviewers. The reviewers judged the overall quality of the submitted papers, together with their relevance to the AIME conference, originality, impact, technical correctness, methodology, scholarship, and quality of presentation. In addition, the reviewers provided detailed written comments on each paper and stated their confidence in the subject area. One Senior Program Committee member was assigned to each paper, who wrote a meta-review and provided a recommendation to the Organizing Committee. A dedicated expert committee comprising the AIME 2024 general chairs, Joseph Finkelstein and Robert Moskovitch, the Scientific Program Committee chair, Enea Parimbelli, and the doctoral consortium chair, Gregor Štiglic, made the final decisions regarding the AIME 2024 conference program. This process was carried out with virtual meetings starting in February 2024. Each long paper was presented in a 20-minute oral presentation during the conference. Each regular short and demonstration paper was presented in a 10-minute presentation, a poster or a demonstration session. The main conference poster session hosted over 70 poster presentations.

The papers were organized according to their topics in the following main themes: (1) Predictive modelling and disease risk prediction; (2) Natural language processing; (3) Bioinformatics and omics; (4) Wearable devices, sensors, and robotics; (5) Medical image analysis; (6) Data integration and multimodal analysis; (7) Explainable AI.

AIME 2024 had the privilege of hosting three invited keynote speakers: Nicholas Tatonetti, Vice Chair of Operations in the Department of Computational Biomedicine and Associate Director of Computational Oncology at Cedars-Sinai Medical Center, Los Angeles, USA, giving the keynote entitled "AI-driven biomedical discoveries using real-world data," Hayit Greenspan, Co-director of the Artificial Intelligence and Emerging Technologies in Medicine PhD concentration at the Icahn School of Medicine at Mount Sinai, New York, USA, who discussed "AI in Medical Imaging – Steps towards supporting Detection and Monitoring of Disease," and Yuan Luo, Director, Institute for Artificial Intelligence in Medicine, Northwestern University, Chicago, USA, who presented "Translational Science: Breakthroughs and Innovations in the new Age of AI."

The doctoral consortium received 11 PhD proposals that were peer reviewed. AIME 2024 provided an opportunity for six of these PhD students to discuss their research goals, proposed methods, and preliminary results in an oral presentation and for four students to present a poster. A scientific panel, consisting of experienced researchers in the field, provided constructive feedback to the students in an informal atmosphere. The doctoral consortium was chaired by Dr. Gregor Štiglic (University of Maribor, Slovenia).

AIME 2024 invited researchers to submit proposals for workshops and tutorials. Twelve workshops were selected by the Organization Committee. These workshops were: (1) Advances in Generating Real-World Evidence from Real-World Data Using Artificial Intelligence (RWE-AI), (2) AI Applications in Telemedicine and Digital Health (TeleHealth-AI), (3) Artificial Intelligence in Oncology Workshop, (4) Third International Workshop on Artificial Intelligence in Nursing (AINurse-24), (5) AI for Drug Discovery: Development in Pharmaceuticals, Academia, or Jointly in Collaborations, (6) AI for Reliable and Equitable Real World Evidence Generation in Medicine, (7) AI for Primary Care: Electronic Scribes and More, (8) AI and Precision Medicine: Innovations and Applications, (9) Implementing AI in Healthcare: Bridging the Gap Between Technical, Cognitive, and Sociotechnical Perspectives, (10) AI Applications in Public Health and Social Services, (11) Mitigating AI Risk through Ethical Data Science and (12) Veteran Health Administration: AI for Clinical Informatics Solutions. In addition to the workshops, two interactive half-day tutorials were selected: (1) Process Mining for Healthcare: From Theory to Practice with pMineR, and (2) Integrating Geography with AI to Improve Decision Support in Public Health.

We would like to thank everyone who contributed to the AIME 2024 conference. First, we would like to thank the authors of the submitted papers and posters and the members of the Program Committee who helped with conducting a successful review. Thank you to the Senior Program Committee for writing meta-reviews and the members of the Senior Advisory Board for guidance. Thanks are also due to the invited speakers as well as to the organizers of the tutorials, the workshops, and the doctoral consortium panel. Many thanks go to the local Organizing Committee members, who managed all the work making this conference possible: Yves Lussier, Penny Atkins, Yue Zhang,

Fatemeh Shah-Mohammadi, Aref Smiley, Te-Yi Tsai, and C. Mahony Reategui-Rivera. We would like to thank Springer for sponsoring the conference, the AIME society for sponsoring awards and the student travel grants, and the University of Utah, Department of Biomedical Informatics, for all the support in organizing it. Finally, we thank the Springer team for helping us in the final preparation of this LNAI book.

May 2024

Joseph Finkelstein
Robert Moskovitch
Enea Parimbelli

Organization

General Chairs

Joseph Finkelstein	University of Utah, USA
Robert Moskovitch	Ben-Gurion University of the Negev, Israel

Program Committee Chairs

Enea Parimbelli (Scientific Program)	University of Pavia, Italy
Xiaoqian Jiang (Scientific Program)	University of Texas at Houston, USA
Marina Sirota (Scientific Program)	University of California at San Francisco, USA
Yves Lussier (Senior Advisory Board)	University of Utah, USA
Panagiotis Papapetrou (Workshop)	Stockholm University, Sweden
Fadia Shaya (Workshop)	University of Maryland, USA
Aref Smiley (Workshop)	University of Utah, USA
Myra Spiliopoulou (Tutorial)	Otto von Guericke University Magdeburg, Germany
Francesca Vitali (Tutorial)	University of Arizona, USA
Fatemeh Shah-Mohammadi (Tutorial)	University of Utah, USA
Gregor Štiglic (Doctoral Consortium)	University of Maribor, Slovenia
Colleen Kenost (Publicity)	University of Utah, USA

Local Organizing Committee

Yves Lussier (Co-chair)	University of Utah, USA
Joseph Finkelstein (Co-chair)	University of Utah, USA
Penny Atkins	University of Utah, USA
Yue Zhang	University of Utah, USA
Fatemeh Shah-Mohammadi	University of Utah, USA

Senior Advisory Board

Serena Jingchuan Guo	University of Florida, USA
Syed Sibte Raza Abidi	Dalhousie University, Canada
Tomohiro Sawa	Teikyo University, Japan
Xinxin (Katie) Zhu	Yale University, USA
Yu-Chuan (Jack) Li	Taipei Medical University, Taiwan
Zhongming Zhao	University of Texas Health Science Center at Houston, USA

Senior Program Committee

Abu Mosa	University of Missouri, USA
Adiebonye Jumbo	SUNY Downstate Health Sciences University, USA
Allan Tucker	Brunel University, UK
Annette ten Teije	Vrije Universiteit Amsterdam, Netherlands
Arianna Dagliati	University of Pavia, Italy
Barbara Oliboni	University of Verona, Italy
Blaz Zupan	University of Ljubljana, Slovenia
Brad Malin	Vanderbilt University, USA
Carlo Combi	Università degli Studi di Verona, Italy
Christopher Tignanelli	University of Minnesota, USA
Craig Kuziemsky	MacEwan University, Canada
Daniel Capurro	University of Melbourne, Australia
Daniel Otzoy	Central American Health Informatics Network, Guatemala
Ece Uzun	Brown University, USA
Elpida Keravnou-Papailiou	University of Cyprus, Cyprus
Gregor Štiglic	University of Maribor, Slovenia
Hamilton Baker	Medical University of South Carolina, USA
Hua Xu	Yale University, USA
Janna Hastings	University of Zurich, Switzerland
John Holmes	University of Pennsylvania, USA
Jonathan Golob	University of Michigan, USA
Jose M. Juarez	Universidad de Murcia, Spain
Kaifu Chen	Harvard University, USA
Kenrick Cato	University of Pennsylvania, USA
Kun Huang	Regenstrief Institute, USA
Li Shen	University of Pennsylvania, USA
Luca Giancarlo	University of Texas Health Science Center at Houston, USA
Lucia Sacchi	University of Pavia, Italy

Mar Marcos	Universitat Jaume I, Spain
Margarita Sordo	Harvard University, USA
Mark Parkulo	Mayo Clinic, USA
Martin Michalowski	University of Minnesota, USA
Maryam Zolnoori	Columbia University, USA
Michael E. Matheny	Vanderbilt University, USA
Milos Hauskrecht	University of Pittsburgh, USA
Mor Peleg	University of Haifa, Israel
Nada Lavrač	Jožef Stefan Institute, Slovenia
Naleef Fareed	Ohio State University, USA
Nan Liu	Duke-NUS Medical School, Singapore
Niels Peek	University of Manchester, UK
Pedro Henriques Abreu	University of Coimbra, Portugal
Peter Lucas	University of Twente, Netherlands
Rui Zhang	University of Minnesota, USA
Sarath Chandra Janga	Indiana University – Purdue University Indianapolis, USA
Silvana Quaglini	University of Pavia, Italy
Stefania Montani	Università del Piemonte Orientale, Italy
Steve Johnson	University of Minnesota, USA
Syed Sibte Raza Abidi	Dalhousie University, Canada
Szymon Wilk	Poznan University of Technology, Poland
Tianxi Cai	Harvard University, USA
Victor Maojo	Universidad Politécnica de Madrid, Spain
Wenyu Song	Brigham and Women's Hospital, USA
Yanshan Wang	University of Pittsburgh, USA
Young Ji Lee	University of Pittsburgh, USA
Yuval Shahar	Ben-Gurion University of the Negev, Israel

Program Committee

Abdolrahman Peimankar	University of Southern Denmark, Denmark
Abeed Sarker	Emory University, USA
Ahmad P. Tafti	University of Pittsburgh, USA
Aida Kamisalic	University of Maribor, Slovenia
Aize Cao	Meharry Medical College, USA
Aleksander Sadikov	University of Ljubljana, Slovenia
Alessio Bottrighi	Università del Piemonte Orientale, Italy
Alfredo Vellido	Universitat Politècnica de Catalunya, Spain
Allan Tucker	Brunel University, UK
Amin Zollanvari	Nazarbayev University, Kazakhstan

Ana Maria Mendonça	Institute for Systems and Computer Engineering, Technology and Science, Portugal
André Fujita	University of São Paulo, Brazil
Anna Fabijańska	Technical University of Lodz, Poland
Annette ten Teije	Vrije Universiteit Amsterdam, Netherlands
Ardalan Naseri	University of Texas Health Science Center at Houston, USA
Aref Smiley	University of Utah, USA
Arianna Dagliati	University of Pavia, Italy
Arif Harmanci	University of Texas Health Science Center at Houston, USA
Arjen Hommersom	Open University of the Netherlands, Netherlands
Arvind Rao	University of Michigan, USA
Barbara Di Camillo	University of Padova, Italy
Baris Suzek	Muğla Sitki Koçman University, Turkey
Beatriz López	University of Girona, Spain
Begoña Martinez-Salvador	Universitat Jaume I, Spain
Bernard Chen	University of Central Arkansas, USA
Bernardo Canovas Segura	Universidad de Murcia, Spain
Bertha Guijarro-Berdiñas	Universidade da Coruña, Spain
Bhasuran Balu	Florida State University, USA
Bibek Paudel	Stanford University, USA
Binh Vu	Research Institute for Telecommunication and Cooperation, Germany
Brad Malin	Vanderbilt University, USA
Brian Locke	University of Utah, USA
Carlo Combi	Università degli Studi di Verona, Italy
Carmela Comito	Institute for High Performance Computing and Networking, Italy
Carson Leung	University of Manitoba, Canada
Chao Cheng	Baylor College of Medicine, USA
Charles Kahn	University of Pennsylvania, USA
Chuming Chen	University of Delaware, USA
Chung-Sheng Li	PwC Information, UK
Cihan Bilge Kayasandık	Istanbul Medipol University, Turkey
Cuncong Zhong	University of Kansas, USA
Daniela Dauria	Free University of Bozen-Bolzano, Italy
Daniela Ferreira-Santos	University of Porto, Portugal
Daniele Pala	University of Pavia, Italy
Denis Newman-Griffis	University of Sheffield, UK
Dvir Aran	Technion – Israel Institute of Technology, Israel
Eleni Kaldoudi	Democritus University of Thrace, Greece

Eneida Mendonca	University of Cincinnati, USA
Erez Shalom	Ben-Gurion University of the Negev, Israel
Fangxiang Wu	University of Saskatchewan, Canada
Fatemeh Shah-Mohammadi	University of Utah, USA
Fernando Luís Barroso Da Silva	University of São Paulo, Brazil
Filip Jagodzinski	Western Washington University, USA
Fleur Mougin	Université de Bordeaux, France
Floriano Scioscia	Polytechnic University of Bari, Italy
Gayo Diallo	Université de Bordeaux, France
Georg Dorffner	Medical University of Vienna, Austria
Giltae Song	Pusan National University, South Korea
Giorgio Leonardi	Università del Piemonte Orientale, Italy
Giorgio Vinciguerra	University of Pisa, Italy
Giovanna Rosone	University of Pisa, Italy
Giovanni Delnevo	University of Bologna, Italy
Giuseppe Agapito	University Magna Graecia of Catanzaro, Italy
Guo Yi	University of Florida, USA
Hajer Baazoui	CY Cergy Paris University, France
Haohan Wang	University of Illinois at Urbana-Champaign, USA
Haridimos Kondylakis	Institute of Computer Science, Greece
Henrik Boström	KTH Royal Institute of Technology, Sweden
Hossein Estiri	Harvard University, USA
Hu Huang	University of Minnesota, USA
Hua Xu	University of Texas Health Science Center at Houston, USA
Iacopo Vagliano	Amsterdam Universitair Medische Centra, Netherlands
Ioanna Miliou	Stockholm University, Sweden
Isabel Sassoon	Brunel University, UK
Isabelle Bichindaritz	State University of New York at Oswego, USA
Jaebum Kim	Konkuk University, South Korea
Jake Chen	University of Alabama at Birmingham, USA
Jan Egger	Graz University of Technology, Austria
Jędrzej Potoniec	Poznan University of Technology, Poland
Je-Keun Rhee	Soongsil University, South Korea
Jens Weber	University of Victoria, Canada
Jerzy Błaszczyński	Poznan University of Technology, Poland
Jesualdo Tomás Fernández-Breis	Universidad de Murcia, Spain
Jie Xu	University of Florida, USA
Jim Zheng	University of Texas Health Science Center at Houston, USA
Jingwen Yan	Indiana University, USA

Joao Setubal	University of São Paulo, Brazil
Joe Song	New Mexico State University, USA
John Holmes	University of Pennsylvania, USA
Jordan Wyrwa	Children's Hospital Colorado, USA
Joseph Romano	University of Pennsylvania, USA
Juexin Wang	Indiana University, USA
Junbai Wang	University of Oslo, Norway
K. S. M. Tozammel Hossain	University of North Texas, USA
Kaifu Chen	Harvard University, USA
Kenrick Cato	University of Pennsylvania, USA
Kenta Nakai	University of Tokyo, Japan
Kerstin Denecke	Bern University of Applied Sciences, Switzerland
Kun-Mao Chao	National Taiwan University, Taiwan
Kyu-Baek Hwang	Soongsil University, South Korea
Laura Moss	University of Aberdeen, UK
Leng Han	Indiana University – Purdue University Indianapolis, USA
Licong Cui	University of Texas Health Science Center at Houston, USA
Lina Sulieman	Vanderbilt University, USA
Liqing Zhang	Virginia Polytechnic Institute and State University, USA
Lisiane Pruinelli	University of Florida, USA
Logan Pierce	University of California at San Francisco, USA
Loris Nanni	University of Padua, Italy
Luca Denti	Università degli Studi di Milano-Bicocca, Italy
Luca Piovesan	Università del Piemonte Orientale, Italy
Luis Rueda	University of Windsor, Canada
Luke Rasmussen	Northwestern University, USA
Manuel Campos	Universidad de Murcia, Spain
Manuel Striani	Università del Piemonte Orientale, Italy
Mar Marcos	Universitat Jaume I, Spain
Marco Spruit	Leiden University, Netherlands
Marcos L. P. Bueno	Eindhoven University of Technology, Netherlands
Martin Chapman	King's College London, UK
Martin Michalowski	University of Minnesota, USA
Maryam Zolnoori	Columbia University, USA
Mei Liu	University of Florida, USA
Mengdi Huai	Iowa State University, USA
Michael Ignaz Schumacher	University of Applied Sciences Western Switzerland, Switzerland
Milos Hauskrecht	University of Pittsburgh, USA

Ming-Yuan Chih	University of Kentucky, USA
Mohamed Nounou	Texas A&M University at Qatar, Qatar
Mor Peleg	University of Haifa, Israel
Morihiro Hayashida	Matsue College, Japan
Nada Lavrac	Jožef Stefan Institute, Slovenia
Nadav Rappoprt	Ben-Gurion University of the Negev, Israel
Nadia Pisanti	University of Pisa, Italy
Nansu Zong	Mayo Clinic, USA
Natalia Grabar	Université de Lille, France
Neil Smalheiser	University of Illinois at Chicago, USA
Nevo Itzhak	Ben-Gurion University of the Negev, Israel
Nguyen Quoc Khanh Le	Taipei Medical University, Taiwan
Niels Peek	University of Manchester, UK
Nisha Puthiyedth	Thompson Rivers University, Canada
Ognjen Arandjelovic	University of St Andrews, Scotland
Panagiotis Papapetrou	Stockholm University, Sweden
Pedro Furtado	University of Coimbra, Portugal
Pedro Larranaga	University of Madrid, Spain
Pedro Pereira Rodrigues	University of Porto, Portugal
Pierre Zweigenbaum	Université Paris-Saclay, France
Primoz Kocbek	University of Maribor, Slovenia
Qianqian Song	University of Florida, USA
Rafał Jóźwiak	Warsaw University of Technology, Poland
Ravi Janardan	University of Minnesota, USA
Reda Alhajj	University of Calgary, Canada
Ricardo Cardoso Pereira	University of Coimbra, Portugal
Rita Casadio	University of Bologna, Italy
Ronald Piscotty	Oakland University, USA
Rui Yin	University of Florida, USA
Rui Zhang	University of Minnesota, USA
Ruisheng Wang	Harvard University, USA
Ruoyu Wang	University of Texas Health Science Center at Houston, USA
Ryan Urbanowicz	University of Pennsylvania, USA
Saeed Salem	Qatar University, Qatar
Samina Abidi	Dalhousie University, Canada
Sanja Avramovic	George Mason University, USA
Satya Sahoo	Case Western Reserve University, USA
Sejung Yang	Yonsei University, South Korea
Seungyoon Nam	Gachon University, South Korea
Shanmughavel Piramanayagam	Bharathiar University, India

Shubo Tian	National Center for Biotechnology Information, USA
Shu-Kay Ng	Griffith University, Australia
Silvana Quaglini	University of Pavia, Italy
Silvia Miksch	Vienna University of Technology, Austria
Simona E. Rombo	Università degli Studi di Palermo, Italy
Simone Ciccolella	University of Milano-Bicocca, Italy
Sing-Hoi Sze	Texas A&M University at Qatar, Qatar
Siru Liu	Vanderbilt University, USA
Soon Chun	City University of New York, USA
Stefania Montani	Università del Piemonte Orientale, Italy
Szymon Wilk	Poznan University of Technology, Poland
Terri Elizabeth Workman	George Washington University, USA
Tian Shubo	National Center for Biotechnology Information, USA
Tiffany Callahan	Columbia University, USA
Valerio Guarrasi	Università Campus Bio-Medico di Roma, Italy
Vasudevan Jagannathan	3M M*Modal, USA
Vipina K. Keloth	Yale University, USA
W. Jim Zheng	University of Texas Health Science Center at Houston, USA
Wenjun Lin	Algoma University, Canada
Xia Ning	Ohio State University, USA
Xiaobo Zhou	University of Texas Health Science Center at Houston, USA
Xiaoming Zeng	University of North Carolina at Chapel Hill, USA
Xiaoyi Raymond Gao	Ohio State University, USA
Xiayuan Huang	University of Wisconsin–Madison, USA
Xing He	University of Florida, USA
Xinghua Shi	Temple University, USA
Xingquan Zhu	Florida Atlantic University, USA
Xu Jie	University of Florida, USA
Xuan Guo	University of North Texas, USA
Yanshan Wang	University of Pittsburgh, USA
Yasir Tarabichi	MetroHealth System, USA
Yejin Kim	University of Texas Health Science Center at Houston, USA
Yi Guo	University of Florida, USA
Ying Li	Regeneron Pharmaceuticals, USA
Yiqing Shen	Johns Hopkins University, USA
Younghee Lee	University of Utah, USA
Yu Zhang	Lehigh University, USA

Yuan An	Drexel University, USA
Yuanxi Fu	University of Illinois at Urbana-Champaign, USA
Yue Zhang	University of Utah, USA
Zhao Li	University of Texas Health Science Center at Houston, USA
Zhe He	Florida State University, USA
Zhengxing Huang	Zhejiang University, China
Zhiyu Wan	Vanderbilt University Medical Center, USA

Contents – Part II

Explainable AI

Contents – Part I

Wearable Devices, Sensors, and Robotics

Medical Imaging Analysis

3T to 7T Whole Brain + Skull MRI Translation with Densely Engineered U-Net Network

Aryan Kalluvila[1,2](✉) and Matthew S. Rosen[1]

[1] Athinoula Martinos Center for Biomedical Imaging, Charleston, MA 08544, USA
aryankalluvila2027@u.northwestern.edu
[2] Northwestern University, 2313 Sheridan Road, Evanston, IL 60201, USA

Abstract. The emergence of 7T MRI scanners offers groundbreaking potential in medical imaging, providing ultra-high resolution compared to standard 3T MRI machines. However, widespread adoption faces challenges due to high costs and safety concerns. Our research addresses this by enhancing 3T MRI image quality to rival 7T scanners, particularly in cases where skull stripping is difficult. We propose a dense U-Net algorithm, surpassing previous deep learning methods, achieving an average PSNR of 23.8638 ± 2.3706 and SSIM of 0.7525 ± 0.0646. Validation on a paired MRI dataset of 3 patients scanned in both 3T and 7T machines showed promising results. Neuroradiologists ranked our tool highest in overall image quality, anatomical structure, and diagnostic confidence compared to two others on the Likert scale. Our proposed method could facilitate broader access to advanced imaging capabilities in clinical settings.

Keywords: Magnetic Resonance Imaging (MRI) · Deep Learning (DL) · CNNs (Convolutional Neural Networks) · Artificial Intelligence (AI)

1 Introduction

Ultra-high field MRI, operating typically at 7 T, revolutionizes brain imaging with superior resolution and sensitivity to anatomical structures and functional activity. Though only around 30 institutions possess 7T machines, their imaging quality far surpasses the more common 3T scanners. While 3T MRI is more accessible and less costly, its images lack the rich contrast and resolution of 7T scans.

The enhanced resolution of 7T MRI enables visualization of intricate neural structures and microstructural changes crucial for understanding brain function. Deep learning, particularly convolutional neural networks (CNNs), enhances MRI image resolution. Studies by Ahmad et al. [1] and Gao et al. [2] demonstrate CNNs' efficacy in improving spatial resolution and image quality. Generative adversarial networks (GANs) further enhance resolution, as shown by Ahmad et al. [1] in brain image enhancement. Transfer learning, as evidenced by Mathivanan et al. [3], employs transfer CNNs to leverage tumor detection, improving diagnostic accuracy.

This study introduces a U-Net architecture tailored for enhancing 3T MRI resolution to match 7T standards. Our innovation lies in this architecture's design and its successful

J. Finkelstein et al. (Eds.): AIME 2024, LNAI 14845, pp. 3–8, 2024.
https://doi.org/10.1007/978-3-031-66535-6_1

application to upscale brain images [4]. Leveraging 7T imaging promises to advance medical imaging capabilities significantly. While deep learning shows promise in image enhancement, challenges such as dataset availability and generalization persist.

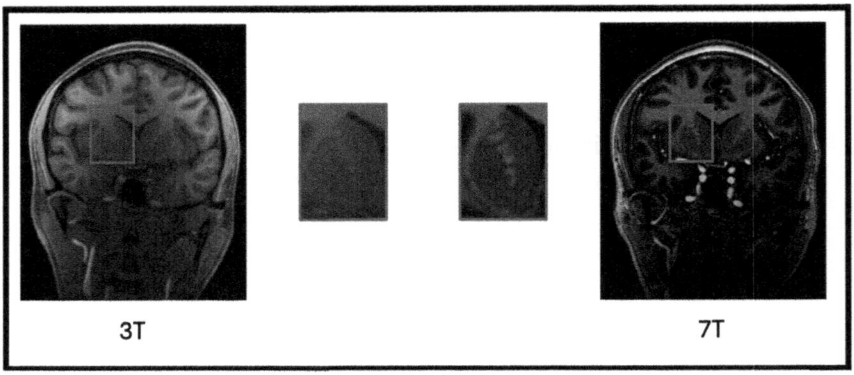

Fig. 1. A Sample 3T vs 7T Patient to Highlight Difference In Image Resolution

The enhanced resolution offered by 7 T (7T) magnetic resonance imaging (MRI) compared to the more prevalent 3 T (3T) scanners is of paramount significance in advancing our understanding of brain structure and function. Figure 1 highlights the difference in resolution of the two scanners, showing a zoomed in version for higher clarity differentiation. The higher field strength of 7T MRI facilitates a finer level of anatomical detail and spatial precision, enabling researchers and clinicians to visualize intricate neural structures and subtle microstructural changes that may be elusive at lower field strengths. This improved resolution holds immense potential in revealing finer cortical layers, small anatomical features, and nuanced tissue boundaries, which are pivotal for accurate anatomical localization and characterization of brain regions. For instance, Isaacs et al. compares manual brain structure delineation between clinical 3T optimized isotropic, and standard 7T MRI scans, revealing superior performance of 7T in visualizing small structures important for clinical localization, suggesting its advantage over 3T in a clinical setting [5].

2 Proposed Methodology

To elevate the resolution of 3T magnetic resonance imaging (MRI) scans to rival that of 7T MRI scans, we introduce a simple mathematical framework mirroring MRI acquisition. Unlike conventional methods assuming blurring and down-sampling, our approach closely aligns with the actual low-resolution MRI acquisition. We begin with the low-resolution 3T MRI image (**y**) and aim to estimate the high-resolution 7T MRI image (**x**). Introducing a mapping function (M) that simulates the MRI acquisition process, we express the low-resolution image (**y**) as the result of applying M to the high-resolution image (**x**).

$$y = M(x) \tag{1}$$

Our simple approach integrates a neural network to approximate the inverse of M. During training, the network learns the relationship between high and low-resolution MRI images, focusing on matrix sizes of 256×308. By assimilating the disparity (**d**) between observed and 3T low-resolution images, the network refines its parameters iteratively, generating refined high-resolution estimates (**x**):

$$d = y - M(x) \tag{2}$$

In this study, the U-Net architecture, particularly the Dense U-Net variant, emerges as superior for MRI scan super-resolution, effectively handling the data's complexities and promoting feature reuse through densely connected convolutional layers. Integration of VGG blocks within Dense U-Net enhances capture of intricate features, vital for accurate reconstruction, while SE blocks further improve adaptability and relevance of feature representation within the network (Fig. 2).

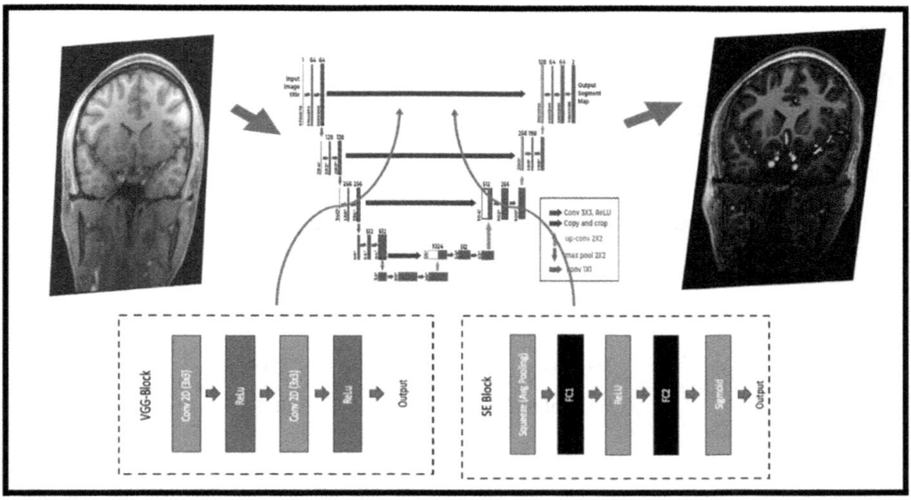

Fig. 2. Dense U-Net Architecture Algorithm Pipeline, incorporating VGG blocks and Self Attention Mechanisms (SE Blocks) [7]

With respect to the dataset, paired T1-weighted and T2-weighted MRI scans from10 healthy subjects at both 3T and 7T strengths were obtained from the public dataset released by Chen et al. [6]. Despite 7T MRI's superior resolution and signal-to-noise ratio, its high costs limit widespread use. The dataset facilitates algorithm benchmarking for synthesizing 7T MR images from 3T MR scans. We used 600 2D slices for training and 400 for testing. All testing was completed on PyTorch systems with NVIDIA GPUs. We used the Adam optimizer with a learning rate of 1e-4 (Fig. 3).

3 Clinical Evaluation and Results

Two board certified neuroradiologists (B.J., J.H) blindly ranked and scored a series of 3T, VDSR, SRCNN, Dense U-Net reconstructed, and 7T MRI scans. They were asked to score the scans with respect to overall image quality on the Likert scale. These scans were

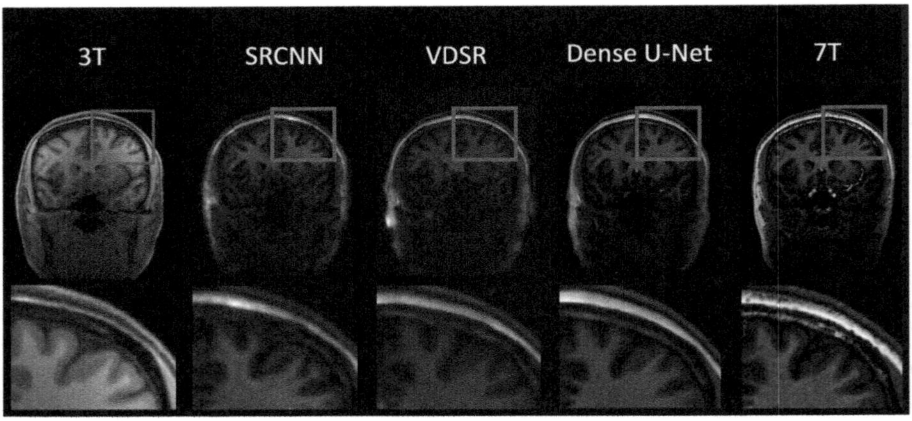

Fig. 3. Qualitative Analysis of Dense U-Net against state of the art with full skull and brain

randomly selected from the testing dataset. The radiologists were calibrated beforehand by looking at several score 1 (3T images) and score 5 (7T images) in order to accurately assess the images.

Table 1. Radiologist Scores and Statistical Analysis

Architecture	Likert Score (1–5)	Tukey Analysis
VDSR (cite)	1.5 ± 0.527	0.01 <
SRCNN (cite)	2 ± 1.0	0.01 <
Dense U-Net	**3.5 ± 0.52**	**0.01 <**

Table 2. Quantitative Analysis via Peak Signal to Noise (PSNR) and Structural Similarity (SSIM)

Architecture	PSNR	SSIM
VDSR	20.7338 ± 2.1507	0.7275 ± 0.05
SRCNN	24.137 ± 2.4137	0.606 ± 0.05
Dense U-Net	**23.8638 ± 2.3706**	**0.7525 ± 0.06**

To qualitatively access the image quality further, a visual output map was generated which shows the output of the Dense U-Net against the other architectures tested. These architectures proved to not reconstruct the images accurately or even over smooth in cortical regions.

To quantitatively access the Dense U-Net, we leveraged PSNR and SSIM metrics. Compared to the other state of the art architectures, the Dense U-net performs competitively, outperforming all of the algorithms with respect to SSIM (the stronger metric for image translation) and places close to the SRCNN for VDSR (Table 1 and 2).

4 Discussion and Conclusion

In this study, we propose a robust methodology for enhancing the resolution of 3T MRI scans to match the quality of 7T MRI scans, particularly focusing on cases where skull stripping is challenging. Our approach employs a dense U-Net algorithm, surpassing previous deep learning methods, achieving promising results with an average PSNR of 23.8638 ± 2.3706 and SSIM of 0.7525 ± 0.0646. Validation on a paired MRI dataset of 3T and 7T scans demonstrates the efficacy of our method. Moreover, clinical evaluation by neuroradiologists ranked our tool highest in overall image quality, anatomical structure, and diagnostic confidence compared to alternative methods.

Our proposed method addresses the limitations of 3T MRI scans by leveraging deep learning techniques and a novel mathematical framework mirroring the MRI acquisition process. The integration of a neural network, particularly the dense U-Net architecture, proves effective in handling the intricacies of MRI data and enhancing spatial resolution. Furthermore, the use of VGG and SE blocks within the DenseUNet architecture enhances feature extraction and reconstruction, leading to improved image quality.

Quantitative analysis using PSNR and SSIM metrics confirms the competitiveness of our method compared to state-of-the-art algorithms. These results underscore the potential of our approach in facilitating broader access to advanced imaging capabilities in clinical settings. By bridging the resolution gap between 3T and 7T MRI scans, our method holds promise in enhancing diagnostic accuracy and medical insight derived from MRI imaging.

Acknowledgements. The authors would like to thank Dr. Benjamin Pulli (B.J.) and Dr. Jeremy Heit (J.H.) for serving as the blind radiologist reviewers for this study.

Disclosure of Interests. The authors have no conflicts of interest.

References

1. Ahmad, W., Ali, H., Shah, Z., et al.: A new generative adversarial network for medical images super resolution. Sci. Rep. **12**, 9533 (2022). https://doi.org/10.1038/s41598-022-13658-4
2. Gao, Y., Li, H., Dong, J., Feng, G.:A deep convolutional network for medical image super-resolution. In: 2017 Chinese Automation Congress (CAC), Jinan, China, pp. 5310–5315 (2017).https://doi.org/10.1109/CAC.2017.8243724
3. Mathivanan, S.K., Sonaimuthu, S., Murugesan, S., et al.: Employing deep learning and transfer learning for accurate brain tumor detection. Sci. Rep. **14**, 7232 (2024). https://doi.org/10.1038/s41598-024-57970-7

4. Kalluvila, A.: Super-resolution of brain MRI via U-Net architecture. Int. J. Adv. Comput. Sci. Appl. (IJACSA), **14**(5) (2023). https://doi.org/10.14569/IJACSA.2023.0140503
5. Isaacs, B.R., et al.: 3 versus 7 Tesla magnetic resonance imaging for parcellations of subcortical brain structures in clinical settings. PLoS ONE **15**(11), e0236208 (2020). https://doi.org/10.1371/journal.pone.0236208.PMID:33232325;PMCID:PMC7685480
6. Chen, X., Qu, L., Xie, Y., et al.: A paired dataset of T1- and T2-weighted MRI at 3 Tesla and 7 Tesla. Sci Data **10**, 489 (2023). https://doi.org/10.1038/s41597-023-02400-y
7. GeeksforGeeks. U-Net Architecture explained (2023). https://www.geeksforgeeks.org/u-net-architecture-explained/

A Sparse Convolutional Autoencoder for Joint Feature Extraction and Clustering of Metastatic Prostate Cancer Images

Zhijun Chen[1], Erolcan Sayar[2], Haoyue Zhang[1], Helen Richards[2], Lucas Liu[2], Baris Turkbey[1], Michael Haffner[2], and Stephanie Harmon[1](✉)

[1] Molecular Imaging Branch, National Cancer Institute, NIH, Bethesda, MD 20892, USA
{zhijun.chen,harry.zhang,ismail.turkbey,stephanie.harmon}@nih.gov

[2] Data Science Integrated Research Center, Fred Hutchinson Cancer Center, Seattle, WA 98109, USA
{esayar,hrichar2,jliu6,mhaffner}@fredhutch.org

Abstract. Metastatic prostate cancer images contain rich and complex information about cellular features. However, due to high level pathogenomic diversity and lack of clinically-validated morphological characteristics, these images are often unlabeled or weakly labeled. Extracting meaningful features from these images using deep learning method is a challenging task. In this paper, we present an unsupervised sparse convolutional autoencoder (SCAE) model to tackle this problem. The imposed sparsity constraints enable the model to perform clustering during feature extraction. The experiments with 856 cores images show that the identified image feature clusters have distinctive cellular feature distributions. By fine-tuning the pretrained encoder module, our proposed model achieves comparable or superior performance to other benchmark models on molecular tumor subtype detection and mutation prediction, suggesting its potential for other downstream supervised tasks.

Keywords: convolutional autoencoder · sparsity · metastatic prostate cancer

1 Introduction

Advanced metastatic prostate cancer (mPC) is characterized by substantial biological heterogeneity [5]. Unlike the Gleason Grading System in localized prostate cancer, there is currently no established framework for the pathological assessment of mPC. Prior studies have reported a wide spectrum of morphologies present in mPC, spanning from commonly observed adenocarcinoma patterns of localized prostate cancer to poorly differentiated morphologies [16]. These poorly differentiated morphologies may be associated with distinct expression profiles, such as neuroendocrine-like differentiation [1,10]; however, the mapping of all morphologies to underlying molecular signatures is an open area of

J. Finkelstein et al. (Eds.): AIME 2024, LNAI 14845, pp. 9–20, 2024.
https://doi.org/10.1007/978-3-031-66535-6_2

interest in the field. A major challenge in developing a clinical framework for reliably characterizing mPC samples is the presence of both inter- and intra-tumor heterogeneity, where metastatic sites can present with multiple morphologies. Therefore, it is desirable to explore automated methods for assessing the entire spectrum of morphological patterns.

Given the challenges in developing robust ground-truth labels for supervised training in mPC, developing an unsupervised approach for learning feature representations across unlabeled data is an attractive solution. Convolutional Autoencoders (CAE) have shown promise in the digital pathology realm for learning compact feature representations that are useful for future downstream tasks [7,13,18]. However, standard CAE is prone to overfitting and the learned features are hard to interpret. In this work, we propose a sparse CAE (SCAE) by introducing sparsity constraints to the output of the encoder. These constraints can regularize the latent space and force the network to discard irrelevant information of the images. Our design allows the network to simultaneously extract image features and perform clustering. The pretrained encoder can be fine-tuned for other supervised learning tasks.

2 Methods

2.1 Dataset

Samples were obtained from rapid autopsy specimens acquired from patients who died of metastatic castration resistant PC and who had provided written informed consent as per the aegis of the Prostate Cancer Donor Program at the University of Washington. The Institutional Review Board of the University of Washington approved this study. Methods for tissue sampling in this cohort have been described in [15]. In total, 411 metastatic sites from 81 patients were included in this study. These metastatic sites were processed into tissue microarrays (TMA), corresponding to 856 cores (1–5 cores/sample). Metastatic site (i.e. bone, lymph node, lung, liver, prostate, etc.) was recorded for all samples.

In a subset of 52 patients, further molecular characterization was completed on core- and patient-level, as described in [9]. This resulted in core-level characterization of immunohistochemical signatures for PC relevant markers including synaptophysin, INSM1, ASCL1, AR, NKX3.1, SOX2, CK8, DLL3, YAP1, EZH2, KI67, FOXA2, HOXB13, ALK, PSMA, and FOXA1. Cores were phenotypically characterized by combined IHC interpretation as positive/negative for Androgen Receptor (AR) expression and positive/negative for neuroendocrine-like (NE) expression. On the patient-level, genomic alterations in AR, CHD1, PTEN, RB1, TP53, and BRAC2 were available for analysis.

All TMA images were scanned on Ventana DP200 scanner at either 20$\times$ or 40$\times$ magnification. Segmentation of cancer regions within each TMA core was completed using a previously developed algorithm. Briefly, a DeepLabV3 segmentation model was trained on 33540 tiles from 121 specimens from two centers, achieving dice performance 0.78. Nuclei segmentation was completed using HoverNet [3]. The dataset was split into the train (60%), validation (10%)

and test (30%) sets using the RandomSplit [11] Python package. The split was stratified on the patient-level, balancing frequency of IHC expression signatures (high/low) across all cores and genomic alterations (altered/intact) across all patients. All cores were tiled at 250 × 250 pixels at 40× magnification. Tiles used in experiments must include more than 10% cancer area by segmentation algorithm, more than 1 cancerous nuclei by HoverNet algorithm and less than 45% white space. In total, 215104 tiles from 451 cores in 26 patients are used for training, 32099 tiles from 49 cores in 5 patients for validation and 157289 tiles from 242 cores in 21 patients for testing.

2.2 Sparse Autoencoder

Our autoencoder consists of two parts: an encoder network f_E and a decoder network g_D. The encoder f_E contains several convolutional, pooling and linear layers. Nonlinear GELU activation and batch normalization are applied between layers. The softmax function is applied to the output of the last linear layer to produce a probability vector of length n. Let $\boldsymbol{X}$ be the input image and $\boldsymbol{p} \in \mathbb{R}^n$ be the corresponding probability, i.e., $\boldsymbol{p} = f_E(\boldsymbol{X})$. Then the feature vector $\boldsymbol{y}$ of this image is obtained by matrix-vector product $\boldsymbol{y} = \boldsymbol{D}\boldsymbol{p}$, where $\boldsymbol{D} \in \mathbb{R}^{m \times n}$ is a non-trainable fixed dictionary matrix whose columns $\{\boldsymbol{d}_1, \ldots, \boldsymbol{d}_n\}$ are i.i.d. sampled from uniform distribution over the unit sphere in $\mathbb{R}^m$ (see Appendix A.1 for rationale behind this design). The decoder g_D is a deconvolutional neural network which reverse the feature extraction process. It uses the image feature $\boldsymbol{y}$ to reconstruct the input image. The details of the network structure is shown in Table 1 and Table 2.

Table 1. Layers in the encoder f_E.

Layer	Kernel, Stride	Output size
Input	–, –	$250^2 \times 3$
Convolution	$3 \times 3, 1$	$248^2 \times 16$
Convolution	$3 \times 3, 1$	$248^2 \times 32$
Avg Pooling	$2 \times 2, 2$	$124^2 \times 32$
Convolution	$3 \times 3, 1$	$124^2 \times 64$
Avg Pooling	$2 \times 2, 2$	$62^2 \times 64$
Convolution	$3 \times 3, 1$	$62^2 \times 128$
Avg Pooling	$2 \times 2, 2$	$31^2 \times 128$
Convolution	$3 \times 3, 1$	$31^2 \times 320$
Convolution	$3 \times 3, 1$	$31^2 \times 240$
Convolution	$3 \times 3, 1$	$31^2 \times 150$
Convolution	$3 \times 3, 2$	$16^2 \times 80$
Convolution	$3 \times 3, 1$	$16^2 \times 40$
Convolution	$3 \times 3, 1$	$16^2 \times 15$
Linear	–, –	n

Table 2. Layers in the decoder g_D.

Layer	Kernel, Stride	Output size
Input	–, –	m
Linear	–, –	$16^2 \times 15$
Deconvolution	$3 \times 3, 1$	$16^2 \times 40$
Deconvolution	$3 \times 3, 1$	$16^2 \times 80$
Deconvolution	$3 \times 3, 2$	$31^2 \times 150$
Deconvolution	$3 \times 3, 1$	$31^2 \times 240$
Deconvolution	$3 \times 3, 1$	$31^2 \times 320$
Deconcolution	$3 \times 3, 1$	$31^2 \times 128$
Deconvolution	$3 \times 3, 2$	$62^2 \times 128$
Deconvolution	$3 \times 3, 1$	$62^2 \times 64$
Deconvolution	$3 \times 3, 2$	$124^2 \times 64$
Deconvolution	$3 \times 3, 1$	$124^2 \times 32$
Deconvolution	$3 \times 3, 2$	$248^2 \times 32$
Deconvolution	$3 \times 3, 1$	$248^2 \times 16$
Deconvolution	$3 \times 3, 1$	$250^2 \times 3$

We train our autoencoder to optimize the following loss terms.

- **Image reconstruction loss.** It is defined as the mean-square difference between the input image $\boldsymbol{X}$ and the reconstructed image $\hat{\boldsymbol{X}} = g_D(\boldsymbol{y})$, i.e.,

$$\ell_R = \frac{1}{MNC} \sum_{i=1}^{M} \sum_{j=1}^{N} \sum_{k=1}^{C} (\boldsymbol{X}_{ijk} - \hat{\boldsymbol{X}}_{ijk})^2,$$

  where M, N and C are the width, height and the number of channels of the image respectively.
- **Sparsity loss.** To prevent overfitting, we introduce the entropy loss ℓ_E of the probability $\boldsymbol{p} = (p_1, \ldots, p_n)$ in the loss function

$$\ell_E = - \sum_{i=1}^{n} p_i \log(p_i).$$

  Optimizing ℓ_E favors p_i to be sparse.
- **Dirichlet loss.** Minimizing the ℓ_E alone may result in network using only a subset of columns of $\boldsymbol{D}$ for feature construction. To address this issue, we introduce the Dirichlet loss ℓ_D over a mini-batch. Suppose that we have a mini-batch of probability vectors $\{\boldsymbol{p}^j\}_{j=1}^{B}$, ℓ_D is defined as

$$\ell_D = - \sum_{i=1}^{n} \alpha_i \log \left(\frac{1}{B} \sum_{j=1}^{B} p_i^j \right),$$

  where α_i are some positive parameters. Optimizing ℓ_D forces the model to use different columns of $\boldsymbol{D}$ to produce different features.

The loss function we optimize is the weighted sum of these three terms $\ell = \lambda_1 \cdot \ell_R + \lambda_2 \cdot \ell_E + \lambda_3 \cdot \ell_D$, where $\lambda_1, \lambda_2, \lambda_3 > 0$. Optimizing the sparsity loss ℓ_E and the Dirichlet loss ℓ_D together favors the clustering of features (see Appendix A.1).

2.3 Downstream Analysis

We use the trained encoder f_E to extract features of tiles for downstream analysis. We first investigate the clustering of tile features using k-means and visualize the result using t-SNE. Then we check the cellular features distributions in each cluster identified by k-means, and compare them with cellular features distributions of tiles from cores marked by molecular phenotype based on androgen receptor (AR) and neuroendocrine marker (NE) expression by IHC.

For supervised tasks, we test our encoder for AR/NE detection and mutation prediction. We create a 2-layer fully connected neural network h which takes tiles features from the encoder and outputs probabilities of the corresponding tile contains AR/NE or genomic alterations. Note that our dataset is weakly labeled (AR/NE labels at core-level, genomic alteration labels at patient-level). At core-level, we have 62% AR+ cores and 32% NE+ cores. At patient-level, we have 54% AR alterations, 4% CHD1 alterations, 43% PTEN alterations, 33%

RB1 alterations, 45% TP53 alterations and 9% BRCA2 alterations. These labels are not mutually exclusive. During supervised training, we assigned all tiles from the same core (patient) the same label as the core (patient). The cross-entropy was used as the loss function.

Our model was compared with commonly used models ResNet50 [6] and DenseNet161 [8]. We measure the model performance using the area under the curve (AUC; higher is better) and expected calibration error [4] (ECE; lower is better). ECE is small when the model's estimated probabilities match the actual probabilities of the ground truth distribution. All results reported were obtained using images not seen by the model before, i.e., images from testing dataset.

3 Results and Discussion

Our sparse autoencoder is first trained by optimizing loss function defined in Sect. 2.2. Data augmentations such as flip and rotation are utilized. We set the length n of the output probability vector to be 3500 and length m of the feature to be 1600. For the weights of loss terms, we fix $\lambda_1 = 1$ and start with large λ_2. We gradually decrease it until the convergent value of ℓ_R is below 0.01. For λ_3 we found that a small value is sufficient to ensure the ℓ_D is close to that of a uniform distribution. The final setting is $\lambda_1 = 1, \lambda_2 = 0.0073$ and $\lambda_3 = 10^{-5}$. The model was trained for 120 epochs with Adam optimizer using learning rate $1.5 \cdot 10^{-4}$, exponential decay rate 0.98 and batch size 100. The model weights that achieve the lowest loss on validation set is stored.

3.1 Tile Clustering

Each tile image is encoded into a sparse probability vector by the encoder f_E. For similar images, their corresponding probability vectors tend to have similar support. That is, the positions of their non-zero elements are similar. This then results in the clustering of feature vectors. We randomly select 2000 tiles from the testing set and extract their features using our trained SCAE. Then we perform k-means clustering with these 1600-dimensional features vectors, and use t-SNE dimension reduction to visualize the result. With our hyperparameter setting, the sparse autoencoder creates 9 clusters. The t-SNE plot is shown in Fig. 1a, and some tile examples in each cluster are shown in Fig. 1b.

To further investigate these clusters, we use Cellpose [17] to extract CellDiv features [12] from tiles and study their distributions in each clusters. The kernel density estimations (KDE) of some selected cellular features are shown in the first line of Fig. 2. We can see that cluster 3 has the most distinctive distribution in these features. We then select tiles which are from cores marked as AR+/NE- or AR-/NE+ based on IHC profiling, and estimate the feature distributions in these AR+ tiles and NE+ tiles. The KDE plots are shown in the second line of Fig. 2. Note that the distributions in cluster 3 resemble those in NE+ tiles. This indicates that our SCAE has the potential to distinguish AR+ and NE+ tiles. Separation of AR+ cores and NE+ cores was also observed when we model the core as a subspace spanned by its tile features (see Appendix A.3).

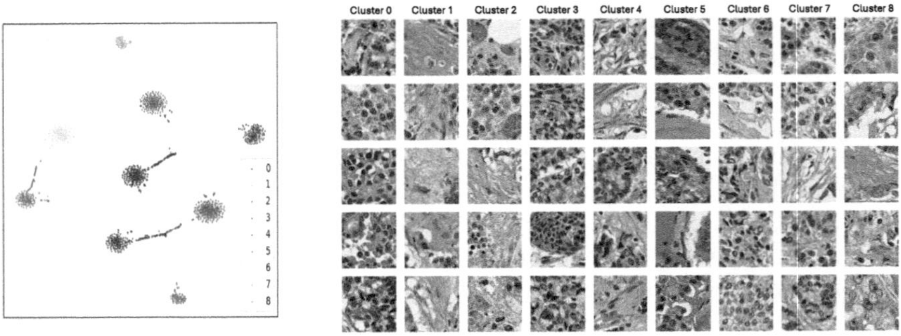

(a) (b)

Fig. 1. a: t-SNE embedding of tile features. K-means clustering result on these features is indicated by different colors. **b:** Tile examples from each cluster. Each column corresponds to one cluster. Clusters appear to correspond to nuclear density (i.e. Cluster 1 vs Cluster 3), nuclear shape (i.e. Cluster 0 vs Cluster 6), and stromal/microenvironment features (Cluster 5).

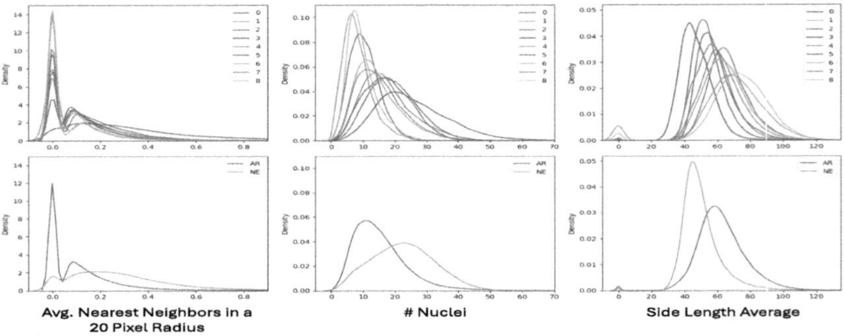

Fig. 2. Selected cellular feature distributions in each cluster, AR+ tiles and NE+ tiles.

3.2 AR/NE Detection

All models (our pretrained encoder, ResNet50 and DenseNet161) are trained for 15 epochs on the training set using batch-size 100 and Adam optimizer with learning rate 10^{-4}. Each model's best performance on testing set at tile-level is summarized in Table 3. In terms of AUC, our model SCAE is comparable to that of pretrained ResNet50 and DenseNet161. However, our model has much smaller ECE. The corresponding receiver operating characteristic (ROC) curves of each model is shown in Fig. 3.

We further evaluate the model performance at core-level. For each core, we take the average of probabilties assigned to its tiles by our model to obtain probabilities of the core being AR+ and NE+. The model performances at core-level is summarized in Table 4. The ROC curves of each model is shown in Fig. 4. The AUC of each model is generally better at core-level than at tile-level. Our

Table 3. Evaluation of models on AR/NE detection at tile level. * indicates the model is pre-trained on ImageNet [14].

Model	AUC(AR/NE)	ECE(AR/NE)
SCAE	0.732/0.843	0.148/0.082
ResNet50	0.691/0.822	0.229/0.203
DenseNet161	0.681/0.828	0.254/0.194
ResNet50*	0.712/0.823	0.224/0.179
DenseNet161*	0.740/0.846	0.224/0.168

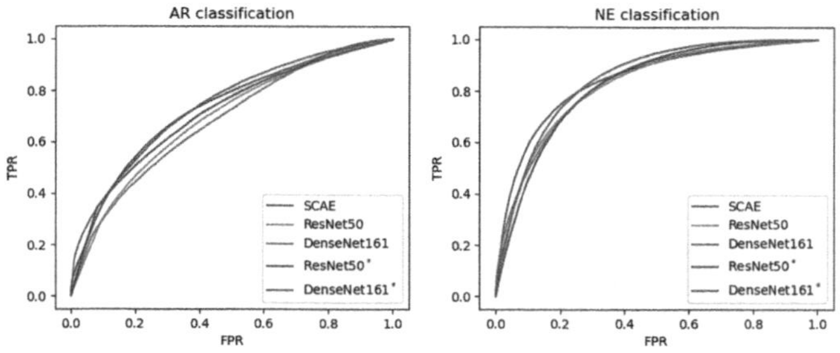

Fig. 3. ROC curves of models for AR/NE detection at tile level.

model's AUC is slightly worse than pretrained DenseNet161, but the our model has much simpler structure with substantially fewer parameters. The output of our model is also compressible. Some AR/NE detection maps of our model are shown in Fig. 5.

Table 4. Evaluation of models on AR/NE detection at core level.

Model	AUC(AR/NE)	ECE(AR/NE)
SCAE	0.742/0.894	0.070/0.089
ResNet50	0.701/0.892	0.133/0.099
DenseNet161	0.647/0.893	0.128/0.084
ResNet50*	0.724/0.907	0.101/0.102
DenseNet161*	0.738/0.918	0.089/0.083

3.3 Mutation Prediction

All models are trained for 10 epochs on training set for predicting PTEN, AR, RB1, CHD1, TP53 and BRCA2 alteration. We use batch-size 100 and Adam optimizer with learning rate 10^{-4}. The best performance of all models were recorded. All models' performances on CHD1, BRCA2 and TP53 alteration are

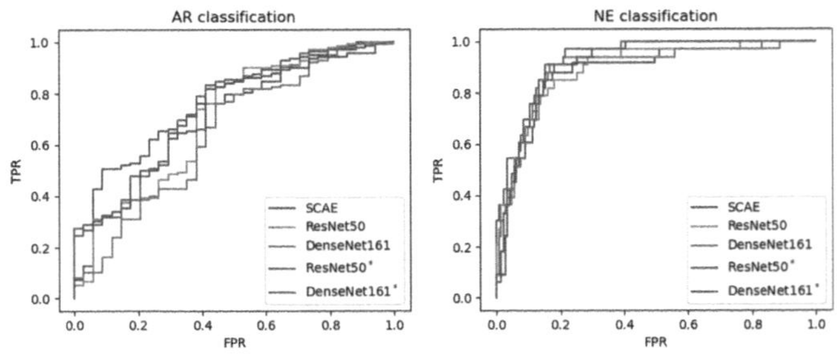

Fig. 4. ROC curves of models for AR/NE detection at core level.

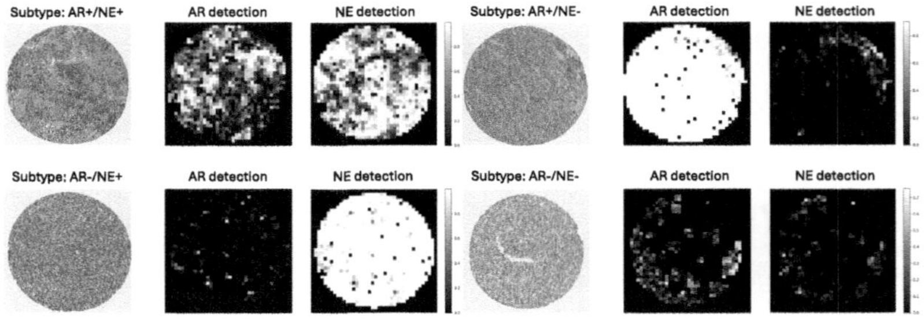

Fig. 5. Examples of correct AR/NE detection. Even though the prediction at core level is correct, intra-tumor heterogeneity may impact results at tile level.

bad (all models have AUC around 0.5). This is possibly due to imbalanced dataset. For PTEN, AR and RB1, models' AUCs and ECEs are summarized in Table 5. Similar to AR/NE detection, our model perform slightly worse than pretrained models, but has lower ECE. The ROC curves are shown in Fig. 6.

Table 5. Evaluation of models on PTEN/AR/RB1 alteration prediction at tile level.

Model	AUC(PTEN/AR/RB1)	ECE(PTEN/AR/RB1)
SCAE	0.604/0.744/0.758	0.182/0.213/0.130
ResNet50	0.606/0.677/0.757	0.279/0.305/0.183
DenseNet161	0.605/0.676/0.794	0.288/0.197/0.185
ResNet50*	0.619/0.710/0.766	0.299/0.202/0.196
DenseNet161*	0.616/0.704/0.795	0.304/0.284/0.200

The distribution of the models' output probabilities for testing tiles are shown by violin plots in Appendix A.4. The ResNet50 and DenseNet161 tend to output probabilities close to 0 or 1 in predictions. This leads to their large ECE shown in Table 5.

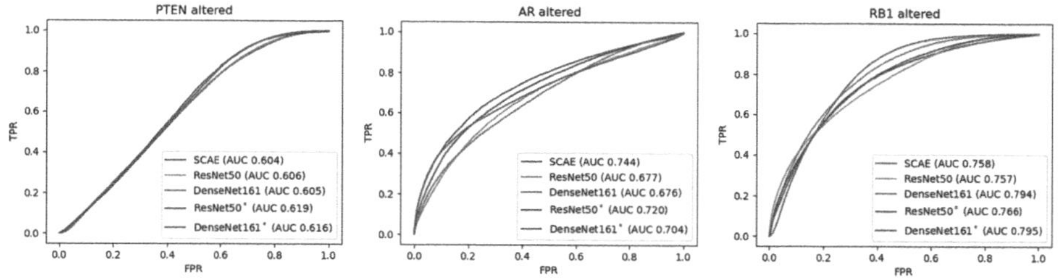

Fig. 6. ROC curves of models for PTEN/AR/RB1 alteration prediction at tile level.

4 Conclusion

In this work, we demonstrate that a sparsity-constrained CAE can robustly identify unique clusters of histomorphological features in metastatic prostate cancer. These features show association to clinically-relevant molecular subtypes of metastatic prostate cancer and demonstrates good quality performance for supervised tasks of predicting core-level molecular subtype and patient-level genomic alterations, performining equivalently to or better than standard supervised CNN-based approaches. Future work expanding to various tissue types and sources is ongoing.

Acknowledgments. This project is in part supported by the Intramural Research Program of the Center for Cancer Research, NCI. The project is additionally supported by DoD CRMP Prostate Cancer Research Program award. The content of this publication does not necessarily reflect the views or policies of the Department of Health and Human Services, nor does mention of trade names, commercial products, or organizations imply endorsement by the U.S. Government.

A Appendix

A.1 Random dictionary

The dictionary $\boldsymbol{D} \in \mathbb{R}^{m \times n}$ in our model is randomly initialized and fixed during training. The columns $\{\boldsymbol{d}_1, \ldots, \boldsymbol{d}_n\}$ are i.i.d. and follow uniform distribution over unit sphere in $\mathbb{R}^m$. This initialization ensures that inner products $\boldsymbol{d}_i^T \boldsymbol{d}_j, i \neq j$ are all small with high probability and columns of $\boldsymbol{D}$ are "almost orthonormal". With high probability, the random matrix $\boldsymbol{D}$ has *restricted isometry property*. More precisely, for some integer $s \leq n$, $\delta \in (0, 1)$, we have the following inequalities hold with high probability

$$(1 - \delta)\|\boldsymbol{x}\|_2^2 \leq \|\boldsymbol{D}\boldsymbol{x}\|_2^2 \leq (1 + \delta)\|\boldsymbol{x}\|_2^2$$

for all $\boldsymbol{x} \in \mathbb{R}^n$ with at most s non-zero elements. This property is essential to the clustering result shown in Sect. 3.1.

A.2 Dirichlet loss

For Dirichlet loss ℓ_D, if all α_i are positive, then minimizing ℓ_D forces the model to use all columns of the dictionary $\boldsymbol{D}$. In particular, if all α_i are positive and equal to some constant, then ℓ_D is equivalent to the (scaled) Kullback-Leibler divergence between the uniform multinomial distribution and a multinomial distribution with parameter $\left(\frac{1}{B}\sum_{j=1}^{B} p_i^j\right)_{i=1}^{n}$.

Optimizing ℓ_E and ℓ_D together favors clustering of features. A similar idea can be found in [2]. One major difference between our methods and theirs is that we use the entropy-based sparsity loss while they use the Gini impurity index. We found in our experiment that the entropy loss works better than Gini index.

A.3 Subspace modeling

The core image in our dataset is of size 5000×5000. It is computationally infeasible to feed this core image into our model. Instead of encoding the whole core image using the encoder, we divide the core image into tiles, encode these tiles into feature vectors and associate these vectors with a low-dimensional subspace. This subspace is then used as the representation of the core image. Specifically, a core image is divided into tiles of size 250×250. After encoding these tiles into feature vectors denoted as $\{\boldsymbol{z}_1, \dots, \boldsymbol{z}_K\} \subset \mathbb{R}^m$, we find a subspace that approximately capture the spatial distribution of these features by calculating the singular values decomposition of the feature matrix $\boldsymbol{Z} = [\boldsymbol{z}_1, \dots, \boldsymbol{z}_K] \in \mathbb{R}^{m \times K}$. Suppose that $\{s_1, \dots, s_K\}$ and $\{\boldsymbol{u}_1, \dots, \boldsymbol{u}_K\}$ are singular values (in descending order) and left singular vectors of $\boldsymbol{Z}$ respectively, then the subspace is determined by the span of $\{\boldsymbol{u}_1, \dots, \boldsymbol{u}_{i^*}\}$, where $i^* = \min\{i \mid \sum_{j=1}^{i} s_j \geq 0.95 \cdot \sum_{j=1}^{K} s_j\}$. The dimension i^* is not fixed and it depends on the the number of tiles in the core and the spatial distribution of tile features. Finally, we use Grassmann metric [19] to calculate the distance between subspaces and this distance is treated as the similarity score between corresponding cores. Based on mutual distances between cores, we can exam the spatial distribution of AR+/NE- and AR-/NE+ cores. The t-SNE plot is shown in Fig. 7.

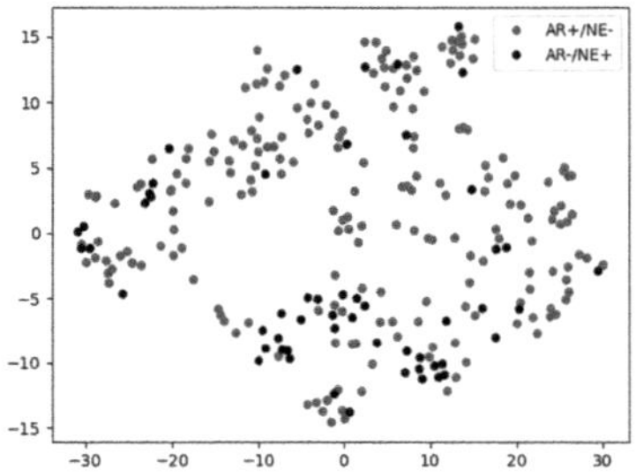

Fig. 7. t-SNE visualization of core distribution

A.4 Distributions of models' output probabilities for mutation prediction

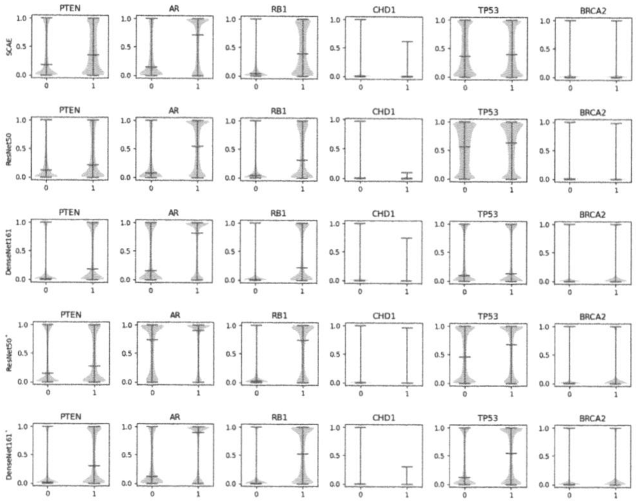

Fig. 8. Distributions of models; predictions on test tiles. The medians are indicated by horizontal bars.

References

1. Beltran, H., et al.: Divergent clonal evolution of castration-resistant neuroendocrine prostate cancer. Nat. Med. **22**(3), 298–305 (2016)
2. Boubekki, A., Kampffmeyer, M., Brefeld, U., Jenssen, R.: Joint optimization of an autoencoder for clustering and embedding. Mach. Learn. **110**(7), 1901–1937 (2021)
3. Graham, S., Vu, Q.D., Raza, S.E.A., Azam, A., Tsang, Y.W., Kwak, J.T., Rajpoot, N.: Hover-net: simultaneous segmentation and classification of nuclei in multi-tissue histology images. Med. Image Anal. **58**, 101563 (2019)
4. Guo, C., Pleiss, G., Sun, Y., Weinberger, K.Q.: On calibration of modern neural networks. In: International Conference on Machine Learning, pp. 1321–1330. PMLR (2017)
5. Haffner, M.C., et al.: Genomic and phenotypic heterogeneity in prostate cancer. Nat. Rev. Urol. **18**(2), 79–92 (2021)
6. He, K., Zhang, X., Ren, S., Sun, J.: Deep residual learning for image recognition. In: Proceedings of the IEEE Conference on Computer Vision and Pattern Recognition, pp. 770–778 (2016)
7. Hou, L., et al.: Sparse autoencoder for unsupervised nucleus detection and representation in histopathology images. Pattern Recogn. **86**, 188–200 (2019)
8. Huang, G., Liu, Z., Van Der Maaten, L., Weinberger, K.Q.: Densely connected convolutional networks. In: Proceedings of the IEEE Conference on Computer Vision and Pattern Recognition, pp. 4700–4708 (2017)
9. Kumar, A., et al.: Substantial interindividual and limited intraindividual genomic diversity among tumors from men with metastatic prostate cancer. Nat. Med. **22**(4), 369–378 (2016)

10. Labrecque, M.P., et al.: Molecular profiling stratifies diverse phenotypes of treatment-refractory metastatic castration-resistant prostate cancer. J. Clin. Investig. **129**(10), 4492–4505 (2019)
11. Lay, N.: Randomsplit (2023). https://github.com/nslay/RandomSplit
12. Lu, C., et al.: A prognostic model for overall survival of patients with early-stage non-small cell lung cancer: a multicentre, retrospective study. Lancet Digital Health **2**(11), e594–e606 (2020)
13. Roy, M., et al.: Convolutional autoencoder based model histocae for segmentation of viable tumor regions in liver whole-slide images. Sci. Rep. **11**(1), 139 (2021)
14. Russakovsky, O., et al.: Imagenet large scale visual recognition challenge. Int. J. Comput. Vision **115**, 211–252 (2015)
15. Sayar, E., et al.: Reversible epigenetic alterations mediate psma expression heterogeneity in advanced metastatic prostate cancer. JCI Insight **8**(7) (2023)
16. Shah, R.B., et al.: Androgen-independent prostate cancer is a heterogeneous group of diseases: lessons from a rapid autopsy program. Can. Res. **64**(24), 9209–9216 (2004)
17. Stringer, C., Wang, T., Michaelos, M., Pachitariu, M.: Cellpose: a generalist algorithm for cellular segmentation. Nat. Methods **18**(1), 100–106 (2021)
18. Xu, J., et al.: Stacked sparse autoencoder (SSAE) for nuclei detection on breast cancer histopathology images. IEEE Trans. Med. Imaging **35**(1), 119–130 (2015)
19. Ye, K., Lim, L.H.: Schubert varieties and distances between subspaces of different dimensions. SIAM J. Matrix Anal. Appl. **37**(3), 1176–1197 (2016)

AI in Neuro-Oncology: Predicting EGFR Amplification in Glioblastoma from Whole Slide Images Using Weakly Supervised Deep Learning

Homay Danaei Mehr[1,2](✉), Imran Noorani[3,4], Priyanka Rana[1], Antonio Di Ieva[2], and Sidong Liu[1,2]

[1] Centre for Health Informatics, Australian Institute of Health Innovation, Macquarie University, Sydney, NSW 2109, Australia
homay.danaeimehr@hdr.mq.edu.au

[2] Computational NeuroSurgery (CNS) Lab, Macquarie Medical School, Macquarie University, Sydney, NSW 2109, Australia

[3] Department of Neuromuscular Diseases, The Francis Crick Institute and University College London, London NW1 1AT, UK

[4] Department of Neurosurgery, National Hospital for Neurology and Neurosurgery, London SO16 6AQ, UK

Abstract. Glioblastoma (GBM), the most aggressive type of brain cancer, is notorious for its resistance to treatments due to its rapid growth and high degree of heterogeneity. Epidermal Growth Factor Receptor (EGFR) is an important diagnostic, prognostic, and therapeutic biomarker in GBM. The current gold standard diagnosis of EGFR detection relies on immunohistochemistry (IHC), Fluorescence in Situ Hybridization (FISH), and sequencing analysis of surgical samples with a turnaround time of several days to weeks, not to mention the laboratory costs related to the analyses. To improve the efficiency and cost-effectiveness of EGFR detection, we propose an alternative solution based on Whole Slide Image (WSI) and artificial intelligence (AI) methods, specifically multiple instance learning (MIL) frameworks. In this study, we conducted a comprehensive analysis of two public datasets, TCGA and CPTAC, with three MIL frameworks, Clustering Constrained Attention Multiple Instance Learning (CLAM), Double–Tier Feature Distillation Multiple Instance Learning (DTFD), and Transformer-based Correlated MIL for WSI Classification (TransMIL), to predict the EGFR status in GBM. With evaluation using 5–fold cross-validation, the CLAM model achieved 0.736 $\pm$ 0.181 and 0.799 $\pm$ 0.164 area under the curve (AUC) on TCGA and CPTAC, respectively, outperforming the state-of-the-art model (AUC of 0.71 $\pm$ 0.031). This study demonstrates the effectiveness of MIL models in the classification of EGFR biomarkers with a high potential to shift the paradigm from the current gold standard to the more efficient and cheaper WSI and AI workflows.

Keywords: Multiple Instance Learning · Weakly Supervised Deep Learning · Glioblastoma · EGFR biomarker · Whole Slide Images

J. Finkelstein et al. (Eds.): AIME 2024, LNAI 14845, pp. 21–29, 2024.
https://doi.org/10.1007/978-3-031-66535-6_3

1 Introduction

Glioblastoma (GBM), the most aggressive brain tumor, is notorious for its resistance to traditional treatments due to its rapid growth and high degree of heterogeneity. Epidermal Growth Factor Receptor (EGFR) is a protein found on the surface of some cells in the body, which is critical in regulating cell growth, survival, proliferation, and differentiation. EGFR plays an important role in the development and progression of around 50% of GBMs and is an important diagnostic and prognostic biomarker, as well as a therapeutic target in GBM [1–3], with several clinical trials ongoing targeting EGFR amplifications. Indeed, the EGFR amplification is diagnostic for Isocitrate DehydroGenase1(IDH1)-wildtype glioblastoma [4, 5].

The gold standard for EGFR detection in the context of cancer diagnostics involves a combination of methods, including immunohistochemistry (IHC), Fluorescence in Situ Hybridization (FISH), and next-generation sequencing. Each of these methods has its advantages and limitations, regarding sensitivity, specificity, the type of sample, and the specific information needed (the overexpression of EGFR protein, EGFR gene amplification, or gene mutations). However, the turnaround time of these analyses is in the order of several days to weeks, their cost can be prohibitive and many of these techniques are not readily accessible to many clinical laboratories.

A few studies attempted to use artificial intelligence (AI) in EGFR detection, offering the benefits of increased cost-effectiveness and higher efficiency. These studies have explored histopathology images in EGFR status prediction for patients with lung cancer [6] and colon cancer [7], but this approach has not been extensively applied to GBM. Nasrallah et al. [8] recently developed the Cryosection Histopathology Assessment and Review Machine (CHARM), an AI model (a vision transformer-based [9] model), using three distinct datasets (i.e., The Cancer Genome Atlas (TCGA), Brigham and Women's Hospital (BWH), and The University of Pennsylvania (UPenn)) to identify GBM biomarkers, including EGFR. The CHARM model achieved an AUC of 0.71 ± 0.031 in the classification of EGFR. However, this model is based on frozen slides, aiming to provide real-time cancer diagnosis during surgery, rather than maximizing the detection performance.

Considering the advantage of hematoxylin and eosin (H&E) stained whole slide images (WSI) over IHC, FISH, and sequencing in EGFR detection, this study aims to develop an alternative method for EGFR prediction, which is based on the state-of-the-art multiple instance learning (MIL) frameworks (i.e., Clustering Constrained Attention Multiple Instance Learning (CLAM) [10], Double–Tier Feature Distillation Multiple Instance Learning (DTFD) [11], and Transformer-based Correlated Multiple Instance Learning for WSI Classification (TransMIL) [12]) and high-quality Formalin-Fixed Paraffin-Embedded (FFPE) WSIs. In this study, we evaluated the proposed pipeline with two large public datasets and three MIL frameworks. Our pipeline outperformed the state-of-the-art method, demonstrating the effectiveness of FFPE-WSI and MIL frameworks in predicting EGFR status in GBM.

2 Methodology

2.1 Datasets

In this study, we used two public GBM WSI datasets with EGFR amplification and mutation annotations, including The TCGA and Clinical Proteomic Tumor Analysis Consortium (CPTAC) from the National Cancer Institute Centre for Cancer Genomics (https://www.cancer.gov/ccg/). The TCGA dataset comprises 329 WSIs, with 217 EGFR-positive cases and 112 EGFR-negative cases. The sizes of these images are up to 35495 × 36929 pixels. The CPTAC dataset includes 217 WSIs, with 99 EGFR-positive cases and 118 EGFR-negative cases. The sizes of these images are up to 23903 × 50374 pixels. WSIs in both datasets are stained using H&E. EGFR amplification was defined as a copy number greater than or equal to 4 [13].

2.2 Data Preprocessing

The analysis of WSIs faces unique challenges due to their large size and high resolution. Consequently, the expansive size of the WSIs may result in the infeasibility of training deep learning models [14]. Patch-based approaches, which divide each WSI into numerous smaller patches, have been commonly used [15]. Following CLAM's framework, the preprocessing of the WSIs includes tissue segmentation using Otsu's method [16] in which the background is eliminated. Each tissue then undergoes a patching process in which the pixels of tissues are grouped into smaller patches (224 × 224). The feature extraction process involves applying the ResNet-50 feature encoder [17] (pre-trained on the ImageNet dataset) to patches (Fig. 1).

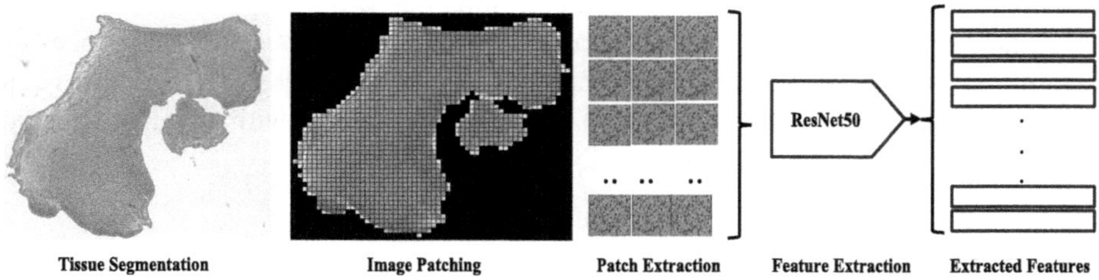

Fig. 1. WSI pre-processing pipeline.

2.3 Multiple Instance Learning (MIL) Frameworks

Hands-on annotating patches by an expert pathologist is a labor-intensive as well as costly process. Generating patches from the WSI can result in abundant numbers of patches. Therefore, disease-positive tissue sections may locate in a minor fraction of the entire sample. This makes the patch-based supervised learning methods unfeasible. Weakly supervised learning methods using the MIL approach have gained popularity, due to their ability to leverage slide-level annotations while still achieving robust performance in image classification tasks [15, 18]. In WSI classification through the MIL, each WSI is considered a "bag" containing various instances (patches) extracted from the slide. Labels of these instance–level patches are unknown. The label of each bag is determined collectively based on the information contained within the patches. Hence, the label of each bag is determined as positive if it contains at least one positive instance otherwise the label is negative [19].

This study compares the performance of three state-of-the-art MIL frameworks, i.e., CLAM [10], DTFD [11], and TransMIL [12]. Each framework represents a different instance embedding strategy. The CLAM is a weakly supervised MIL framework that uses an **attention mechanism and patch-level clustering** to infer a bag-level label [10]. The DTFD uses two tiers to group the patches through the concept of **pseudo–bags**. Tier 1 weighs each patch in the pseudo–bags using the attention mechanism and then a feature distillation strategy (i.e., Maximum-Minimum Selection (MaxMinS), Maximum selection (MaxS), Aggregated Feature Selection (AFS), and Maximum Attention Score Selection (MAS)), to derive a feature vector for each pseudo–bag. Tier 2 refines the feature representations of a parent bag for the final classification [11]. TransMIL, based on the **transformer model**, uses a self-attention mechanism to focus on the relevant patches that inform the label for the entire slide. Additionally, it incorporates the proposed pyramid position encoding generator to encode spatial information [12].

The following are the adjusted hyperparameters for the three models in this study. For the CLAM model, we adjusted the smooth SVM loss function for patch-level classification and the cross-entropy loss function for slide-level classification. With a focus on a single predominant class due to the dataset's two-label structure, we adjusted the training model based on a single batch CLAM model. It was trained with an Adam optimizer, a weight decay of 0.0000001, and a learning rate of 0.0002 over 200 epochs. The DTFD model was trained for 200 epochs using the cross-entropy loss function and Adam optimizer. For the DTFD model, the learning rate was adjusted to 0.000001. The scheduler adjusted the learning rate by a learning rate ratio of 0.2, which reduces the learning rate to 20% of its value after every 100 epochs. The TransMIL model was configured with a learning rate of 0.0002, a Lookahead optimizer with a weight decay of 0.00001, and a cross-entropy loss function, over 200 epochs.

2.4 Performance Evaluation

To evaluate the three models' performance, we employed 5-fold cross-validation on each of the two datasets (TCGA and CPTAC), with a split ratio of 80% for training, 10% for validation, and 10% for testing in each fold. The results are derived from the mean and standard deviation of performance metrics obtained from the 5-fold cross-validation. For external validation purposes, in all three models, we assigned 85% of the TCGA dataset for training and 15% for internal validation. Subsequently, we allocated the whole CPTAC as an unseen test set. Similarly, we utilized 85% of the CPTAC dataset for training and 15% for internal validation and used the whole TCGA as an unseen test set. Furthermore, Area under the curve (AUC), Sensitivity (true positive rate), and Specificity (true negative rate) have been used as the metrics for performance evaluation.

3 Results

We conducted a comprehensive performance comparison to predict GMB EGFR biomarkers among three state-of-the-art MIL–based models and the transformer-based CHARM model [10], which is the only available study in the literature for GMB EGFR classification. Table 1 illustrates the classification results of all three models trained on TCGA and CPTAC datasets using 5-fold cross-validation. It is observed that the CLAM model with AUC of 0.736 ± 0.181(on TCGA) and 0.799 ± 0.164 (on CPTAC) performed the best. It is obvious that the attention mechanism in the CLAM model significantly enhances performance by focusing on more class–relevant features. Furthermore, the CLAM model's high Sensitivity values using 5-fold cross-validation on TCGA (Sensitivity: 0.741 ± 0.194) and on CPTAC (Sensitivity: 0.759 ± 0.268) indicate its high performance in identifying a high rate of positive cases.

We conducted a comparative analysis between our models and the CHARM model [8]. The CHARM model was trained based on a random selection of three distinct datasets i.e., TCGA (different from our used dataset), BWH, and UPenn, including 2334 cryosection WSIs taking advantage of the slide diversity and abundant data for training.

The results demonstrate that the CLAM model, achieving higher AUCs, outperformed the CHARM model which obtained an AUC of 0.71 ± 0.031. Similarly, the DTFD (MaxS) model with an AUC of 0.727 ± 0.201, also surpassed the CHARM model.

Furthermore, the CHARM model uses cryosection-based datasets suffering from uneven tissue thickness that can affect performance. In contrast, the WSI datasets in this study are based on FFPE slides.

Table 1. Five-fold cross-validation performance evaluation

Dataset	Model	AUC	Specificity	Sensitivity
TCGA	**CLAM**	**0.736 ± 0.181**	0. 545 ± 0.351	0.741 ± 0.194
	DTFD (AFS)	0.706 ± 0.0755	0.959 ± 0.6	0.512 ± 0.113
	TransMIL	0.688 ± 0.172	0.627 ± 0.419	0.548 ± 0.604
CPTAC	**CLAM**	**0.799 ± 0.164**	0.6618 ± 0.277	0.759 ± 0.268
	DTFD (MaxS)	**0.727 ± 0.201**	0.805 ± 0.176	0.595 ± 0.254
	TransMIL	0.694 ± 0.176	0	1
TCGA/BWH/UPenn	CHARM	0.71 ± 0.031	N/A	N/A

For external validation, a reciprocal comparison approach was utilized with the TCGA dataset for training and validation, and the CPTAC dataset for an unseen test set. Subsequently, the roles of the datasets were reversed (Table 2). The results demonstrate that the DTFD (AFS) trained by the TCGA (including 329 WSI) dataset and tested by an unseen CPTAC dataset (including 217 WSI) with AUC = 0.714 outperformed the other two MIL-based models. Moreover, DTFD's high Specificity (0.928) indicates that it could identify a high rate of negative cases.

On the other hand, when CPTAC (including 217 WSI) dataset was utilized for training and TCGA (including 329 WSI) was used for the unseen test set, the performance of all three models decreased. It demonstrates a strong correlation between the model's performance and the size of the training set. Using larger datasets like TCGA improved model performance. However, the model's performance decreased when using a limited dataset like CPTAC for training. In other words, it demonstrates that models trained on larger datasets exhibit better generalization capabilities.

Based on all the experiments conducted, it was observed that the TransMIL model, which uses the transformer architecture, gained lower performance than other models. The CHARM model, based on transformers, did not perform as well as the CLAM and DTFD models. This indicates that attention mechanisms are more effective in identifying the most relevant class-specific features over transformer-based models.

Table 2. Performance evaluation on an unseen test set

Train Set	Test Set	Model	AUC	Specificity	Sensitivity
TCGA	CPTAC	CLAM	0.707	0.711	0.575
		DTFD (AFS)	**0.714**	0.928	0.5
		TransMIL	0.692	0.262	0.946
CPTAC	TCGA	CLAM	0.639	0.776	0.414
		DTFD (AFS)	0.672	0.899	0.444
		TransMIL	0.623	0.262	0.946

The CLAM model generates slide-level heatmaps by identifying and aggregating regions of WSI with high diagnostic significance (high attention score, red regions) while discounting regions with low diagnostic importance (low attention score, blue regions) associated with the prediction. Since the CLAM model using 5-fold cross-validation outperformed the other two MIL-based models as well as the CHARM model, following the CLAM's framework we generated the attention scores that are visualized as heatmaps (Fig. 2).

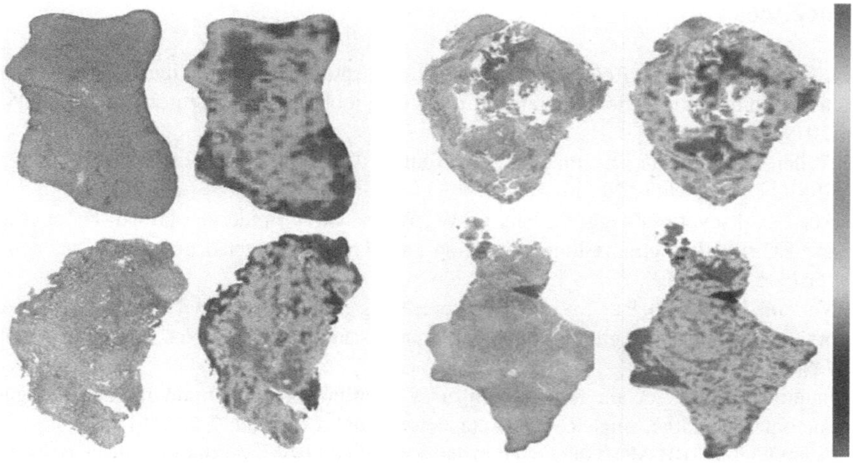

Fig. 2. Examples of heatmaps of EGFR-positive and negative cases.

4 Conclusion

In this work, we conducted a comprehensive comparison among three superior MIL-based models (CLAM, DTFD, and TransMIL) and the vision transformer-based CHARM model to predict GBM EGFR amplification biomarkers. The experimental results showed that the CLAM model using 5-fold cross-validation on both TCGA and CPTAC datasets was able to successfully predict EGFR amplification from H&E slides. Notably, it not only outperformed the other two models but also exceeded the performance of the CHARM model. Upon extensive evaluation using an unseen CPTAC dataset, the DTFD (AFS) trained on the TCGA dataset surpassed the other two MIL-based models. Results demonstrate the attention mechanism's ability to enhance performance in the CLAM and DTFD models, by focusing on class-specific features.

In our future work, we aim to propose our MIL–based framework to improve results regarding the methodological gaps present in the current models. Eventually, our proposed model will serve as a complementary analytic tool to be added to the neuro-oncology diagnostic armamentarium.

Disclosure of Interests. The authors have no competing interests to declare that are relevant to the content of this article.

References

1. Lassman, A.B., et al.: Epidermal growth factor receptor (EGFR) amplification rates observed in screening patients for randomized trials in glioblastoma. J. Neurooncol **144**, 205–210 (2019)
2. Weller, M., Le Rhun, E., Preusser, M., Tonn, J.-C., Roth, P.: How we treat glioblastoma. ESMO Open **4**, e000520 (2019)
3. An, Z., Aksoy, O., Zheng, T., Fan, Q.-W., Weiss, W.A.: Epidermal growth factor receptor and EGFRvIII in glioblastoma: signaling pathways and targeted therapies. Oncogene **37**, 1561–1575 (2018)
4. Noorani, I., Mischel, P.S., Swanton, C.: Leveraging extrachromosomal DNA to fine-tune trials of targeted therapy for glioblastoma: opportunities and challenges. Nat. Rev. Clin. Oncol. **19**, 733–743 (2022)
5. Jaunmuktane, Z., et al.: Methylation array profiling of adult brain tumours: diagnostic outcomes in a large, single centre. Acta Neuropathol. Commun. **7**, 24 (2019)
6. Ren, Q., et al.: IIB-MIL: Integrated instance-level and bag-level multiple instances learning with label disambiguation for pathological image analysis. In: Greenspan, H., et al. (eds.) International Conference on Medical Image Computing and Computer-Assisted Intervention, pp. 560–569 Cham: Springer Nature, Switzerland (2023). https://doi.org/10.1007/978-3-031-43987-2
7. Teichmann, M., Aichert, A., Bohnenberger, H., Ströbel, P., Heimann, T.: End-to-end learning for image-based detection of molecular alterations in digital pathology. In: Wang, L., Dou, Q., Fletcher, P. T., Speidel, S., Li, S. (eds.) Medical Image Computing and Computer Assisted Intervention – MICCAI 2022. MICCAI 2022. Lecture Notes in Computer Science, vol. 13432, pp. 88–98. Springer, Cham, Switzerland (2022)
8. Nasrallah, M.P., et al.: Machine learning for cryosection pathology predicts the 2021 WHO classification of glioma. Med **4**, 526-540.e4 (2023)
9. Dosovitskiy, A., et al.: An image is worth 16×16 words: Transformers for image recognition at scale (2020). arXiv preprint arXiv:2010.11929
10. Lu, M.Y., et al.: Data-efficient and weakly supervised computational pathology on whole-slide images. Nat Biomed Eng **5**, 555–570 (2021)
11. Zhang, H., et al.: DTFD-MIL: double-tier feature distillation multiple instance learning for histopathology whole slide image classification. In: 2022 IEEE/CVF Conference on Computer Vision and Pattern Recognition (CVPR), pp. 18780–18790. IEEE (2022). https://doi.org/10.1109/CVPR52688.2022.01824
12. Shao, Z., et al.: TransMIL: Transformer based Correlated Multiple Instance Learning for Whole Slide Image Classication. Neural Information Processing Systems (2021)
13. Lopez-Gines, C., et al.: New pattern of EGFR amplification in glioblastoma and the relationship of gene copy number with gene expression profile. Mod. Pathol. **23**, 856–865 (2010)
14. Tellez, D., Litjens, G., van der Laak, J., Ciompi, F.: Neural image compression for gigapixel histopathology image analysis. IEEE PAMI **43**, 567–578 (2021)
15. Campanella, G., et al.: Clinical-grade computational pathology using weakly supervised deep learning on whole slide images. Nat. Med. **25**, 1301–1309 (2019)
16. Otsu, N.: A threshold selection method from gray-level histograms. IEEE Trans. Syst. Man Cybern. **9**, 62–66 (1979)
17. He, K., Zhang, X., Ren, S., Sun, J.: Deep residual learning for image recognition. In: 2016 IEEE Conference on Computer Vision and Pattern Recognition (CVPR), pp. 770–778. IEEE (2016). https://doi.org/10.1109/CVPR.2016.90

18. Hashimoto, N., et al.: Multi-scale domain-adversarial multiple-instance CNN for cancer subtype classification with unannotated histopathological images. In: Proceedings of the Conference on Computer Vision and Pattern Recognition (IEEE/CVF), pp. 3852–3861. IEEE (2020). https://doi.org/10.1109/CVPR42600.2020.00391
19. Wang, S., et al.: RMDL: Recalibrated multi-instance deep learning for whole slide gastric image classification. Med. Image Anal. **58**, 101549 (2019)

An Exploration of Diabetic Foot Osteomyelitis X-ray Data for Deep Learning Applications

Brandon Abela[1](✉), Martin Masek[1], Jumana Abu-Khalaf[1], David Suter[1], and Ashu Gupta[2]

[1] Edith Cowan University, Joondalup, WA 6027, Australia
babela@our.ecu.edu.au, b.abela@ecu.edu.au
[2] Fiona Stanely Hospital, Murdoch, WA 6150, Australia

Abstract. The features of Diabetic FootOsteomyelitis (DFO) in medical X-rays are subtle. If more salient features are correlated with the DFO class, the model may become overdependent on these salient features and underperform when they are absent. In this paper, we identify DFO correlated features present in a DFO X-ray dataset that become a source of class differentiation for DL models. The dataset contains 882 dorsoplantar foot X-rays with DFO in the forefoot. To isolate distinct information in the original X-rays, six sub-image sets were created from the original X-ray images. A CNN model was then trained on each sub-image set. In order to identify the impact of correlated features, saliency maps were used in conjunction with the corresponding reports to identify which features of the image were activating the model during classification. The highest performing model achieved an accuracy of 76.8% which suggests the model is proficient at classifying DFO, however, the saliency map outputs indicate that classification is being assisted by features correlated with DFO, such as amputations and ulcers. The results of this investigation suggest that when developing X-ray datasets for DFO or other similar diseases, the effect of co-morbid abnormalities needs to be taken into consideration.

Keywords: Deep Learning · Osteomyelitis · Medical Imaging

1 Introduction

Approximately 1 in 10 adults, or 537 million people, are estimated to have diabetes [1]. Projections indicate that this will increase to 643 million by 2030 and increase further to 783 million by 2045 [1]. When left untreated, diabetes results in hyperglycemia (elevated blood glucose levels), which progressively impairs various internal systems, leading to a spectrum of health issues [2]. Among these complications, Diabetic Foot Osteomyelitis (DFO) (bone infection) is a significant concern and serves as the central focus of this paper.

This project was funded by the Future Health Research and Innovation Fund scheme through the Research Translation Projects 2021 Program.

J. Finkelstein et al. (Eds.): AIME 2024, LNAI 14845, pp. 30–39, 2024.
https://doi.org/10.1007/978-3-031-66535-6_4

DFO occurs when Diabetic Foot Ulcers (DFU) become infected, leading to exposure of underlying tissue [3]. This allows bacteria to penetrate deeper into the bone. If left untreated, the infected bone may need to be amputated which has been shown to subsequently increase patient mortality. A meta-analysis [4] found that across 28 studies and 17,325 subjects, mortality rates after minor amputations were: 3.5% at 1 month, 20% at 1 year, 28% at 3 years, 44.1% at 5 years, 51.3% at 6–7 years, and 58.5% at 8 to 9 years.

Detecting DFO features in an X-ray is challenging. Radiologists perform at an average sensitivity of 68.91% [5]. This difficulty is due in part to the subtle features and slow presentation, with features only becoming visible to radiologists 7–10 days after infection [6]. Improving detection of DFO from X-rays will improve patient outcomes since X-rays are quick, cheap, and accessible. These factors contribute to a prompt diagnosis which helps prevent severe complications [7]. A potential method of improving detection using X-rays is to leverage Deep Learning (DL) models to assist radiologists.

DL models have demonstrated effectiveness in detecting various health related concerns, such as fractures [8], brain tumors [9], and pneumonia [10]. When training DL models for a new task, the question of whether the results can be trusted should be explored. This is an important question to ask since generalization of performance despite a distribution shift is required for clinical adoption of such technology [11]. Distribution shifts primarily affect models that rely on input features indirectly related to the target class when making a prediction. For example, Winkler et al. [12] trained a model to detect skin melanoma but found their model was sensitive to surgical skins markers. Oakden-Rayner et al. [13] trained a model to detect pneumothorax but found that model performance relied on the presence of chest tubes. And Zech et al. [14] trained a model to detect pneumonia but found their model was able to separate the positive and negative class based on hospital imaging procedures.

Two methods for exploring how different factors within an image impact classification are ablation studies [15] and saliency maps [16]. These techniques allow for the analysis of the model's decision-making process. In DL research, ablation refers to selectively removing individual components from a model, such as specific layers, filters, or input features, for the purpose of assessing the impact of that component [15]. For example, a study [17] used deep learning models to predict clinical management of skin lesions using a combination of clinical images, thermoscopic images, and patient metadata. An ablation study was conducted to evaluate the contribution of each of these input components and found that patient metadata had the most impactful contribution to model performance.

In DL research, saliency maps can be used to visualize which regions of an input contributed most significantly to the final output [16]. These maps can be combined with existing knowledge about the data, for instance, the disease's location, to gain insights into whether the model is effectively focusing on relevant regions. For example, DeGrave et al. [18] used saliency maps to identify that their model, trained to detect Covid-19, had learned to associate lateral markers and shoulder positioning with the positive class. This was indicated by activation of these regions.

The present study employs both ablation and saliency map techniques to investigate the performance of a DL model classifying foot X-rays containing DFO in the forefoot. The goal is to determine whether this data is affected by potential confounding factors that should be considered in the curation of future models or X-ray datasets. The main contributions of the current study are:

1. The identification of potentially confounding features in DFO X-rays that may impact DL model performance.
2. An evaluation of a CNN model's performance when applied to a DFO image dataset that has been developed without considering co-morbid abnormalities.

2 Data

The X-ray images used in the current study were obtained from three hospitals located in the south metropolitan region of Perth, Australia. A total of 16,056 X-ray examinations that contained reports explicitly stating that the patient was diabetic, or queried osteomyelitis, were obtained.

From these 16,056 exams, those that report the presence of DFO in the forefoot were extracted resulting in 533 positive cases. The forefoot was selected due to it being the most common region for DFO. The study was further constrained to a single X-ray view per case, the dorsoplantar view. The dorsoplantar views were extracted from exams, where available, and screened to verify the appearance of the forefoot and its concordance with DICOM header information, resulting in 471 positive cases. With these images, a set of negative cases balanced on age and gender were obtained from the original 16,056 exams using the same method of filtering, resulting in a total of 942 X-ray images – 471 positives and 471 negatives. An example of the original X-ray can be found in Fig. 1.

Lastly, 250 of these images were used to train a UNet [19] model to segment the foot from the background. This is an important ablation due to the background containing orientation information (i.e. L, R) that may influence model performance. The UNet model was able to successfully segment the majority of the images, however, several images were not accurately segmented. After removing and rebalancing the data, the final data count was 882; 441 positive cases and 441 negative cases. All left foot X-rays were flipped along the vertical axis to ensure that orientation information did not affect model performance.

3 Methods

To investigate the influence of features present in DFO X-rays, multiple sub-image sets focusing on separate regions of the original X-rays were created. Each set was created to isolate specific features so that we could examine how those features influence model performance. The following sets were created:

1. **Complete Foot + B (with background):** This set is the original DICOM image padded along the shortest axis with 0's to create a square image. The images were then resized to 224x224 to match the dimension requirements of a standard VGG16 model. Padding was provided before resizing to retain image dimensions.
2. **Complete Foot -B (without background):** This set is obtained using the same process as the complete foot + B set, except this set used the UNet mask to convert the X-ray background into 0's.

3. **Forefoot + B:** This set is created by cropping the top half of the foot using the x and y coordinates of the UNet masks, padding the short axis, and resizing the images to 224x224. This set focuses on the regions with the features of interest (according to reports).
4. **Forefoot -B:** This set is obtained using the same process as the forefoot + B set, except this set used the UNet mask to convert the background to 0's.
5. **Hindfoot + B:** This set is obtained using the same process as the forefoot + B set except with the bottom half of the foot. This region of the foot does not contain any of the known DFO features and is created to see if there are any unknown patient characteristics that distinguish the two classes from each other.
6. **Background Patches:** This set is created by taking 64x64 crops from the background and resizing them to 224x224. These crops were taken from various regions of the background and do not contain any part of the foot. This subset was created to see if there are any differences in imaging settings when radiologists suspect DFO.

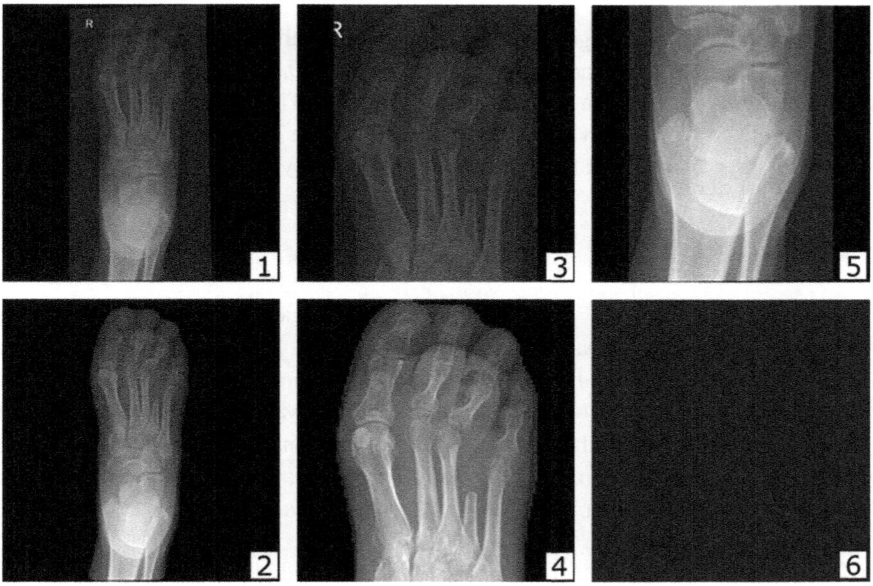

Fig. 1. An example for each image set. Each image has a corresponding description provided in the list above, indicated by the number label.

Each of these six sub-image sets were used to train a VGG16 model [20]. Each model was trained with a batch size of 8, a learning rate of 4×10^{-7}, a dropout of 0.5, patience of 50 epochs, and utilized the Adam optimizer [21]. For each sub-image set, a 10-fold cross-validation was performed with an 85/15 train/test split. The data splitting process used a stratified method, ensuring an even distribution of positive and negative samples in the test split and each of the 10 folds.

Next, reports for both the positive and negative exams were manually examined, and the most common abnormalities were counted and used to create a correlation matrix. This matrix was used to identify co-morbid abnormalities.

Lastly, the highest performing fold for the forefoot + B input was explored using saliency maps in conjunction with the X-rays corresponding reports. The reports indicate the location of abnormalities, where this information was used with the saliency maps to assess if the correct features were being activated during a prediction.

4 Evaluation

The corresponding model for each sub-image was evaluated using four metrics, accuracy, precision, recall, and F1. Precision is the quality of the correct predictions, recall is the percentage of positive cases correctly identified and is sometimes referred to as sensitivity, and F1 is the harmonic mean between precision and recall. The correlations in the correlation matrix were calculated using Pearson's r. A description of these metrics is provided in [22].

Grad-CAM was used to generate saliency maps for the qualitative evaluation [16]. Grad-CAM produces a saliency map overlay by first obtaining the output gradients for each of the feature maps of a specific layer of interest. Then a global average for each map is calculated which provides each feature map with a weight. A weighted sum of the feature maps is then used to produce the final saliency map.

5 Results

The performance of each sub-image is shown in Table 1 and Fig. 2. Both forefoot models achieved the highest performance with equivalent results. Since the features of DFO are in these sets, these results contrasted with the complete foot results suggest that focusing on the region of interest improves performance. The equivalency of both forefoot results also suggests that background information does not influence performance. However, for the complete foot, the background information provides a minor performance increase. This may be due to the orientation information being partly removed for the forefoot images, which can be seen in Fig. 1.

The model trained with the hindfoot images produced non-trivial results. This suggests that there are features within the hindfoot salient enough for a DL model to use as a source of class differentiation. The model trained with the background patches also produced non-trivial results suggesting a minor influence for imaging features. This also suggests the results for the hindfoot are primarily due to imaging features. This is discussed further in Sect. 6.

The most common abnormalities reported in the dataset have been presented in the correlation matrix in Fig. 3. Of these abnormalities, amputations, ulcers, and calcifications had the highest correlations with DFO. These correlations were 0.361 for amputations, 0.276 for ulcers, and -0.547 for calcifications. DFO's correlations can be used to understand the distribution of these abnormalities between the two classes, with a negative correlation indicating a higher proportion in the negative class, and vice versa.

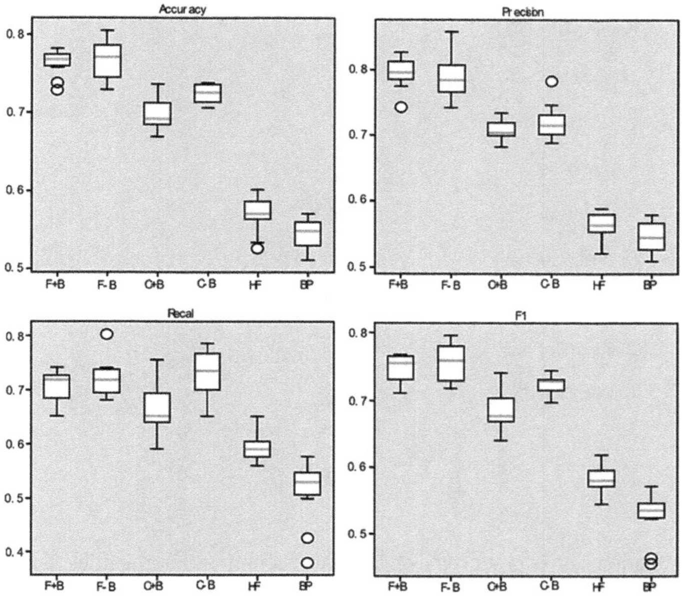

Fig. 2. Box plot figures showing the performance of each of the six image sets. White circles represent extreme fold values that deviate greatly from the other folds (outliers).

Table 1. A performance comparison of all image sets across all performance metrics. The forefoot images gave the highest performance which suggests that focusing on DFO features improves performance.

Input	Accuracy	Precision	Recall	F1
Forefoot -B	0.763 ± 0.017	0.795 ± 0.024	0.706 ± 0.033	0.746 ± 0.020
Forefoot + B	0.768 ± 0.025	0.788 ± 0.031	0.729 ± 0.041	0.757 ± 0.028
Complete Foot -B	0.698 ± 0.021	0.708 ± 0.014	0.667 ± 0.044	0.686 ± 0.029
Complete Foot + B	0.724 ± 0.011	0.720 ± 0.026	0.730 ± 0.043	0.724 ± 0.014
Hindfoot + B	0.569 ± 0.023	0.562 ± 0.022	0.598 ± 0.030	0.579 ± 0.022
Background Patch	0.545 ± 0.020	0.545 ± 0.023	0.511 ± 0.058	0.525 ± 0.036

For the forefoot -B model, the correlations between co-morbid abnormalities and positive predictions were calculated. The most notable finding from this was that the models' positive predictions had a greater correlation with amputations ($r = 0.539$) than is observed in the ground truth ($r = 0.361$). Amputations were also found to have the same correlation with positive predictions as DFO (amputations $r = 0.539$ vs DFO $r = 0.538$). The correlation of ulcers is equivalent for model predictions ($r = 0.289$) and the ground truth ($r = 0.276$).

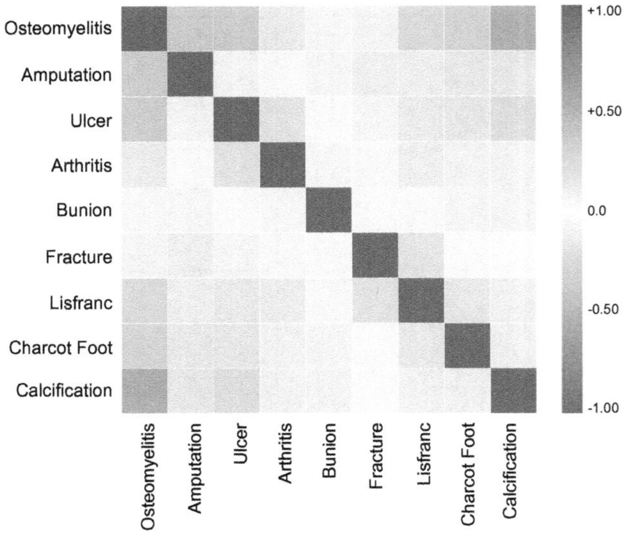

Fig. 3. A correlation matrix produced using the most common abnormalities found in the current study's diabetic foot X-ray dataset.

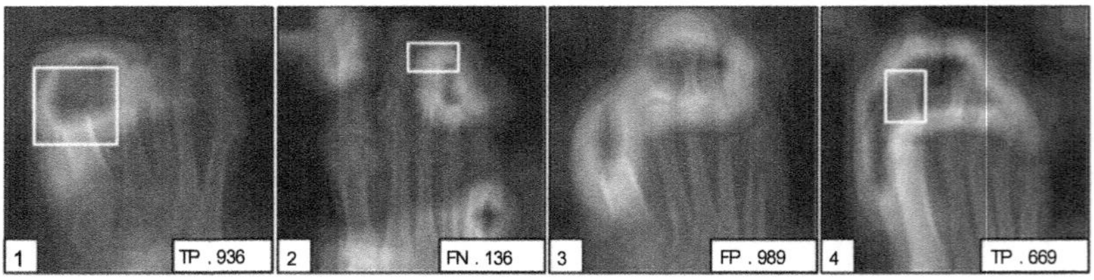

Fig. 4. Salience maps for an individual forefoot -B model fold. The white boxes indicate DFO in DFO positive examples.

Four saliency maps from an individual forefoot -B cross validation fold have been provided in Fig. 4. Saliency map 1 is the true positive with the highest confidence (0.936). This saliency map appears to correctly localize the region with DFO; however, a closer look shows activation primarily in the region of the missing toe.

Saliency map 2 is the false negative with the highest confidence (0.136). Various regions are being activated including the soft tissue ulcer beside the DFO. Saliency map 3 is the false positive with highest confidence (0.989) and activates across the top of the digits with an amputation. Lastly, saliency map 4 shows a positive classification with no amputations, which correctly activates the section with periosteal reaction, a symptom of DFO, however this region is also experiencing soft tissue swelling along the red activation region due to an ulcer. The confidence of this positive prediction (0.669) is noticeably lower than saliency map 1 and 3 which contain amputations.

6 Discussion and Conclusion

The current study investigated the features of a DFO X-ray dataset using a trained VGG16 model combined with ablation studies and Grad-CAM analysis. The key focus of the study was to identify what features are most influential in the classification of DFO.

We found that by cropping out the forefoot (region with DFO) and using it as input, model performance was notably enhanced as reported in Table 1, which suggests the model is using DFO features when classifying X-rays. However, the qualitative evaluation showed high activation for ulcers and amputations which suggests that the model may be engaging in shortcut learning using these abnormalities. The results of the correlation analysis also showed amputations and ulcers as the abnormalities most correlated with DFO, which is consistent with the qualitative analysis. Furthermore, when each abnormalities correlation is calculated for model predictions, amputations were found to be just as correlated with the positive predictions as DFO (amputations $r = 0.539$ vs DFO $r = 0.538$) and 0.178 higher than the correlation observed in the dataset. Overall, these results suggest the model is engaging in shortcut learning using amputations, and to a lesser extent ulcers. What this means is that while performance using the standard evaluation procedure indicates a performant model, the model is expected to fail to maintain its performance in a clinical setting due to a shift in the feature distribution disallowing the effective use of learned shortcuts.

To address the potential for imaging procedures to effect model performance, we investigated whether the model could separate the classes based on crops of the background. The results indicated that the model could not use imaging procedures to differentiate classes. The rationale for using background crops was that we would be able to evaluate imaging features separate from patient features. However, background crops do not contain many of the feature vectors through which imaging features are expressed. For example, adjustment in kilovolt peak increases penetration of material (i.e. bone) [23], which can only be accurately expressed when there is material to penetrate. A comparison of the results obtained for the background crops and hindfoot input shows a small increase in performance for the hindfoot. Though minor, this increase could be due to imaging features having a better medium through which they can be expressed. It is also possible that this minor increase is the result of some undefined patient features.

The findings of the current study suggest that when developing X-ray datasets for DFO or other similar diseases, the effect of co-morbid abnormalities needs to be taken into consideration. Specifically, abnormalities that present a large shift in X-ray appearance, since these likely overshadow the changes seen in more subtle diseases, such as DFO. This could be achieved through pair matching, however, matching all potential confounders is very limiting in a domain where data is already limited [24]. Therefore, future work should focus on developing a process through which the influence of identified confounders is confirmed. This would allow matching to be constrained to confounders confirmed to affect model performance. Another potential solution is to take a patch-based approach, where a dataset is created using patches of bone infected with DFO. This would remove the influence of amputations and other abnormalities that could confound model performance.

References

1. Saeedi, P., et al.: Global and regional diabetes prevalence estimates for 2019 and projections for 2030 and 2045: Results from the International Diabetes Federation Diabetes Atlas, 9th edition. Diabetes Research and Clinical Practice. 157, 107843 (2019)
2. Amiel, S.A.: The consequences of hypoglycaemia. Diabetologia **64**, 963–970 (2021)
3. Derebaşınlıoğlu, H., Aksoy, O.: Evaluation of osteomyelitis in diabetic foot ulcers with exposed bone. Wound Practice and Research. 31 (2023)
4. Yammine, K., Hayek, F., Assi, C.: A meta-analysis of mortality after minor amputation among patients with diabetes and/or peripheral vascular disease. J. Vasc. Surg. **72**, 2197–2207 (2020)
5. Llewellyn, A., Kraft, J., Holton, C., Harden, M., Simmonds, M.: Imaging for detection of osteomyelitis in people with diabetic foot ulcers: a systematic review and meta-analysis. Eur. J. Radiol. **131**, 109215 (2020)
6. Leone, A., et al.: The role of serial radiographs to diagnose diabetic foot bone infection. Mediterranean Journal of Hematology and Infectious Diseases. 14 (2022)
7. Matheson, E.M., Bragg, S.W., Blackwelder, R.S.: Diabetes-related foot infections: diagnosis and treatment. Am. Fam. Physician **104**, 386–394 (2021)
8. Inoue, T., et al.: Automated fracture screening using an object detection algorithm on whole-body trauma computed tomography. Scientific Reports. 12 (2022)
9. Khan, M., et al.: Accurate brain tumor detection using deep convolutional neural network. Comput. Struct. Biotechnol. J. **20**, 4733–4745 (2022)
10. Szepesi, P., Szilágyi, L.: Detection of pneumonia using convolutional neural networks and Deep Learning. Biocybernetics Biomed. Eng. **42**, 1012–1022 (2022)
11. Kelly, C.J., Karthikesalingam, A., Suleyman, M., Corrado, G., King, D.: Key challenges for delivering clinical impact with Artificial Intelligence. BMC Med. **17** (2019)
12. Winkler, J.K., et al.: Association between surgical skin markings in dermoscopic images and diagnostic performance of a deep learning convolutional neural network for melanoma recognition. JAMA Dermatol. **155**, 1135 (2019)
13. Oakden-Rayner, L., Dunnmon, J., Carneiro, G., Re, C.: Hidden stratification causes clinically meaningful failures in machine learning for medical imaging. In: Proceedings of the ACM Conference on Health, Inference, and Learning (2020)
14. Zech, J.R., Badgeley, M.A., Liu, M., Costa, A.B., Titano, J.J., Oermann, E.K.: Variable generalization performance of a deep learning model to detect pneumonia in chest radiographs: A cross-sectional study. PLOS Med. **15** (2018)
15. Sheikholeslami, S., Meister, M., Wang, T., Payberah, A.H., Vlassov, V., Dowling, J.: AutoAblation. In: Proceedings of the 1st Workshop on Machine Learning and Systems (2021)
16. Selvaraju, R.R., Cogswell, M., Das, A., Vedantam, R., Parikh, D., Batra, D.: Grad-cam: Visual explanations from deep networks via gradient-based localization. In: 2017 IEEE International Conference on Computer Vision (ICCV) (2017)
17. Abhishek, K., Kawahara, J., Hamarneh, G.: Predicting the clinical management of skin lesions using deep learning. Sci. Rep. **11** (2021)
18. DeGrave, A.J., Janizek, J.D., Lee, S.-I.: AI for radiographic COVID-19 detection selects shortcuts over signal. Nature Mach. Intell. **3**, 610–619 (2021)
19. Ronneberger, O., Fischer, P., Brox, T.: U-Net: Convolutional Networks for Biomedical Image Segmentation. Lecture Notes in Computer Science, pp. 234–241 (2015)
20. Simonyan, K., Zisserman, A.: Very Deep Convolutional Networks for Large-Scale Image Recognition. arXiv (2014)
21. Kingma, D.P., Ba, J.: Adam: A Method for Stochastic Optimization. arXiv (2014)
22. Pagano, M., Gauvreau, K.: Principles of Biostatistics. Duxbury, Pacific Grove, CA (2000)

23. Samei, E., Dobbins, J.T., Lo, J.Y., Tornai, M.P.: A framework for optimising the radiographic technique in digital X-ray imaging. Radiat. Prot. Dosimetry. **114**, 220–229 (2005)
24. Chae, J., Kim, J.: An investigation of transfer learning approaches to overcome limited labeled data in medical image analysis. Appl. Sci. **13**, 8671 (2023)

Automated Detection and Characterization of Small Cell Lung Cancer Liver Metastases on CT

Sophia Ty[1], Fahmida Haque[1], Parth Desai[2], Nobuyuki Takahashi[2,3], Usamah Chaudhary[1], Benjamin Simon[1], Peter L. Choyke[1], Anish Thomas[2], Baris Turkbey[1], and Stephanie A. Harmon[1](✉)

[1] Artificial Intelligence Resource, NCI, NIH, Bethesda, USA
stephanie.harmon@nih.gov
[2] Developmental Therapeutics Branch, NCI, NIH, Bethesda, USA
[3] Department of Medical Oncology, National Cancer Center Hospital East, Kashiwa, Japan

Abstract. Small cell lung cancer patients with metastatic disease often present with liver lesions that can be detected in computed tomography (CT). These lesions have highly diverse phenotypes resulting from heterogenous expression of various neuroendocrine (NE) genes. In this work, we present a two-step machine learning framework to automatically detect liver lesions in CT scans using a 3D segmentation model followed by radiomics-based analysis to stratify the patients based on their NE tumor profile. Our liver lesion detection model achieved a lesion sensitivity greater than 66% and up to 83% for two datasets. The radiomics-based neuroendocrine phenotype classifier can distinguish patients based on their NE expression profile (NE vs. non-NE) and achieved an F1 score of 85%.

Keywords: Small cell lung cancer · gene expression · tumor classification

1 Introduction

Small cell lung cancer (SCLC) is a highly aggressive form of lung cancer strongly associated with smoking. Lung cancer as a whole presents a major global burden and is the leading cause of cancer-related death in the United States [1], with SCLC accounting for 13% of cases [2]. SCLC is marked by ubiquitous presence of TP53, RB1 loss of function events, and high chromosomal instability [3, 4] that lead to its highly metastatic nature [5]. It commonly spreads to numerous locations: hepatic metastases are seen in 21–27% patients at presentation and 69% at autopsy [6], making the liver the most common site of metastases after mediastinal lymph nodes. Liver metastasis is also an independent marker of poor prognosis [7, 8]. Thus, automated identification of SCLC lesions on liver imaging provides a great deal of utility from diagnosis to treatment management. SCLC demonstrates a high level of heterogeneity, manifesting under various transcriptional subgroupings with therapeutic implications [9]. Underscoring these transcriptional groups is the heterogeneous expression of various neuroendocrine (NE)

J. Finkelstein et al. (Eds.): AIME 2024, LNAI 14845, pp. 40–44, 2024.
https://doi.org/10.1007/978-3-031-66535-6_5

genes which can further sub-stratify patients based on NE or non-NE tumor presentation [10, 11]. Herein, we propose a machine learning framework for the automated segmentation of the liver and cancerous lesions followed by classification of molecular subtypes on computed tomography (CT) scans. We first implemented a 3D segmentation model able to simultaneously detect the liver and lesions within the organ. Then, a radiomics-based classifier is used to distinguish patients' tumor profile as positive or negative for harboring NE-like phenotype based on expression profiling.

2 Methods

2.1 Dataset

Our work utilizes two datasets: (1) the Liver Tumor Segmentation Challenge (LiTS), a public dataset containing 131 CT scans [12], henceforth referred to as the LiTS dataset, and (2) an 86 patient CT dataset retrospectively obtained from patients at our local institution, henceforth referred to as the mSCLC dataset. The LiTS dataset included expert liver and lesion segmentations with both primary and secondary liver tumors. Scans were obtained from multiple clinical centers around the globe. This research study was conducted retrospectively using human subject data made available in open access by [12]. Ethical approval was not required as confirmed by the license attached with the open access data.

The mSCLC dataset contains 52 patients and 233 scans with CT-visible liver lesions confirmed through radiology reports and successfully matched to gene-based NE scores. Within these, 50 scans underwent ground truth lesion segmentation by an experienced radiologist (>10 yrs) using ITK-Snap, representing at least one scan per patient for those with biopsy confirmed SCLC that can be annotated manually. Remaining scans were labeled based on outputs from the segmentation algorithm. This single institution study was approved by the local institutional review boards from the cohort (ClinicalTrials.gov Identifier: NCT02769962, NCT03554473, NCT02487095, NCT02484404, NCT02146170). Written informed consent was obtained from all patients. All patients underwent tissue (FFPE) or blood sampling (cfDNA) for RNA-Seq and WES profiling. Each scanning timepoint was matched to the nearest profiling date to determine NE vs non-NE phenotype groups. Methods for the score-based gene signatures have been described previously [11].

2.2 Segmentation Algorithm

Liver lesions in CT images were segmented using three different models: (1) a 3D UNet, (2) a 3D SegResNet, and (3) a 3D nnUNet model. The UNet and SegResNet models were built using the Medical Open Network for AI (MONAI) platform (version 1.3.0), a PyTorch-based open-source framework for deep learning in medical imaging [13]. Preprocessing and hyperparameter selection were performed to identify the best configuration for each model. The following data transformations were applied during training: foreground cropping, variable CT windowing, and random cropping by labels with a sampling ratio of 4:1:3 for the background, liver and lesion with 12 samples taken per image. Both models were trained using the LiTS dataset (train/dev/test split of 104/17/10) with a learning rate of $1e^{-4}$, optimized using an adaptive moment estimator and lesion level DICE loss for 1500 epochs; Inference was conducted using a sliding window technique. The 3D nnUNet, a self-configuring deep learning model, was implemented using the built-in *3d_fullres* five-fold cross-validation [14]. For this regimen, the LiTS training (n = 104) and development (n = 17) sets were merged. All models were trained solely using LiTS; Model performance was assessed on independent test sets from both data sources (LiTS: n = 10, mSCLC: n = 50). The best model architecture was selected and retrained using combined data with LiTS splits conserved and mSCLC stratified at the patient level (train-dev/test: 65/21). Lesion annotations from this model was used for subsequent radiomics analysis and tumor phenotype classification.

2.3 Classification Algorithm

SCLC metastatic liver lesions were characterized by extracting qualitative imaging features from lesion contours provided by the segmentation model. CT image feature extraction was performed using PyRadiomics (v3.0.1) resulting in 107 features representing first-order statistics, shape (2D and 3D), and texture. The number of liver lesions for each scan was determined using connected-components-3d (v3.12.1) and served as an additional image feature providing a total of 108 features. A subset of imaging characteristics correlated with tumor phenotype were selected using LASSO regression (scikit-learn v1.2.2) on mSCLC scans with matched gene-based NE scores and corresponding lesion contours successfully obtained from the segmentation model (n = 227).

Radiomics-based stratification of tumor phenotype was conducted using three classification algorithms: (1) Logistic Regression (scikit-learn v1.2.2), (2) Random Forest (scikit-learn v1.2.2), and (3) XGBoost (v2.0.3). All models incorporated the selected subset of imaging features derived from LASSO regression as explanatory variables for binary classification of tumors as NE or non-NE phenotype. Each classifier was trained with and without class-based weights derived from the data distribution of the training set. All classifiers were evaluated using a hold-out test with a training/test set split of 177/50. Model accuracy, recall, precision, and F1 scores were calculated and compared across all models to assess performance. The best model was selected using the F1 score as the primary performance criterion.

3 Results and Discussion

3.1 Liver Lesion Detection

Among the three segmentation models implemented for liver lesion detection, the nnUNet provided the most robust and accurate method for lesion identification. When assessed on independent test groups from each dataset, the nnUNet model outperformed the UNet and SegResNet models. The *3d_fullres* nnUNet model had a median lesion DICE of 0.607 and 0.750, lesion sensitivity of 0.539 and 0.813, and PPV of 0.99 and 1.0 for the mSCLC and LiTS test sets respectively. For both cohorts, the nnUNet model has 0–1 false positives per scan. For a complete comparison of model performance, see

Table 1. Upon selection of the nnUNet as the best segmentation model, a *3d_fullres* nnUNet was retrained with combined LiTS and mSCLC dataset. This model achieved a median DICE of 0.771 and 0.640 with sensitivities of 0.826 and 0.667 for the LiTS test set (n = 10) and mSCLC test set (n = 12) respectively. For both datasets, the model had a PPV of 1.0 and 0 false positives per scan. For all mSCLC scans matched with gene-based NE scores with confirmed liver lesions, the nnUNet model provided automated segmentations. Of these scans (n = 233), the nnUNet model correctly detected the presence of tumors for 227 cases.

Table 1. Summary of Lesion Segmentation Model Performance.

	UNet		SegResNet		nnUNet	
Metric	LiTS	mSCLC	LiTS	mSCLC	LiTS	mSCLC
Median DICE	0.439	0.297	0.417	0.418	0.750	0.607
Sensitivity	0.395	0.244	0.662	0.447	0.813	0.539
PPV	0.971	0.975	0.962	0.987	1.0	0.998
FP/scan	0.1 (0–1)	0.1 (0–1)	0.2 (0–1)	0.1 (0–1)	0	0.02 (0–1)

3.2 SCLC Liver Lesion Characterization and Phenotype Classification

The LASSO regression identified twenty-seven imaging features correlated with NE-related tumor phenotypes ($R^2 = 0.43$). Among these, shape characteristics account for nine out of the twenty-seven correlated features, appearing to be the most dominant radiomic feature subtype found in the subset. The top five imaging characteristics determined during feature selection were: major axis length, number of lesions, minor axis length, maximum 2D diameter row, and the maximum 3D diameter. Among all the models, the random forest classifier trained with class-based weights had the best performance with an F1 Score of 85%. A summary of the classification performance from all models is presented in Table 2.

Table 2. Summary of mSCLC NE Phenotype Classification.

Training	Classifier	Accuracy	Precision	Recall	F1 Score
Unweighted Classes	Logistic Regression	0.72	0.76	0.92	0.83
	Random Forest	0.74	0.80	0.89	0.83
	XGBoost	0.72	0.80	0.84	0.82
Weighted Classes	Logistic Regression	0.70	0.77	0.87	0.81
	Random Forest	0.76	0.79	0.92	0.85
	XGBoost	0.72	0.83	0.78	0.81

4 Conclusion

Molecular subtypes of SCLC metastatic liver lesions can be identified using a two-step machine learning framework that automatically segments liver lesions and subsequently classifies the tumor using radiomics features associated with gene-based neuroendocrine expression profiles. This approach offers a pathway to determine NE transcriptional subgroups from CT images of metastatic liver lesions and may support physicians during clinical decision-making and treatment planning.

Disclosure of Interests. All authors have no competing interests to declare.

References

1. Horner, M.J., et al.: SEER Cancer Statistics Review, 1975–2006, National Cancer Institute, Bethesda, MD. https://seer.cancer.gov/csr/1975_2006/, based on November 2008 SEER data submission, posted to the SEER web site (2009)
2. Van Meerbeeck, J.P., Fennell, D.A., De Ruysscher, D.K.: Small-cell lung cancer. The Lancet **378**(9804), 1741–1755 (2011)
3. Rudin, C.M., Brambilla, E., Faivre-Finn, C., Sage, J.: Small-cell lung cancer. Nat. Rev. Dis. Primer **7**(1), 3 (2021)
4. George, J., et al.: Comprehensive genomic profiles of small cell lung cancer. Nature **524**(7563), 47–53 (2015)
5. Sher, T., Dy, G.K., Adjei, A.A.: Small cell lung cancer. Mayo Clin. Proc. **83**(3), 355–367 (2008)
6. Jackman, D.M., Johnson, B.E.: Small-cell lung cancer. Lancet **366**(9494), 1385–1396 (2005)
7. Cai, H., Wang, H., Li, Z., Lin, J., Yu, J.: The prognostic analysis of different metastatic patterns in extensive-stage small-cell lung cancer patients: a large population-based study. Future Oncol. **14**(14), 1397–1407 (2018)
8. Kagohashi, K., Satoh, H., Ishikawa, H., Ohtsuka, M., Sekizawa, K.: Liver metastasis at the time of initial diagnosis of lung cancer. Med. Oncol. Northwood Lond. Engl. **20**(1), 25–28 (2003)
9. Rudin, C.M., et al.: Molecular subtypes of small cell lung cancer: a synthesis of human and mouse model data. Nat. Rev. Cancer **19**(5), 289–297 (2019)
10. Schroeder, B.A., Thomas, A.: SCLC subtypes and biomarkers of the transformative immunotherapy responses. J. Thorac. Oncol. **18**(9), 1114–1117 (2023)
11. Lissa, D., et al.: Heterogeneity of neuroendocrine transcriptional states in metastatic small cell lung cancers and patient-derived models. Nat. Commun. **13**(1), 2023 (2022)
12. Bilic, P., et al.: The Liver Tumor Segmentation Benchmark (LiTS). Med. Image Anal. **84**, 102680 (2023)
13. "Medical Open Network for Artificial Intelligence (MONAI)". https://monai.io/. Accessed 11 March 2024
14. Isensee, F., Jaeger, P.F., Kohl, S.A.A., Petersen, J., Maier-Hein, K.H.: NnU-Net: a self-configuring method for deep learning-based biomedical image segmentation. Nat. Methods **18**(2), 203–211 (2021)

Content-Based Medical Image Retrieval for Medical Radiology Images

Dario Barać[1], Teo Manojlović[1,2], Mateja Napravnik[1], Franko Hržić[1,2], Mihaela Mamula Saračević[3], Damir Miletić[4], and Ivan Štajduhar[1,2](✉)

[1] Faculty of Engineering, University of Rijeka, 51000 Rijeka, Croatia
ivan.stajduhar@uniri.hr
[2] Center for Artificial Intelligence and Cybersecurity, University of Rijeka, 51000 Rijeka, Croatia
[3] Medico Special Hospital, 51000 Rijeka, Croatia
[4] Clinical Hospital Centre Rijeka, 51000 Rijeka, Croatia

Abstract. In this study, a CBIR system for medical radiology images of various anatomical regions and imaging modalities was designed and implemented. Three descriptor-extraction retrieval methods were evaluated: BoVW, HOG, and CNN. Optimal configuration for each method was selected empirically. We used a dataset consisting of 38,000 2D images, originating from CHC Rijeka, Croatia. Retrieval with ImageNet-pretrained CNNs outperformed both BoVW and HOG retrieval. Re-ranking retrieved images using a different method improved results in almost all cases. Out of all methods compared, CNN retrieval (ShuffleNet combined with average pooling) with HOG re-ranking performed best. Our experiments can be recreated using the code available at https://github.com/dbarac/cbmir.

Keywords: Information Retrieval · Re-ranking · Content-Based Image Retrieval · CBIR · Medical Radiology Images · CBMIR · Feature Extraction

1 Introduction

Finding relevant images in large image databases, or image retrieval (IR) can be challenging. The traditional approach is searching based on keywords. This requires manual image annotation, which is time-consuming. Content-based IR (CBIR) is an alternative approach which attempts to index images automatically using visual features such as image texture, shape or colour distribution. Relevant images are retrieved by extracting visual features from a given query image and comparing them with the features of images in the database. The scope of CBIR systems (their purpose and image domain), effects of sensory and semantic gaps, and options for each of the necessary computational steps (image processing, feature extraction, similarity measurement) are discussed in [26].

This paper describes the design of a content-based medical IR (CBMIR) system, that can be used for assistance in medical diagnosis, education and research.

J. Finkelstein et al. (Eds.): AIME 2024, LNAI 14845, pp. 45–59, 2024.
https://doi.org/10.1007/978-3-031-66535-6_6

The results suggest that convolutional neural networks (CNNs) pretrained on ImageNet can be used as feature extractors for CBMIR without additional fine-tuning. Moreover, we show that retrieval result re-ranking can slightly improve overall CBMIR performance.

2 Related Work

A review of approaches for large scale medical IR is made in [13]. The paper describes challenges (system efficiency and accuracy), opportunities in the medical field (computer-aided diagnostics, visual pattern exploration) existing applications, datasets and methods for feature extraction, indexing and search. According to the paper, many medical IR systems use hand-crafted local feature descriptors. Scale-invariant feature transform (SIFT) [14], speeded up robust features (SURF) [3], and local binary patterns (LBPs) [19] are frequently used, often with bag-of-visual-words (BoVW) [25] models for feature quantisation. Global feature descriptors such as GIST [20], HOG [4] or colour histograms are also used for medical IR. The alternative to hand-crafting features is learning useful feature representations directly from the dataset. For example, a convolutional stacked autoencoder has been used for feature learning for brain MR image registration [27].

To avoid comparing the query image feature vector with all feature vectors in the database, hierarchical clustering and vocabulary trees [18] have been used in large scale medical IR systems to improve retrieval efficiency. Another method used for image indexing is feature hashing: feature vectors are hashed into short binary codes, which can be efficiently compared (for finding the nearest neighbours of an image) by calculating the Hamming distance. The hashing functions can be data-independent [10] or learned on a given set of images [6]. The dimensionality of large feature vectors can also be reduced with principal component analysis (PCA), or by pruning redundant words from visual vocabularies with probabilistic latent semantic analysis (PLSA) [5]. After finding the nearest neighbours of the query image in the database, re-ranking (RR) the retrieved images can lead to more accurate results. For example RR can be done by comparing images using different types of features, or by asking the user for relevance feedback. For evaluation of systems, metrics such as precision, recall and mean average precision (mAP) are often used.

For texture features, Fourier transforms, wavelets and Gabor filters have been used in medical radiology applications [1]. Edges, contours and polylines have been used to describe shape features. In [28], the authors used Radon features and barcodes in combination with support vector machines (SVMs) for X-ray IR. In [16], Gabor Wavelets, Gray Level Co-occurrence Matrix (GLCM), Radon Transforms, Orientation Histograms and Fourier Descriptors were used on regions of interest selected by radiologists for the retrieval of spine vertebrae irregularity images. Many CBMIR systems created before 2011 are mentioned in [1].

3 Materials and Methods

We investigate developing a CBMIR system for radiology images. The purpose of the system is category search, i.e. the system should retrieve images from the database which belong to the same class as the query image, w.r.t. the anatomical region in the image (e.g. chest images), the imaging modality (e.g. MR images) or both (e.g. MR images of the head).

The system should take as its input a single query image and the value for the parameter k, which is the number of images the system should retrieve from the database and return to the user. After comparing the query image with the images stored in the system's database by using a selected distance measure for specific feature descriptor types (choices explained in more detail in Appendix B), the system should return the top-k most similar images to the user.

3.1 Dataset

The dataset used in this study consists of 38,000 medical radiology images of 24 anatomical regions and 6 imaging modalities, sampled from a wider data collection which was gathered retrospectively in 2017 from the CHC Rijeka's picture archiving and communication system (PACS) [15,17]. 30,000 images are used for hyperparameter tuning and model training, and the remaining 8,000 images are used for testing. All images are 8-bit grayscale, and have a fixed size of 256×256 px. Other relevant dataset characteristics are briefly explained in Appendix A.

3.2 System Description

We implemented three CBMIR methods: a BoVW model [25] with SIFT descriptors [14], HOG [4], and CNNs [11]. Because these pipelines are fairly standard, for brevity we omit their detailed description from the main text of this paper, and refer the reader to Appendix B for implementation details.

In general (regardless of the method used), the system operates in two phases: offline and online. The offline phase is necessary for extracting and storing latent representations of images from the training subset (the set of images which can be retrieved). In the online phase, the user provides a query image and the system returns similar images. Optionally, retrieved images can be re-ranked using another method. An illustration of both phases is shown in Fig. 1.

The dataset image descriptors computed in the offline phase are loaded into memory at run-time. In the online phase, after computing the query image descriptor, the distance between the query descriptor and all loaded dataset image descriptors is computed. After sorting the dataset images by distance, the top-k most similar image IDs are returned to the user. Since the number of images in the training subset is not too large (30,000), it is possible to do this under 1 second. For larger datasets, more advanced indexing techniques should be used.

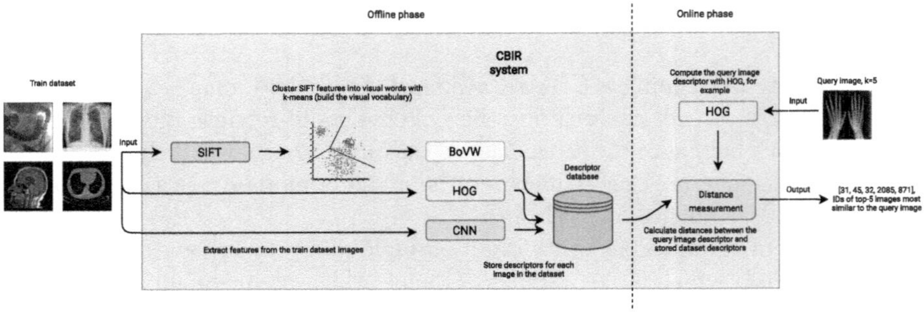

Fig. 1. Our CBMIR system operates in two phases: offline and online. In the offline phase, BoVW, HOG and CNN descriptors for the training images are extracted and stored. Before computing BoVW descriptors, local SIFT features are extracted and clustered using k-means for building the visual vocabulary. In the online phase, the user provides a query image and the system returns similar images. Images are retrieved by measuring the distance between the computed query image descriptor (BoVW, HOG or CNN) and the stored dataset image descriptors.

To find appropriate algorithm hyperparameters for BoVW, HOG, and CNN selection, and compare model performance, a stratified subset of the training data was used for validation. This validation subset consists of 20% of images of each anatomical region (5,991 in total). The validation images were not processed in the offline stage (their descriptors were not stored). We cover the hyperparameter selection in more detail in Sect. 4.1.

3.3 Evaluation Metrics

For evaluating and visually comparing the performance between different methods and algorithm parameters, precision at k ($P@k$) for $k \in \{1, \dots, 10\}$ was calculated by dividing the number of relevant images retrieved for a specific query image by the number of retrieved images k. A retrieved image is said to be relevant if both the imaging modality and the anatomical region of the query image and the retrieved image are the same (as an alternative, only the anatomical region or the imaging modality can be checked). Finally, we calculate the mean precision at k by calculating the $P@k$ for each query image and divide it with the number of query images.

To summarise the scores for multiple $k \in \{1, \dots, k_{max}\}$, mAP was calculated as described in [13].

4 Results and Discussion

Variable names used in this section are self explanatory. We refer the reader interested in a more detailed explanation of the IR pipelines to Appendix B for implementation details.

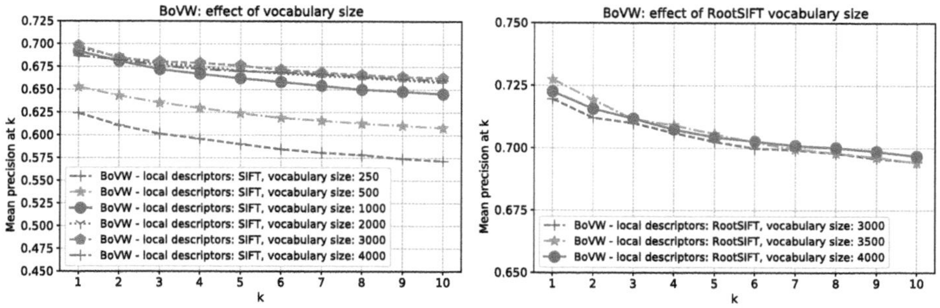

Fig. 2. Performance comparison for BoVW vocabulary sizes: (left) Size over 3, 000 did not improve the performance for BoVW with SIFT descriptors. (right) For smaller k-s, $n_{words} = 3,500$ had the best performance. For larger k-s, $n_{words} = 4,000$ was the best.

4.1 Hyperparameter Selection for Feature Extraction

BoVW. A total of 4, 152, 087 SIFT descriptors were extracted from the training subset. Appropriate visual vocabulary size for BoVW was determined empirically by evaluating the model on the validation subset. Vocabulary sizes of 250, 500, 1, 000, 2, 000, 3, 000, and 4, 000 were compared. For vocabulary learning, k-means was run for 50 iterations on the SIFT descriptors, with 5 retries for every vocabulary size. Increasing the number of iterations did not result in a performance improvement. The comparison can be found in Fig. 2. Vocabulary size (n_{words}) 3, 000 performed best.

After that, RootSIFT [2], a simple modification of SIFT descriptors was tested using the chosen vocabulary size (3, 000). For RootSIFT, the SIFT descriptors are modified first by L_1-normalising them, and then taking the square root for each element in the descriptor. For $n_{words} = 3,000$, using RootSIFT descriptors resulted in a 2.2% performance increase for mean P@1. Since using RootSIFT improved the performance, we tested additional vocabulary sizes (3, 000, 3, 500 and 4, 000). The results are shown in Fig. 2.

HOG. For HOG retrieval, it is necessary to choose values for $size_{cell}$, $size_{block}$, $stride_{block}$, and n_{bins} (the number of bins in each cell histogram). First, we tested 6 combinations of cell size, block size and block stride (overlapping and non-overlapping), with a fixed $n_{bins} = 9$. Best performance was attained for $size_{cell} = 64$, $size_{block} = 128$, and $stride_{block} = 64$ (overlapping blocks). Next, the effect of changing n_{bins} was tested. The results are shown in Fig. 3.

CNN Retrieval. IR performance from CNN-extracted descriptors was compared for: ResNet18 [7], ResNet34 [7], AlexNet [9], SqueezeNet [8] and VGG16 [23]. The models were pretrained on ImageNet, and require 3-channel images as input, so the grayscale channels were stacked before input. SqueezeNet

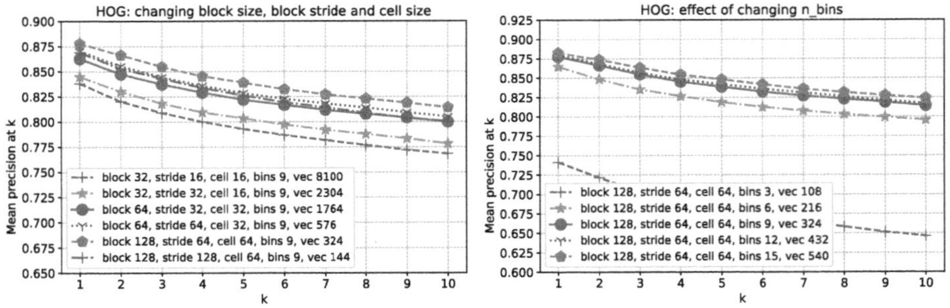

Fig. 3. Performance comparison for different HOG hyperparameter values: (left) Block size, block stride, and cell size, for a fixed $n_{bins} = 9$. HOG descriptor dimensionality is stated in the legend (*vec*). Best performance attained for $size_{cell} = 64$, $size_{block} = 128$, and $stride_{block} = 64$ (overlapping blocks). (right) Histogram sizes. Best performance attained for $n_{bins} = 15$.

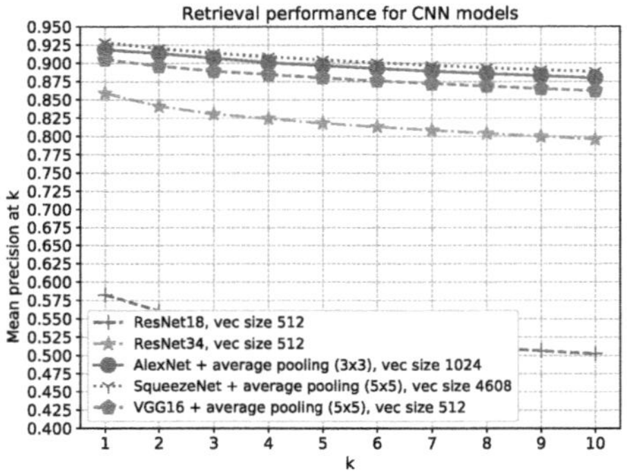

Fig. 4. Performance comparison for different CNN architectures. Output of the final convolutional (or pooling) layer is used as the descriptor. Best performance was attained for SqueezeNet, having a 5×5 average pooling layer on top.

– with a 5×5 average pooling layer to reduce output dimensionality – performed best. The results are shown in Fig. 4.

4.2 Retrieval Result Re-ranking

Images retrieved with one method (e.g. BoVW) can be re-ranked with another method (e.g. HOG). This is done by first retrieving k_{prep} images with the first method, sorting the retrieved images according to the distance of descriptors computed with the second method, and finally returning the top-k results with the lowest distance to the query image descriptor. Result RR was tested for various combinations of IR and RR methods, and k_{prep} values. The results are shown in Fig. 5. In most cases, RR improved model performance. Performance improvement was the greatest for IR with BoVW.

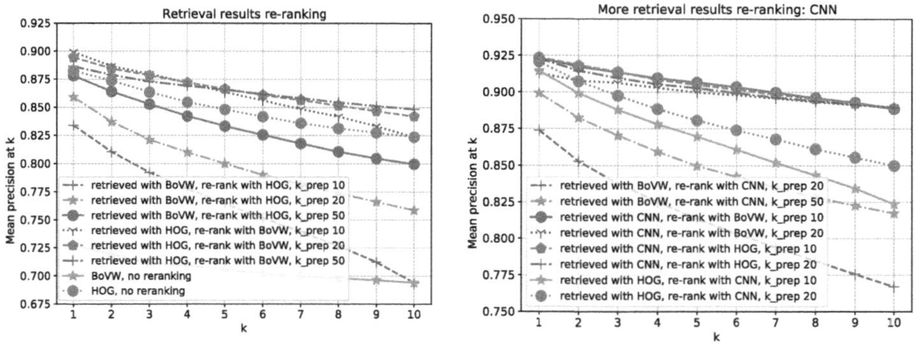

Fig. 5. RR performance comparison: (left) BoVW and HOG, using different k_{prep} values. RR with HOG causes a significant performance increase for images retrieved with BoVW. Higher k_{prep} values work better. HOG with BoVW RR outperforms HOG without RR. (right) RR with CNN as the IR method, or as the RR method. Using CNN for RR images retrieved with BoVW or HOG improves model performance.

4.3 Final Model Evaluation

For the final evaluation on the test subset, for each IR method the configuration that attained the best performance on the validation subset was chosen. The best RR configuration was also included. Optimal hyperparameter values are laid out in Table 1.

The results for the mean P@k ($k = 1, ..., 10$) are shown in Fig. 6. Comparing individual methods, CNN retrieval performed the best, followed by HOG, and finally BoVW. CNN retrieval results were slightly improved with HOG RR for some k values. The mAP scores (for $k_{max} = 10$) were calculated by considering anatomical region, imaging modality, and both for determining the relevance of the retrieved images. The results are shown in Table 1. With respect to imaging modality, CNN retrieval outperformed the rest (mAP$=$ 0.9654); with respect to anatomical region and to both, CNN with HOG RR ranked best (map$=$0.8855, and 0.8835, respectively). Examples of retrieved images for each anatomical region can be found in Appendix C.

Times required in the offline phase and times necessary to retrieve images with each method are shown in Table 2. HOG retrieval is the fastest. Time was measured on an AMD Ryzen 5 2600 @ 3.400 GHz CPU with 16 GB RAM.

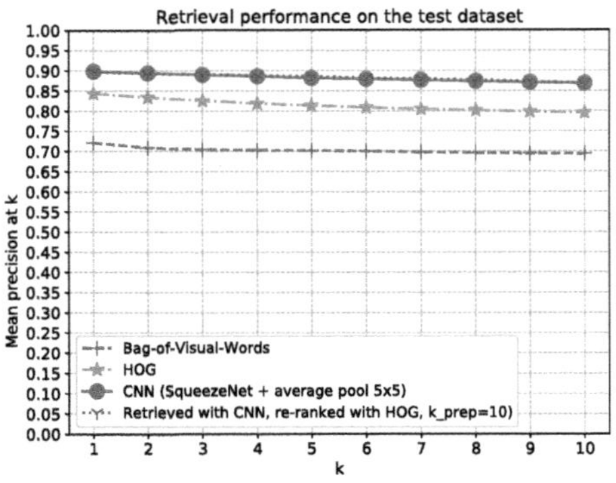

Fig. 6. Performance comparison on the test subset. Retrieved images were considered relevant if both the anatomical region and the imaging modality of the query image and the retrieved image were the same. For top-1 retrieval, SqueezeNet outperformed other models. For higher k values, RR with HOG performed slightly better.

Table 1. Performance comparison on the test subset (mAP), and the optimal configuration used. Best results are emphasised.

Method	mAP[a]	mAP[b]	mAP[c]	Hyperparam. val. of optimal configurations
BoVW	0.7097	0.8576	0.7022	$n_{words} = 4,000$, RootSIFT local features
HOG	0.8249	0.9121	0.8146	$size_{cell} = 64$, $size_{block} = 128$, $stride_{block} = 64$ (overlapping blocks), $n_{bins} = 15$
CNN	0.8845	**0.9654**	0.8821	SqueezeNet + 5×5 average pooling for extracting image descriptors
CNN+RR	**0.8855**	0.9635	**0.8835**	Re-ranking (RR) with HOG, $k_{prep} = 10$

[a] anatomical region, [b] imaging modality, [c] both.

Table 2. Time required for the offline phase for each IR method, and mean IR time per query. The latter was measured by using each test image as the query image.

Method	Offline phase duration	Retrieval time
BoVW	141 min (SIFT 6 min, vocab 121 min, descriptors 14 min)	334 ms
HOG	1 min (extract descriptors)	53 ms
CNN	22 min (extract descriptors)	339 ms
CNN + HOG RR	–	341 ms

5 Conclusion

In this study, a CBMIR system for radiology images of various anatomical regions and imaging modalities was designed and implemented. Three IR methods were evaluated: BoVW, HOG, and CNN. The experiments presented here can be recreated using the code available at https://github.com/dbarac/cbmir.

The configuration for each method was chosen empirically by comparing the results on the validation subset. BoVW was tested with SIFT and RootSIFT local features. The use of RootSIFT descriptors improved the results. For IR with CNN descriptors, five architectures pretrained on the ImageNet dataset were evaluated. ShuffleNet provided the best performance. IR with CNNs outperformed both BoVW and HOG retrieval. The CNN models were not additionally trained. Fine-tuning the CNNs for classifying the anatomical region or modality of radiology images could potentially further improve IR results. Additional architectures could be tested as well.

Retrieving images with one method and then RR them according to another method was also tested. RR improved the results in almost all cases. BoVW retrieval showed the most significant performance increase after RR. Out of all methods evaluated on the test subset, CNN retrieval (ShuffleNet with 5×5 average pooling) with HOG RR performed the best.

Acknowledgements. This work has been supported by the Croatian Science Foundation [grant number IP-2020-02-3770] and by the University of Rijeka [grant number uniri-iskusni-tehnic-23-12].

A Appendix – Dataset

The initial RadiologyNET data collection comprises around 2 million distinct examinations conducted between 2008 and 2017 within the standard clinical procedures at CHC Rijeka. This collection was retrospectively compiled in 2017 and underwent anonymisation during extraction to eliminate any sensitive information. Ethical approval from the relevant Ethics Committee was secured to gather and process the data, with a stipulation that the data remains confidential in its current state. For the work presented here, we are utilising only a small subset sampled from this overall collection. Imaging modality and anatomical region were automatically reconstructed from DICOM data [15,17]. This resulted in a noisy annotation, as can be seen in some examples in Fig. C.1. Overall statistics of the dataset used in this study are presented in Table A.1. No windowing was used for adapting the dynamic range of images [15].

Table A.1. Number (#) of images in the training and test subsets, per: (left) anatomical region; (right) imaging modality.

Anatomical region	#training	#test
Chest	5,123	1,587
Urinary tract	5,000	1,961
Colon	3,672	1,011
Abdomen	3,351	758
Head	3,239	555
Cerebrala	3,062	879
Breast	1,104	432
L-spine	944	65
C-spine	877	128
Knee	505	55
Shoulder	477	68
Heart	369	69
Arm	337	15
Pelvis	306	118
Extremity	292	70
Foot	282	63
Neck	260	59
Ankle	231	16
Hand	202	15
Spine	162	22
Whole body	77	40
Lower extrem	70	5
T-spine	42	12
Extrem	16	6

Imaging modality	#training	#test
Radio fluoroscopy (RF)	5,123	1,587
X-ray angiography (XA)	5,000	1,961
Computed tomography (CT)	3,672	1,011
Magnetic resonance (MR)	3,351	758
Computed radiography (CR)	3,239	555
Nuclear medicine (NM)	3,062	879

B Appendix – Calculating Feature Descriptors and Image Retrieval

B.1 Retrieval with BoVW Descriptors

Local Feature Extraction with SIFT. The first step for using the BoVW model is local feature extraction. In this study, we used SIFT [14]. To detect scale-invariant features, SIFT uses a difference-of-Gaussians (DoG) keypoint detector, which is an approximation of the Laplacian-of-Gaussian, computed with a discrete Gaussian image pyramid. The image pyramid is generated by blurring and downsampling the original image multiple times. Image keypoints are selected by finding local extrema in the DoG image pyramid, and discarding points which are on edges or have low-contrast. Each keypoint has a characteristic scale and orientation, which are used for computing the SIFT descriptor.

The SIFT descriptor is generated by computing gradient directions for pixels in the area surrounding the keypoint, dividing the area into blocks, calculating histograms of gradient directions for each block, and finally, concatenating the histograms. The keypoint scale and orientation are used during these steps to make sure the descriptors are scale and rotation invariant. Detailed steps for SIFT keypoint detection and descriptor computation can be found in [21].

Visual Vocabulary Learning. Since many SIFT features are found in each image (50–350 were found per training subset image), it would be computationally expensive to directly compare features between all images in the dataset and the query image. The BoVW model quantises all local features found in an image, aggregates them and describes the image with a single fixed-length vector. To quantise local SIFT features, it is necessary to first build the visual vocabulary. This is done by clustering all SIFT features extracted from the training subset by running k-means, with $k = n_{words}$ (the size of the vocabulary). The vocabulary is the set of n visual words, i.e. the n most representative SIFT descriptors from the dataset, which are selected by taking the cluster centroids generated by k-means.

BoVW Image Descriptor Computation. After visual vocabulary learning, the BoVW descriptors [25] for each image can be generated. First, each SIFT descriptor extracted from an image is quantised to the closest visual word. This is done by measuring the Euclidean distance between the descriptor and each visual word (k-means cluster centroid), and assigning the ID of the closest visual word to the SIFT descriptor.

The BoVW descriptor is a histogram of visual word occurrences in an image. The dimensionality of the descriptor is the same as the visual vocabulary size (n_{words}). The descriptor is computed by starting with zeros in all positions in the histogram, and then incrementing the position of the assigned visual word ID for each SIFT descriptor. Each image contains only a subset of all visual words, so the BoVW descriptor is a sparse vector.

TF-IDF Descriptor Re-weighting. To improve performance, the BoVW descriptors are re-weighted using term frequency-inverse document frequency (TF-IDF) [12]. The re-weigthing aims to increase the bin magnitude for meaningful words (words which rarely occur), and reduce the bin magnitude of words which are not meaningful. Each histogram bin in the descriptor is re-weighted using the following expression:

$$t_{id} = \frac{n_{id}}{n_d} log \frac{N}{n_i}, \tag{1}$$

where t_{id} is the histogram bin of word i for image d, n_{id} is the number of occurences of word i in image d, n_d is the total number of word occurences in image d, n_i is the number of images that contain word i, and N is the number of images.

Distance Measure. For image retrieval, we use cosine distance measure [24] to compute the distance between the query image descriptor x and the stored descriptors of dataset images y:

$$d_{cos}(\mathbf{x}, \mathbf{y}) = 1 - \frac{\mathbf{x}^T \mathbf{y}}{\|\mathbf{x}\| \|\mathbf{y}\|}. \tag{2}$$

The cosine distance ignores the magnitude between the vectors, and considers the cosine of the angle between them.

B.2 Retrieval with HOG Descriptors

HOG Image Descriptor Computation. The HOG descriptors are computed once globally for the whole image. The first step is computing the gradient direction and magnitude for each pixel in the image. After computing the gradient (the vector of partial x and y image derivatives):

$$\nabla f(x, y) = \begin{bmatrix} g_x \\ g_y \end{bmatrix} = \begin{bmatrix} \frac{\partial f}{\partial x} \\ \frac{\partial f}{\partial y} \end{bmatrix} = \begin{bmatrix} f(x+1, y) - f(x-1, y) \\ f(x, y+1) - f(x, y-1) \end{bmatrix}, \tag{3}$$

the magnitude and direction of the gradient can be calculated as follows:

$$g = \sqrt{g_x^2 + g_y^2}, \tag{4}$$

$$\theta = arctan(g_y / g_x). \tag{5}$$

For computing the descriptor, the image is divided into a grid of equally sized cells, $size_{cell} \times size_{cell}$. For each cell, a histogram of gradient orientations of pixels is computed. The histograms have n_{bins} bins each. Then, individual cell histograms for blocks of 2×2 cells are concatenated into a single vector and normalised. The HOG descriptor is the concatenation of block vectors in each image position. Block positions are determined by sliding the block across the image with stride equal to $stride_{block}$.

Distance Measure. For computing the distance between HOG image descriptors, L_1 distance was used, since it showed better performance than L_2 (Euclidean):

$$d_1(\mathbf{x}, \mathbf{y}) = \|\mathbf{x} - \mathbf{y}\|_1. \tag{6}$$

B.3 Retrieval with CNN Descriptors

CNN Image Descriptor Computation. CNNs [11] are a class of neural networks frequently used for computer vision tasks, such as image classification or object detection. In general, for image classification, a CNN's architecture consist of various blocks of convolutional and pooling layers for feature extraction, and one or more fully connected layers at the top of the network.

For IR, the fully connected layers at the top can be removed. Then, image descriptors can be computed by using images as the network input and taking the flattened output of the final convolutional (or pooling) layer [22].

In this study, several CNN architectures were tested in such a manner. IR performance was tested for ResNet18 [7], ResNet34 [7], AlexNet [9], SqueezeNet [8] and VGG16 [23]. The models are all pretrained on the ImageNet dataset, and were not fine-tuned to our dataset. For some models, an average pooling layer was added at the end to reduce the dimensionality of the final output vector.

Distance Measure. For computing the distance between CNN image descriptors, L_2 (Euclidean) distance was used:

$$d_2(\mathbf{x}, \mathbf{y}) = \|\mathbf{x} - \mathbf{y}\|_2. \tag{7}$$

C Appendix – Example Retrieved Images

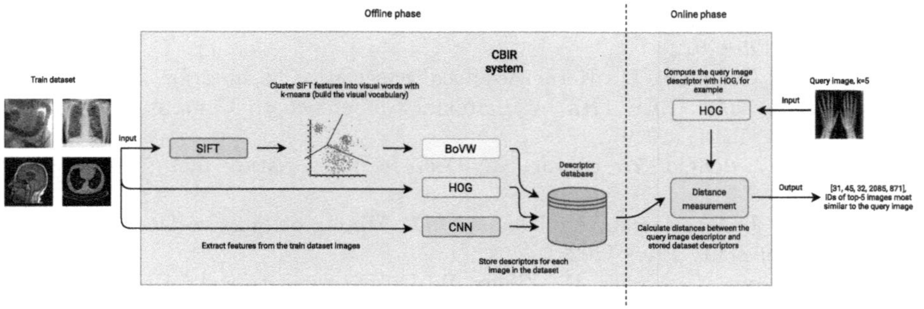

Fig. C.1. One query image was randomly chosen for each anatomical region in the dataset. The leftmost images in both columns (red frame) are the query images. The remaining 5 images in each row are the top-5 most similar images retrieved with the best performing retrieval method (retrieval with SqueezeNet + 5×5 average pooling, RR with HOG, $k_{prep} = 10$). The automated ground truth annotation process resulted in errors in some volume slices examples shown (e.g. NECK, CSPINE) because it was performed on volumes, not individual slices. (Color figure online)

References

1. Akgül, C.B., Rubin, D.L., Napel, S., Beaulieu, C.F., Greenspan, H., Acar, B.: Content-based image retrieval in radiology: current status and future directions. J. Digit. Imaging **24**, 208–222 (2011)
2. Arandjelović, R., Zisserman, A.: Three things everyone should know to improve object retrieval. In: 2012 IEEE Conference on Computer Vision and Pattern Recognition, pp. 2911–2918. IEEE (2012)

3. Bay, H., Tuytelaars, T., Van Gool, L.: SURF: speeded up robust features. In: Leonardis, A., Bischof, H., Pinz, A. (eds.) ECCV 2006, Part I 9. LNCS, vol. 3951, pp. 404–417. Springer, Heidelberg (2006). https://doi.org/10.1007/11744023_32
4. Dalal, N., Triggs, B.: Histograms of oriented gradients for human detection. In: 2005 IEEE Computer Society Conference on Computer Vision and Pattern Recognition, CVPR 2005, vol. 1, pp. 886–893. IEEE (2005)
5. Foncubierta-Rodríguez, A., García Seco de Herrera, A., Müller, H.: Medical image retrieval using bag of meaningful visual words: unsupervised visual vocabulary pruning with PLSA. In: Proceedings of the 1st ACM International Workshop on Multimedia Indexing and Information Retrieval for Healthcare, pp. 75–82 (2013)
6. Gong, Y., Lazebnik, S., Gordo, A., Perronnin, F.: Iterative quantization: a procrustean approach to learning binary codes for large-scale image retrieval. IEEE Trans. Pattern Anal. Mach. Intell. **35**(12), 2916–2929 (2012)
7. He, K., Zhang, X., Ren, S., Sun, J.: Deep residual learning for image recognition. In: Proceedings of the IEEE Conference on Computer Vision and Pattern Recognition, pp. 770–778 (2016)
8. Iandola, F.N., Han, S., Moskewicz, M.W., Ashraf, K., Dally, W.J., Keutzer, K.: SqueezeNet: AlexNet-level accuracy with 50x fewer parameters and $<$0.5 mb model size. arXiv preprint arXiv:1602.07360 (2016)
9. Krizhevsky, A., Sutskever, I., Hinton, G.E.: ImageNet classification with deep convolutional neural networks. In: Advances in Neural Information Processing Systems, vol. 25 (2012)
10. Kulis, B., Grauman, K.: Kernelized locality-sensitive hashing for scalable image search. In: 2009 IEEE 12th International Conference on Computer Vision, pp. 2130–2137. IEEE (2009)
11. LeCun, Y., Bengio, Y., Hinton, G.: Deep learning. nature **521**(7553), 436–444 (2015)
12. Leskovec, J., Rajaraman, A., Ullman, J.D.: Mining of Massive Data Sets. Cambridge University Press (2020)
13. Li, Z., Zhang, X., Müller, H., Zhang, S.: Large-scale retrieval for medical image analytics: a comprehensive review. Med. Image Anal. **43**, 66–84 (2018)
14. Lowe, D.G.: Distinctive image features from scale-invariant keypoints. Int. J. Comput. Vis. **60**, 91–110 (2004). https://doi.org/10.1023/B:VISI.0000029664.99615.94
15. Manojlović, T., Štajduhar, I.: Deep semi-supervised algorithm for learning cluster-oriented representations of medical images using partially observable DICOM tags and images. Diagnostics **11**(10), 1920 (2021)
16. Mustapha, A., Hussain, A., Samad, S.A., Zulkifley, M.A., Diyana Wan Zaki, W.M., Hamid, H.A.: Design and development of a content-based medical image retrieval system for spine vertebrae irregularity. Biomed. Eng. Online **14**(1), 1–24 (2015)
17. Napravnik, M., Hržić, F., Tschauner, S., Štajduhar, I.: Building RadiologyNET: unsupervised annotation of a large-scale multimodal medical database. arXiv preprint arXiv:2308.08517 (2023)
18. Nister, D., Stewenius, H.: Scalable recognition with a vocabulary tree. In: 2006 IEEE Computer Society Conference on Computer Vision and Pattern Recognition, CVPR 2006, vol. 2, pp. 2161–2168. IEEE (2006)
19. Ojala, T., Pietikäinen, M., Harwood, D.: A comparative study of texture measures with classification based on featured distributions. Pattern Recogn. **29**(1), 51–59 (1996)
20. Oliva, A., Torralba, A.: Modeling the shape of the scene: a holistic representation of the spatial envelope. Int. J. Comput. Vis. **42**, 145–175 (2001)

21. Otero, I.R.: Anatomy of the SIFT Method. Ph.D. thesis, École normale supérieure de Cachan-ENS Cachan (2015)
22. Razavian, A.S., Sullivan, J., Carlsson, S., Maki, A.: Visual instance retrieval with deep convolutional networks. ITE Trans. Media Technol. Appl. **4**(3), 251–258 (2016)
23. Simonyan, K., Zisserman, A.: Very deep convolutional networks for large-scale image recognition. arXiv preprint arXiv:1409.1556 (2014)
24. Singhal, A., et al.: Modern information retrieval: a brief overview. IEEE Data Eng. Bull. **24**(4), 35–43 (2001)
25. Sivic, Zisserman: Video Google: a text retrieval approach to object matching in videos. In: Proceedings Ninth IEEE International Conference on Computer Vision, pp. 1470–1477. IEEE (2003)
26. Smeulders, A.W., Worring, M., Santini, S., Gupta, A., Jain, R.: Content-based image retrieval at the end of the early years. IEEE Trans. Pattern Anal. Mach. Intell. **22**(12), 1349–1380 (2000)
27. Wu, G., Kim, M., Wang, Q., Munsell, B.C., Shen, D.: Scalable high-performance image registration framework by unsupervised deep feature representations learning. IEEE Trans. Biomed. Eng. **63**(7), 1505–1516 (2015)
28. Zhu, S., Tizhoosh, H.R.: Radon features and barcodes for medical image retrieval via SVM. In: 2016 International Joint Conference on Neural Networks (IJCNN), pp. 5065–5071. IEEE (2016)

Cross-Modality Synthesis of T1c MRI from Non-contrast Images Using GANs: Implications for Brain Tumor Research

Mehnaz Tabassum[1,2(✉)], Priyanka Rana[1], Eric Suero Molina[2,3], Antonio Di Ieva[2], and Sidong Liu[1,2]

[1] Centre for Health Informatics, Australian Institute of Health Innovation, Macquarie University, Sydney, NSW, Australia
mehnaz.tabassum1@hdr.mq.edu.au

[2] Computational NeuroSurgery (CNS) Lab, Macquarie Medical School, Macquarie University, Sydney, NSW, Australia

[3] Department of Neurosurgery, University Hospital of Münster, Münster, Germany

Abstract. Magnetic Resonance Imaging (MRI) has revolutionized medical diagnostics by providing a multifaceted view of the human body's intricate structures and pathologies. With its different sequences, MRI offers comprehensive information about the structure and function of the brain, providing valuable clues in neuro-oncology. However, post-contrast (gadolinium) T1 sequences (T1c) MRI, which offers unparalleled detail in blood-brain barrier disruption, is sometimes inaccessible due to logistical and clinical constraints (e.g., costs, pregnancy, allergic reaction to the agent, etc.). Our study aims to address this issue by synthesizing T1c MRI images using generative AI methodologies. We have expanded on existing Generative Adversarial Network (GAN) frameworks and addressed the vanishing gradient problem, mode collapse, and overfitting by integrating an adaptive loss function. Our approach has shown notable improvements in image fidelity, as evidenced by the Structural Similarity Index Measure (SSIM), Peak Signal to Noise Ratio (PSNR) metrics, and visual analysis. With our conditional GAN model, we have made significant progress in medical imaging, ensuring cross-modality image synthesizing and advancing diagnostic processes in scenarios where T1c MRI access is limited.

Keywords: Missing modality · Generative adversarial network · Image synthesis · Wasserstein Loss · WGAN · pix2pix

1 Introduction

Magnetic resonance imaging (MRI) utilizes various modalities such as T1, T2, FLAIR, and T1C to collect crucial disease diagnostic information. The choice of MRI modality is pivotal in medical diagnostics, as each type unveils specific tissue characteristics essential for accurate diagnosis. T1-weighted MRI is in-valuable for its high-resolution depiction of anatomical structures, distinguishing between grey and white matter with

J. Finkelstein et al. (Eds.): AIME 2024, LNAI 14845, pp. 60–69, 2024.
https://doi.org/10.1007/978-3-031-66535-6_7

clarity. Complementing this, T2-weighted MRI excels in identifying lesions and inflammations thanks to its sensitivity to fluid content. FLAIR imaging further enhances lesion detection adjacent to the cerebrospinal fluid by effectively suppressing fluid signals, offering a clearer view of pathological changes. Meanwhile, T1c MRI, enhanced with a gadolinium- based contrast, is crucial for delineating tumors and their margins and assessing vascularity, providing comprehensive insights into tumor biology. To clarify, we visualize a case from the BraTS 2023 [14] dataset in Fig. 1, where we can see different contrast MRI sequences. These modalities form a cohesive imaging strategy that significantly improves the diagnostic process. However, acquiring all desirable imaging modalities is sometimes not feasible due to practical constraints, such as potential allergic reactions to the contrast agent, higher costs, longer processing times, image corruption, and lack of availability in certain medical facilities. Hence, a missing modality issue may raise uncertainty for tumor localization and segmentation since each modality contributes unique and comprehensive information [1, 2, 24].

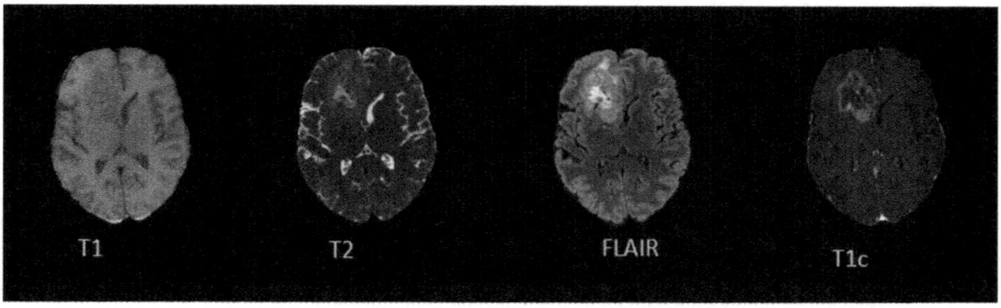

Fig. 1. Illustration of all (four) modalities with different sensitivities to distinct brain regions

Over time, various methods have been developed to address the challenge of missing modalities in MRI scans, transitioning from early synthesis techniques to sophisticated deep-learning models [7, 11, 23]. Early methods focused on generating the absent modality within the training phase, necessitating modality-specific networks that hampered holistic training efficacy [8]. In contrast, newer strategies strive for efficiency by forming a universal model that integrates representations from all available modalities [19], addressing the challenge of missing data more effectively. These approaches have explored creating missing modal images by extracting crucial features and employing methods like modality translation to shared spaces [12], knowledge distillation [3, 6, 21], and feature optimization [16, 20, 25]. In recent years, Generative Adversarial Networks (GANs) [10] have been established for synthetic data generation [4, 5, 17]

[13] to solve the missing modality issue. However, training GANs is complicated due to various challenges, such as the potential for generating imputation noise, difficulties in model convergence, extensive training requirements, and generator instability during training. Additionally, problems like vanishing gradients [10] and mode collapse [13] add layers of complexity, making the practical application of GANs in medical imaging a nuanced task that requires careful consideration and specialized approaches. To cope with these problems, Wasserstein GAN (WGAN) [22] was proposed, which has demonstrated more stable optimization processes and realistic synthetic images [18].

Despite advancements, the quality of synthesized neuroimages had yet to fully meet clinical standards, especially in tasks requiring localization or segmentation of regions of interest (ROI) of tumors. In this study, we incorporate an adaptive loss function that mitigates common issues of the existing GAN, enhancing the reliability of synthesized images. Our application of conditional GANs demonstrates significant improvements in image quality, as evidenced by superior SSIM and PSNR metrics, alongside positive visual evaluations. This advancement presents a promising avenue when lacking T1c MRI images, potentially transforming tumor diagnosis and treatment planning.

2 Methods and Materials

In this study, we employ a GAN framework, adapted from the models in [13], for generating T1c MR images from other modalities (T1, T2, FLAIR). This GAN framework is configured explicitly for conditional image generation, mapping an observed MRI image x and a random noise vector z to an output T1c MR image y, formulated as $G: \{x, z\} \rightarrow y$.

2.1 Dataset

This study's analysis used the publicly available BraTs2023 dataset [14]. A random selection of 360 participants was made, each contributing four types of multi-modal magnetic resonance (MR) images: T1, T1c, T2, and Flair, all featuring a consistent voxel resolution of $240 \times 240 \times 155$. An extensive preprocessing approach was implemented to prepare the dataset. This included standardizing the image formats, aligning the scans to a uniform anatomical framework, and enhancing the image resolution.

2.2 Preprocessing

To address the inherent complexity of 3D MRI volumes in the Neuroimaging Informatics Technology Initiative (NIfTI) file format, thorough preprocessing is essential. This involves alignment and standardizing the modalities to ensure accurate pairing with T1c modalities. This step is critical for the consistency of multi-sequence analysis and subsequent neuroimaging assessments. Our model employs a split-input fusion module to handle different MRI modalities such as T1, T2, and FLAIR skillfully. Initially processing them separately, it extracts distinctive features from each, which are consolidated. A self-attention mechanism meticulously emphasizes the most vital features from the combined data, ensuring that each modality's significant attributes are captured. The fusion of these refined features, when paired with T1C sequences, constitutes the input for our GAN. This advanced approach generates enriched T1C MRI images, providing sharper, more detailed visuals for enhanced diagnostic accuracy. Figure 2 depicts the advanced input fusion model we've integrated as a preprocessing step for our synthesis task.

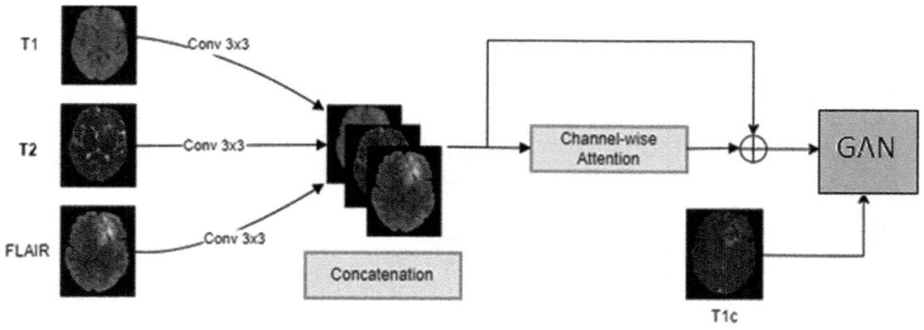

Fig. 2. Input Fusion Model

2.3 Network Architecture

We employed the pix2pix GAN model [13] for image-to-image translation, enabling the generation of T1C image modalities from T1, T2, and FLAIR image modalities. Pix2Pix GAN belongs to a category of conditional GAN (cGAN) [15] and is trained using joint optimization of two networks: the Generator (G) and Discriminator (D). The G network generates synthetic or fake data, while the D network determines whether the generated samples are authentic or fake. Both networks undergo alternating training to maximize the probability of correctly classifying real images and generated samples. The GAN achieves optimal training when the generator can produce data samples as diverse as the original data distribution and deceive the discriminator into perceiving them as real. The adversarial training involves a discriminator *D*, employing a PatchGAN [13] structure, which assesses the authenticity of local patches within the generated images. The adversarial objective function for the generator and discriminator is formalized as follows:

$$L_{cGAN}(G, D) = \mathrm{E}_{x,y}\left[\log D(x, y)\right] + \mathrm{E}_{x,z}[\log(1 - D(x, G(x, z)))]$$

This is complemented by an L1 loss that promotes fidelity to the target image, leading to the composite loss function:

$$L_{composite}(G, D) = L_{cGAN}(G, D) + \lambda \mathrm{E}_{x,y,z}[||y - G(x, z)||_1]$$

Here, x is the input from the source modality, y represents the actual target image, and z is a random vector that introduces diversity in the generated output. λ serves as a balance between the pix2pix GAN and L1 losses. The generator (G) seeks to minimize the overall loss, while the discriminator (D) aims to increase the GAN loss (Fig. 3).

Adaptive Loss Function. We enhanced the pix2pix GAN by revising its discriminator with an additional convolution layer towards the end. This modification leads to a one-dimensional output through a Sigmoid activation, allowing the model to detect continuous data points essential for intricate image analysis. Such advancements boost the model's prediction accuracy and ease the output's interpretability, proving invaluable for complex image generation tasks.

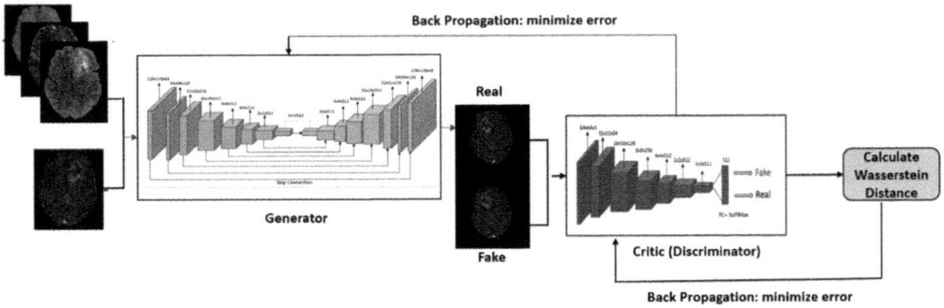

Fig. 3. Schematic diagram of our proposed GAN incorporating Wasserstein Distance to update Generator and Critic Loss

The loss function utilized for GAN training is one of the primary reasons for the unstable training, vanishing gradients (where the discriminator reaches perfect optimization that does not provide any valuable information for the generator to get better), and mode collapse (where the generator produces limited similar samples) [18, 22]. Accordingly, we replace the standard GAN loss function with a Wasserstein Loss function based on Wasserstein distance (W-dis). This measures how easily synthetic and real images can be distinguished, offering functional gradients through the objective function. The Equation for the Wasserstein loss in the context of GANs, where have a generator G aiming to produce data resembling the real data distribution P_r, and a discriminator D (or critic in the context of WGANs) that evaluates the data, can be formulated as follows:

$$L_{WL} = \mathrm{E}_{x \sim P} r[D(x)] - \mathrm{E}_{\hat{x} \sim P} g[D(\hat{x})]$$

where L_{WL} is the Wasserstein loss; $\mathrm{E}_{x \sim P}$ $[D(x)]$ is the expected value of the discriminator's output for real data x drawn from the real data distribution.

P_r; $\mathrm{E}_{\hat{x} \sim P}$ $[D(\hat{x})]$ is the expected value of the discriminator's output for generated data $x\hat{}$ drawn from the generator's data distribution P_g; $D(x)$ is the discriminator's output when evaluating a real data instance x, and $D(x\hat{})$ is the discriminator's output when evaluating a generated data instance $\hat{x}$.

In the original formulation of the WGAN [9], the discriminator (or critic) is not trained to classify instances as real or fake but rather to provide a score that measures the "realness" of the data. The generator aims to maximize this score for its generated instances, thereby minimizing the Wasserstein distance between the real and generated distributions.

2.4 Experimental Setup

We developed our model in Python 3.9.18 using TensorFlow 2.4.1 and employed mini-batch Stochastic Gradient Descent (SGD) alongside the Adam optimization algorithm, setting a learning rate of 0.0002 and $\beta_1 = 0.5$ for momentum. To ensure a fair comparison, the same hyperparameters were maintained across both the baseline pix2pix model and the enhanced pix2pix WGAN variant. The training spanned 115 epochs with single-instance batches and required 250 h on an NVIDIA Tesla V100 GPU.

3 Results

This study employed the pix2pix GAN to generate T1C MRI modality from other MRI modalities. We optimized the performance of this image-to-image translational model by using a new loss function based on W-dis. We used PSNR and SSIM metrics to evaluate the improvements, comparing the outcomes with our refined model and the original pix2pix GAN. This comparison was crucial in validating the effectiveness of our methodology.

3.1 Evaluation Metrics

The PSNR is a standard metric for image quality assessment, which compares the maximum potential power of a signal to the power of corrupting noise. The PSNR of our synthetic images is calculated as:

$$\mathrm{PSNR} = 10.\log_{10}\frac{\mathrm{MAX}^2}{\mathrm{MSE}} \tag{1}$$

where MAX_I represents the maximum possible pixel value of the image, and MSE denotes the mean squared error between the original and synthetic images.

The SSIM measures the perceived quality of digital images and videos. Our computed SSIM values, defined by:

$$\mathrm{SSIM}(x, y) = \frac{(2\mu_x\mu_y + C_1)(2\sigma_{xy} + C_2)}{(\mu_x^2 + \mu_y^2 + C_1)(\sigma_x^2 + \sigma_y^2 + C_2)} \tag{2}$$

where x and y are the original and synthetic images, respectively, indicating a substantial structural congruence.

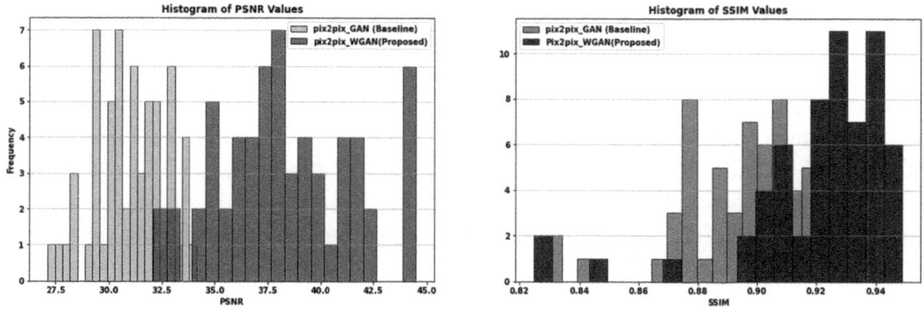

Fig. 4. Comparison Evaluation of SSIM and PSNR

3.2 Quantitative Analysis

Figure 4 illustrates the quantitative findings of the research, showcasing a side-by-side assessment of image synthesis performance between the suggested pix2pix WGAN and the standard pix2pix, as measured by PSNR and SSIM indices. The PSNR values for our method, marked in green, commonly fall within an elevated range of 35 to 45 dB, surpassing the standard model shown in blue. Similarly, the SSIM scores, represented in purple for our method, regularly exceed 0.90, indicating a higher consistency in structural integrity compared to the baseline's more varied results in yellow. These metrics confirm the enhanced capability of our proposed method in synthesizing higher-quality images.

3.3 Qualitative Analysis

Figure 5 illustrates a side-by-side T1c MRI image synthesis comparison using two GAN approaches. The top row, marked as "Baseline," demonstrates the results from our baseline pix2pix GAN, and Directly below, the "Proposed" row reveals the outputs from our pix2pix WGAN model. The bottom row offers the "Ground Truth" - the actual T1c MRI scans for reference. Upon visual inspection, the images generated by our proposed model show a remarkable resemblance to the ground truth, with improved clarity and detail fidelity compared to the baseline GAN-generated images. Notably, the proposed model better captures the contrast and intricate structures within the brain, which are crucial for accurate medical diagnosis. While retaining the general structure, the baseline images exhibit less definition and sharpness, with the potential loss of detail that could be vital in a clinical setting.

4 Discussion

Our proposed pix2pix WGAN model, showcasing enhanced outcomes, marks a significant step forward in synthesizing T1c MR images. Incorporating an adaptive loss function has been pivotal in addressing some of the most persistent challenges in GAN development, such as vanishing gradients, mode collapse, and the risk of overfitting. Our findings, underscored by enhanced PSNR and SSIM metrics, indicate a clear qualitative and quantitative superiority over existing baseline models. Critically, the visual quality of the synthesized images has shown remarkable closeness to actual T1c MRI scans, underscoring the practical value of our model. Such advancements promise to bridge the gap where T1c MRIs are not readily available, thereby supporting the intricate process of brain tumor delineation and subsequent treatment planning. The broader impact of this work is multi-fold. From a clinical perspective, democratizing access to essential diagnostic tools could fundamentally change patient outcomes, especially in under-resourced healthcare environments. Our method's versatility also hints at its potential applicability across different imaging modalities, suggesting a future where AI not only assists but amplifies the capabilities of medical imaging.

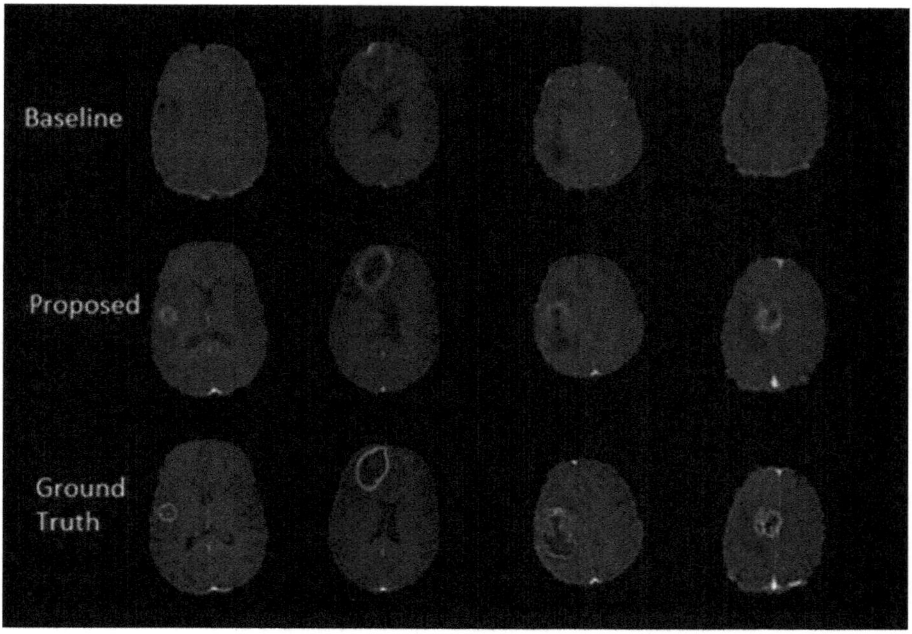

Fig. 5. Visual Representation of the synthesize T1c from Multi-modal MRI's

5 Conclusion and Future Direction

In essence, this study provides more than just a methodological enhancement; it offers a new lens through which we can envision the integration of AI in medical diagnostics. The promising results lay the groundwork for expansive research and development in medical image synthesis, fostering a synergy between AI innovation and clinical expertise.

Evaluating the model's performance in real-world clinical scenarios is imperative. Testing against more extensive and varied datasets will further elucidate the robustness and generalizability of our approach. Moreover, clinical validation is critical to understanding how synthesized images fare in actual diagnostic workflows and their interpretability by medical professionals.

References

1. Azad, R., Khosravi, N., Dehghanmanshadi, M., Cohen-Adad, J., Merhof, D.: Medical image segmentation on mri images with missing modalities: A review. arXiv preprint arXiv:2203.06217 (2022)
2. Azad, R., Khosravi, N., Merhof, D.: Smu-net: Style matching u-net for brain tumor segmentation with missing modalities (2022)
3. Azad, R., Khosravi, N., Merhof, D.: Smu-net: style matching u-net for brain tumor segmentation with missing modalities. In: International Conference on Medical Imaging with Deep Learning, pp. 48–62. PMLR (2022)

4. Cao, B., Zhang, H., Wang, N., Gao, X., Shen, D.: Auto-gan: self-supervised collaborative learning for medical image synthesis. In: Proceedings of the AAAI Conference on Artificial Intelligence, vol. 34, pp. 10486–10493 (2020)
5. Chang, Q., et al.: Multi-modal asyndgan: Learn from distributed medical image data without sharing private information. arXiv preprint arXiv:2012.08604 (2020)
6. Chen, C., Dou, Q., Jin, Y., Liu, Q., Heng, P.A.: Learning with privileged multi- modal knowledge for unimodal segmentation. IEEE Trans. Med. Imagingging **41**(3), 621–632 (2021)
7. Dalmaz, O., Yurt, M., C¸ukur, T.: Resvit: Residual vision transformers for multimodal medical image synthesis. IEEE Trans. Med. Imaging **41**(10), 2598–2614 (2022)
8. Dayarathna, S., Islam, K.T., Uribe, S., Yang, G., Hayat, M., Chen, Z.: Deep learning based synthesis of MRI, CT and pet: review and analysis. Med. Image Anal. **92**, 103046 (2024)
9. Frogner, C., Zhang, C., Mobahi, H., Araya, M., Poggio, T.A.: Learning with a wasserstein loss. Advances in neural information processing systems, 28 (2015)
10. Goodfellow, I., et al.: Generative adversarial nets. Advances in neural information processing systems 27 (2014)
11. Hamghalam, M., Frangi, A.F., Lei, B., Simpson, A.L.: Modality completion via gaussian process prior variational autoencoders for multi-modal glioma segmentation. In: Medical Image Computing and Computer Assisted Intervention–MICCAI 2021: 24th International Conference, Strasbourg, France, September 27–October 1, 2021, Proceedings, Part VII 24, pp. 442–452. Springer (2021)
12. Havaei, M., Guizard, N., Chapados, N., Bengio, Y.: Hemis: Hetero-modal image segmentation. In: Medical Image Computing and Computer-Assisted Intervention– MICCAI 2016: 19th International Conference, Athens, Greece, October 17–21, 2016, Proceedings, Part II 19, pp. 469–477. Springer (2016)
13. Isola, P., Zhu, J.-Y., Zhou, T., Efros, A.A.: Image-to-image translation with conditional adversarial networks. In: Proceedings of the IEEE Conference on Computer Vision and Pattern Recognition, pp. 1125–1134 (2017)
14. Kazerooni, A.F., et al.: The Brain Tumor Segmentation (BraTS) Challenge 2023 (2023). https://arxiv.org/abs/2305.17033. Accessed: 2024-03-10
15. Mirza, M., Osindero, S.: Conditional generative adversarial nets. arXiv preprint arXiv:1411.1784 (2014)
16. Pan, Y., Chen, Y., Shen, D., Xia, Y.: Collaborative image synthesis and disease diagnosis for classification of neurodegenerative disorders with incomplete multi-modal neuroimages. In: Medical Image Computing and Computer Assisted Intervention–MICCAI 2021: 24th International Conference, Strasbourg, France, September 27–October 1, 2021, Proceedings, Part V 24, pp. 480–489. Springer (2021)
17. Qin, Z., Liu, Z., Zhu, P., Ling, W.: Style transfer in conditional gans for cross-modality synthesis of brain magnetic resonance images. Comput. Biol. Med. **148**, 105928 (2022)
18. Rana, P., Sowmya, A., Meijering, E., Song, Y.: Imbalanced cell-cycle classification using wgan-div and mixup. In: 2022 IEEE 19th International Symposium on Biomedical Imaging (ISBI), pp. 1–4. IEEE (2022)
19. Sharma, A., Hamarneh, G.: Missing MRI pulse sequence synthesis using multi- modal generative adversarial network. IEEE Trans. Med. Imaging **39**, 1170–1183 (2019)
20. Sylvain, T., et al.: Cross-modal information maximization for medical imaging: Cmim. arXiv preprint arXiv:2010.10593 (2020)
21. Van Tulder, G., de Bruijne, M.: Why does synthesized data improve multi- sequence classification? In: Medical Image Computing and Computer-Assisted Intervention–MICCAI 2015: 18th International Conference, Munich, Germany, October 5–9, 2015, Proceedings, Part I 18, pp. 531–538. Springer (2015)
22. Wu, J., Huang, Z., Thoma, J., Acharya, D., Van Gool, L.: Wasserstein divergence for gans. In: Proceedings of the European Conference on Computer Vision (ECCV), pp. 653–668 (2018)

23. Zhang, Y., et al.: Modality-aware mutual learning for multi-modal medical image segmentation. In: Medical Image Computing and Computer Assisted Intervention–MICCAI 2021: 24th International Conference, Strasbourg, France, September 27–October 1, 2021, Proceedings, Part I 24, pp. 589–599. Springer (2021)
24. Zhao, Z., Yang, H., Sun, J.: Modality-adaptive feature interaction fornbsp; brain tumor segmentation withnbsp missing modalities. In: Medical Image Computing and Computer Assisted Intervention – MICCAI 2022: 25th International Conference, Singapore, September 18–22, 2022, Proceedings, Part V, pp. 183–192. Springer, Heidelberg (2022)
25. Zhou, T., Canu, S., Vera, P., Ruan, S.: Latent correlation representation learning for brain tumor segmentation with missing MRI modalities. IEEE Trans. Image Process. **30**, 4263–4274 (2021)

Harnessing the Power of Graph Propagation in Lung Nodule Detection

Sudipta Modak[1(✉)], Yash Trivedi[2], Esam Abdel-Raheem[1], and Luis Rueda[2]

[1] Department of ECE, University of Windsor, Windsor, ON, Canada
{modak,eraheem}@uwindsor.ca
[2] School of Computer Science, University of Windsor, Windsor, ON, Canada
{trived22,lrueda}@uwindsor.ca

Abstract. Automatic detection of lung nodules has been a key element in modern medical research in the past decade. By utilizing accurate nodule detection approaches, lung cancer can be treated in its early stages, reducing its mortality rate. Manual detection of nodules is hindered by structures contained in the computed tomography scan images such as veins, bronchioles and lymphatics. Furthermore, nodules are tiny and sometimes have haphazard boundaries making them easy to miss. In this work, we propose a two-step lung nodule detection method that uses a novel hierarchical superpixel merging (HSPM) module to reduce the number of proposed regions of interest and couples it with a new graph neural network namely, the regional downsampled residual superpixel aggregation network (dRes-SPAN) to accurately detect nodules in the lung parenchyma. Experiments on the Lung Image Database Consortium and Image Database Resource Initiative (LIDC-IDRI) dataset show a high improvement in performance in terms of nodule classification and detection while achieving an overall low false positive rate. A significant reduction in training time and trainable parameters has also been observed along with the relaxation in the requirement for heavy annotation compared to the state-of-the-art object detection architectures.

Keywords: Lung Nodule Detection · Superpixel Graph · Graph Neural Networks · Hierarchical Merging · Biomedical Image Processing

1 Introduction

Lung nodules are the primary reason for developing malignant tumors in the lungs that can result in death. Currently, the survival rate for lung cancer is only 21% which is the lowest amongst all types of cancers [18]. A lung nodule is a circular opacity that can have well-defined or chaotic boundaries measuring up to 3 cm in diameter [19]. Low-dose computed tomography (LDCT) is usually used to screen the interior of the lungs and performs better at providing longitudinal and cross-sectional views of the nodules [18]. It is seen from clinical studies that the survival rates of patients for various types of lung cancers increase

J. Finkelstein et al. (Eds.): AIME 2024, LNAI 14845, pp. 70–80, 2024.
https://doi.org/10.1007/978-3-031-66535-6_8

if nodules are detected early [15]. Manual diagnosis of lung nodules however is not only time-consuming for radiologists but also labor-intensive [18]. Most nodules occupy a very small area within a CT scan image and are easy to miss during diagnosis due to the presence of artifacts such as veins, bronchioles, and lymphatics [11].

To aid radiologists with the detection of pulmonary nodules and to reduce the diagnosis time, computer-aided diagnosis (CAD) systems usually employ convolutional neural networks (CNN) to detect nodule candidates within 2D CT scan images. The main objective of any region-based nodule detection algorithm is to propose bounding boxes for different structures contained within the lung parenchyma and assign a confidence score to these boxes highlighting the regions of interest (ROI) in the image. This method is similar to the Regions with CNN features (RCNN) architecture for object detection presented in [8]. The boxes with the highest scores are then selected and further processed to find the approximate positions of the nodules. This mechanism is known as the localization of nodules in the image [14]. In this work, a fast two-step method of lung nodule detection is presented that can effectively detect and localize nodules in CT scan images using a variant of graph neural networks and superpixel graphs.

The main contributions of this paper are i) A novel graph neural network-based encoder network, namely, downsampled residual superpixel aggregation network (dRes-SPAN), that can learn intrinsic features from superpixel graphs to effectively classify regions as nodules or background; ii) An edge formation mechanism that transforms an image into a superpixel graph; iii) A new method of region proposal, namely, hierarchical superpixel merging (HSPM) that uses the unsupervised superpixel generation method, superpixels extracted via energy-driven sampling (SEEDs) [4], hierarchical graph merging and multilevel thresholding [20] to reduce the number of proposed regions that must be classified by the trained network; iv) A comparison with the state-of-the-art lung nodule detection algorithms in the medical field on the Lung Image Database Consortium and Image Database Resource Initiative (LIDC-IDRI) [1] dataset.

2 Related Works

In the last decade, several deep learning-based computer-aided diagnosis (CAD) systems have been proposed for lung nodule detection. These algorithms mainly focus on the localization of nodules in 2D CT scan images. As most nodules are tiny and occupy a very small area in an image, localization of nodules is an important task for proper diagnosis [14]. For accurate localization of nodules, regional convolutional neural networks (RCNN) [8] and you only look once (YOLO) [22] are usually modified and employed through transfer learning. The RCNN architecture is a two-stage object detection network that has a region proposal network [25] built-in to propose regions. Furthermore, RCNN and its variants presented in [6,7] usually employ a CNN to train on patches acquired from the training images. On the other hand, YOLO [22] is a real-time object detection algorithm that utilizes a grid-based CNN which classifies the smaller grids on an image by its object detection criteria. While these architectures are well-suited for the detection of daily life objects, they have to be retrained on medical data to detect nodules. Most medical datasets have low data contained in them and this data is often not properly annotated [13]. This hindrance limits the training capability of deep neural networks on these datasets resulting in overfitting, improper localization, and a high false positive rate [13]. Therefore, to counter this limitation, architectures such as the ones presented in [15,21] modify these detection algorithms heavily to account for a high number of false positives and negatives. Ji *et al.* in their work in [12], mentions that YOLOv5s is not able to classify structures accurately in the lung parenchyma by using its backbone CNN module. Therefore the work proposes several modifications such as Atrous Spatial Pyramid Pooling and a Contextual Transformer module to make YOLOv5s suitable for nodule detection.

Other architectures that are not dependent on traditional CNN encoders used in YOLO and RCNN have begun to emerge in the last few years. The work in [23] presents a novel multi-encoder-based self-distilled network (MEDS-Net) that utilizes bidirectional maximum intensity projections and feature fusion to detect nodules. The architecture employs the aid of multi-slices to improve the detection efficiency of nodules and achieves a lower false positive rate in detection. Some methods such as the one presented in [16,17] have also developed 3D ROI detection for lung nodules using 3D CNNs. Even though these methods show a high detection sensitivity, low false positive rates, and better generalization capability than their 2D CNN counterparts, the number of trainable parameters for these architectures is huge, resulting in a high training time.

Traditionally, graph neural networks (GNN) have not been used for image classification as they can not directly process grid-like structured data [9]. However, in recent times, a few methods have been developed such as the ones presented in [3,10] that assisted GNNs in capturing intrinsic features from images that can be used for image classification. Therefore, harnessing the power of graph propagation and superpixel merging we propose a two-step lung nodule detection method that shortens the training time tremendously while achieving a high performance in detection.

3 Methodology

3.1 Downsampled Residual Superpixel Aggregation Network

Superpixel Generation: In this stage, the SEEDs [4] algorithm is utilized to generate superpixels from training patches. The training patches used are 128 × 128 in size which have been extracted from nodule images from the CT scan records present in the LIDC-IDRI database [1]. Through experimentation, it has been found that the number of superpixels generated by the SEEDs [4] for any 2D CT scan image is 4^N making it perfect for graph classification tasks. For experimentation, the number of superpixels generated for each training patch is taken as,

$$S = 2 \times X, \tag{1}$$

where X is the x-dimension of the patch and S is the number of superpixels generated.

Superpixel Graph Formation: For a given patch once superpixels are generated, the Euclidean distances between these superpixels are calculated using the following equation,

$$L_{(c,d)} = \sqrt{w \times (I_c - I_d)^2 + (j_c - j_d)^2 + (k_c - k_d)^2}\,, \tag{2}$$

where c and d are the superpixel numbers and the variables, and j and k represent the locations of the centroid of each superpixel in the patch. The variable, I represents the average grayscale intensity of a superpixel and w represents the weight associated with the average intensity. Next, the nearest-neighbor algorithm is used to identify only eight nearest neighbors for each superpixel. These neighbors are then stored in the connectivity list, $\mathbf{B}$. Next, an adjacency matrix, $\mathbf{A}$, with dimensions $S \times S$ is created and initialized with zeros. From the list, $\mathbf{B}$, the connectivity is taken one at a time, and a 1 is added to the positions in matrix $\mathbf{A}$ that represent the connection between two connected superpixels. The resultant matrix, $\mathbf{A}$ will be symmetric. For the feature matrix, $\mathbf{F}$, only the average grayscale intensities, I of each superpixel are included.

Graph-Based Network for Training. An undirected graph G can be expressed in the form $G = (V, E)$ where V represents the vertices or nodes and E represents the edges. For the superpixel graph that has been generated for a given patch, a new message-passing function is developed. The proposed message-passing function is divided into two parts, an unweighted aggregation function H, that accumulates messages from neighbors of each superpixel, and the weighted update function $\mathbf{U}$ that updates the features of each node and transforms the feature vectors to a higher dimensional space. The aggregation function, H function is given by,

$$H_u = \Sigma_{e=1}^{S} \Sigma_{v=1}^{S} A_e \times D_v^{-1} \times F_v + F_u \tag{3}$$

where u represents the node number for which the aggregation is being performed. The variable v represents the neighbor number of the node u, and e represents the element number in the vth row of the adjacency matrix $\mathbf{A}$. Elements of the vth row of the adjacency matrix $\mathbf{A}$ is represented by A, D is the degree normalization value for neighbor v, and F is the feature vector of the neighbor v. The values, H_u are then stored in matrix $\mathbf{H}$. After the aggregation of features, the algorithm moves to its update function where trainable weights are used along with an activation function. The update equation is performed using the formula,

$$\mathbf{U} = \lambda(\Sigma_{m=1}^{z} \mathbf{H} \otimes \mathbf{W}_m) \tag{4}$$

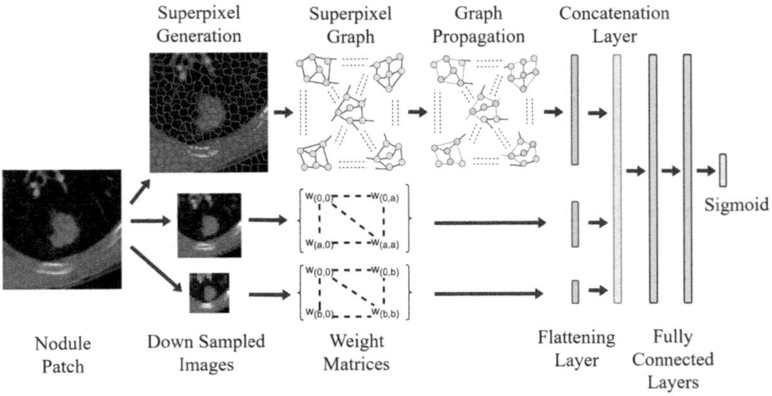

Fig. 1. Block diagram for the training of the proposed dRes-SPAN.

where the operator, $\otimes$ constitutes the matrix multiplication of matrices $\mathbf{H}$ and $\mathbf{W}$ that represent the feature aggregation matrix and trainable weight matrix, respectively. The number of weight matrices involved in the transformation is given by z, and λ constitutes the Gaussian Error Linear Unit activation function. The Eqs. (3) and (4) account for the message-passing operation of a single graphical layer. This network is called the superpixel aggregation network (SPAN). Furthermore, inspired by residual networks presented in [6], we add downsampled residues of patches into our architecture to enhance its generalization capability. This method consists of two max-pooling layers that transform the patch into its coarser versions. These images are then coupled with separate trainable weight matrices where these matrices undergo matrix multiplication with the coarsened images. We call this method downsampled residue (dRes) generation. By utilizing two layers of SPAN, two dRes images at different levels of downsampling, batch normalization, flattening, concatenation, two fully connected layers, and binary cross-entropy loss, the model can update the model weights to learn specific properties from nodule patches. The network is displayed in Fig. 1.

3.2 Hierarchical Superpixel Merging Region Proposal Network

The images are preprocessed to segment the lungs and reduce any noise contained in them. The lung parenchyma is segmented using the lung segmentation algorithm described in [19]. Consequently, for noise removal, the images are filtered using a median filter. Next, the SEEDs [4] algorithm is used to generate superpixels by using Eq. (1). Equation (2) is then used to calculate the distances between superpixels, while the mean grayscale intensities of the superpixels are used to decide if two superpixels should be merged to form a cluster. If two superpixels have an edge between them and the intensity difference between these superpixels is below a threshold T then these superpixels will be merged.

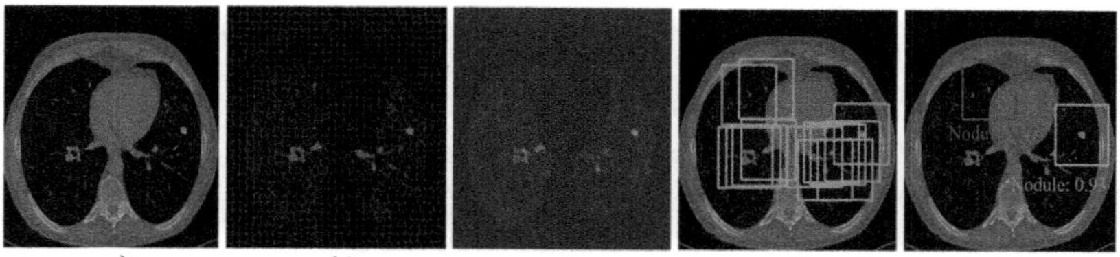

Fig. 2. Region Proposal Network and Inference a) CT scan slice b) Segmented lung and Superpixel generation c) Cluster formation d) Thresholding and Bounding box Generation e) Classification and Overlapping Removal.

Once the merging phase ends, we end up with a few disjoint clusters in the image as shown in Fig. 2c). These clusters are then passed through two static thresholds α and β to filter out giant clusters. The logic is such that if the number of pixels contained within a particular cluster is between the values of α and β then this cluster will be sent to the trained classification network, dRes-SPAN, for classification. A bounding box which is 128×128 is then drawn around each cluster centers cropping the region. These cropped boxes are then fed to the trained dRes-SPAN to obtain the probability of having a nodule. The boxes with probabilities above 0.5 are then selected. Now, there might be more than one box in the same region that contains a nodule due to multiple clusters being detected close to each other (as shown in Fig. 2). We define a threshold Γ that is used to select the best box containing a nodule. The probabilities that are assigned to each box by dRes-SPAN are taken and the box with the highest value is selected. If the Euclidean distance between the center of this box and an adjacent box is below the value of Γ then the box with the lower probability is removed. The entire process is summarized in Fig. 2.

4 Experimental Results and Analysis

4.1 Dataset Description

The dataset used for the experimentation has been collected from the LIDC-IDRI [1] database. This database consists of 1,018 CT scan images [18] and the nodule count is 2,669 that are at least 3 mm in diameter. The data from this database has been analyzed and annotated by four expert radiologists [18]. We have only considered solid and part-solid nodules that have been annotated by at least two radiologists for training and testing the proposed architecture. The dataset is divided into training/validation and test subsets where we use the data from the first 800 records for training and validation while the rest are used for testing. The training set is enhanced by using data augmentation that involves mirroring, rotation, and zooming in by 20%.

4.2 Evaluation Metrics and Implementation Details

The evaluation of our proposed architecture is done on common classification and nodule detection metrics. The metrics used for evaluating the learning capability of the dRes-SPAN are accuracy (ACC), true positive rate (TPR), Precision ($PREC$), and false positive rate (FPR). Furthermore, the TPR and FPR are used to evaluate the region of convergence (ROC) and the area under curve (AUC) [24]. The equations for the metrics are given by,

$$ACC = \frac{TP+TN}{TP+FP+FN+TN}, \quad (5)$$

$$TPR = \frac{TP}{TP+FN}, \quad (6)$$

$$PREC = \frac{TP}{TP+FP}, \quad (7)$$

$$FPR = \frac{FP}{FP+TN}, \quad (8)$$

where TP, FP, TN, and FN signify true positives, False Positives, True Negatives, and False Negatives, respectively. To evaluate the performance of the detection, the FP/scan metric is used [21]. This metric counts the total number of patches wrongly detected when a CT scan image is processed slice by slice.

All experiments have been conducted using an I7-13700, 2.1 GHz system with 32 GB RAM and NVIDIA RTX A2000 12 GB GPU. The learning rates, weight decay, loss function, and the optimizer that is used for all methods included in the comparison are taken as specified by Aydoghmishi *et al.* in [2] for training large networks on a small group of images. Furthermore, all models are trained for 100 epochs using a batch size of 256. The fully connected layers of traditional encoder models such as VGGNet and ResNet have been reduced to only two layers similar to the proposed dRes-SPAN model.

4.3 Results and Discussion

The experiments for evaluating the proposed method have been conducted in two stages: i) Evaluation of the proposed encoder architecture on nodule classification in terms of ACC, TPR, $PREC$, FPR, time per epoch (T/epoch), and trainable parameters, along with ROC-AUC analysis; ii) Evaluation of the proposed detection framework in terms of TPR versus FP/scan. For each evaluation, a comparison with the state-of-the-art architectures is also included. From Table 1, it can be seen that the proposed dRes-SPAN achieves higher performance in all respects than traditional CNN encoders. Furthermore, the proposed method shows a significant decrease in training time compared to very large networks making it highly suitable for nodule detection tasks. From Fig. 3c), it can be seen that the model harbors good generalization capability with an area under curve of 0.9909 for predicting nodules. It takes approximately 95 epochs for the proposed model to converge showing a classification accuracy of 95.85% on the training set. On the test set, the proposed model shows an accuracy of 95.35% after 100 epochs of training.

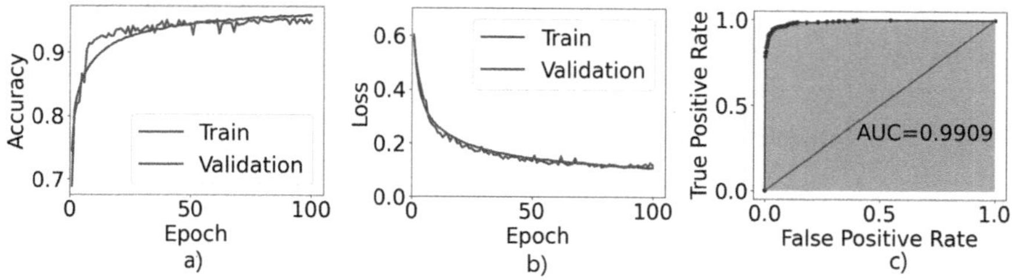

Fig. 3. Training results for SPAN a) Accuracy analysis b) Loss reduction analysis c) Generalization capability.

Table 1. Evaluation of Metrics for the distinction of nodules from the background using dRes-SPAN compared to CNN encoders for traditional object detection methods on the test set after 100 epochs of training. Here, **Params** represent trainable parameters.

Model	*ACC* (%)	*TPR* (%)	*PREC* (%)	*FPR* (%)	T/epoch (s)	Params
VGGNet [7]	90.74	91.03	89.34	10.51	686	17.94×10^6
ResNet [6]	91.57	92.31	94.82	6.40	228	36.37×10^6
dRes-SPAN	95.35	94.95	96.07	4.12	54	1.02×10^6

Similarly, for the second phase of evaluation, we have picked a few state-of-the-art nodule detection architectures in the field that are evaluated on 2D CT scan images. The results and comparison are shown in Table 2. From the table, it can be seen that even though the detection sensitivity for the proposed HSPM + dRes-SPAN is marginally better and the FP/scan is lower than the method presented in [21], the proposed method shows faster inference compared to the method due to its efficiency in generating lesser number of region propositions. Furthermore, the YOLO architecture requires a high amount of annotations which come in the form of seven annotations per patch in an image, and that can mount up to thousands of images in a CT scan that need seven annotations each. Compared to YOLO, the proposed architecture only requires two annotations for every nodule in a CT scan making it much lighter and suitable for medical image diagnosis.

Table 2. Comparison of detection sensitivity versus average false positive per scan for various state-of-the-art nodule detection architectures. Here, the term, Annotations, refers to the amount of annotated data needed to train an architecture and Speed refers to the inference speed to process the entire CT scan slice by slice.

Method	TPR (%)	FP/scan	Annotation	Speed
Mean-shift + Faster RCNN [21]	94.31	**2.7**	Light	Slow
YOLOv3 [5]	89.43	6.8	Heavy	Fast
HSPM + dRes-SPAN (proposed)	**94.43**	3.1	Light	Fast

5 Conclusion

The architecture presented in this paper utilizes the power of graph propagation, residual learning, and superpixel clustering to produce a powerful lung nodule detection method. It can be seen from the experimentations that by selecting a suitable superpixel generation method and using a Euclidean distance-based edge formation technique, superpixels can be used along with graphical networks to learn powerful intrinsic features. Furthermore, results from our experiments show a significant reduction in training time and trainable parameters while achieving a better overall performance in nodule detection compared to several available architectures in the field. For future works, the architecture will be trained and tested on more open-access datasets to improve its robustness. Consequently, the dRes-SPAN will be used in nodule segmentation where it will be trained to segment nodules with high boundary adherence from 2D CT scan images. Furthermore, the proposed architecture will also be used in nodule malignancy prediction and grading.

Acknowledgments. This work has been funded by the Council of Canada (NSERC) Discovery Grants (RGPIN-2018-05523) and (RGPIN-2019-04696), and the University of Windsor, Office of Research Services and Innovation.

References

1. Armato, S.G., III., et al.: The lung image database consortium (LIDC) and image database resource initiative (IDRI): a completed reference database of lung nodules on CT scans. Med. Phys. **38**(2), 915–931 (2011). https://doi.org/10.1118/1.3528204
2. Aydoghmishi, F.M., Modak, S., Abdel-Raheem, E., Rueda, L.: Examining the performance of melanoma classification using superpixel segmentation: a comparative analysis. In: 2023 International Conference on Microelectronics (ICM), pp. 113–118. IEEE (2023). https://doi.org/10.1109/ICM60448.2023.10378939
3. Bae, J.H., et al.: Superpixel image classification with graph convolutional neural networks based on learnable positional embedding. Appl. Sci. **12**(18), 9176 (2022). https://doi.org/10.3390/app12189176

4. Van den Bergh, M., Boix, X., Roig, G., Van Gool, L.: SEEDS: superpixels extracted via energy-driven sampling. Int. J. Comput. Vis. **111**, 298–314 (2015). https://doi.org/10.1007/s11263-014-0744-2
5. Bu, Z., et al.: Lung nodule detection based on YOLOv3 deep learning with limited datasets. Mol. Cell. Biomech. **19**(1), 17–28 (2022). https://doi.org/10.32604/mcb.2022.018318
6. Cheng, B., Wei, Y., Shi, H., Feris, R., Xiong, J., Huang, T.: Revisiting RCNN: on awakening the classification power of faster RCNN. In: Ferrari, V., Hebert, M., Sminchisescu, C., Weiss, Y. (eds.) ECCV 2018. LNCS, vol. 11219, pp. 473–490. Springer, Cham (2018). https://doi.org/10.1007/978-3-030-01267-0_28
7. Girshick, R.: Fast R-CNN. In: Proceedings of the IEEE International Conference on Computer Vision, pp. 1440–1448 (2015)
8. Girshick, R., Donahue, J., Darrell, T., Malik, J.: Rich feature hierarchies for accurate object detection and semantic segmentation. In: Proceedings of the IEEE Conference on Computer Vision and Pattern Recognition, pp. 580–587 (2014)
9. Han, K., Wang, Y., Guo, J., Tang, Y., Wu, E.: Vision GNN: An image is worth graph of nodes. In: Advances in Neural Information Processing Systems, vol. 35, pp. 8291–8303 (2022)
10. Han, Y., Wang, P., Kundu, S., Ding, Y., Wang, Z.: Vision HGNN: an image is more than a graph of nodes. In: Proceedings of the IEEE/CVF International Conference on Computer Vision, pp. 19878–19888 (2023)
11. Hovinga, M., Sprengers, R., Kauczor, H.-U., Schaefer-Prokop, C.: CT imaging of interstitial lung diseases. In: Schoepf, U.J.J., Meinel, F.G.G. (eds.) Multidetector-Row CT of the Thorax. MR, pp. 105–130. Springer, Cham (2016). https://doi.org/10.1007/978-3-319-30355-0_7
12. Ji, Z., et al.: Lung nodule detection in medical images based on improved YOLOv5s. IEEE Access (2023). https://doi.org/10.1109/ACCESS.2023.3296530
13. Kim, M., et al.: Deep learning in medical imaging. Neurospine **16**(4), 657 (2019). https://doi.org/10.14245/ns.1938396.198
14. Li, R., Xiao, C., Huang, Y., Hassan, H., Huang, B.: Deep learning applications in computed tomography images for pulmonary nodule detection and diagnosis: a review. Diagnostics **12**(2), 298 (2022). https://doi.org/10.3390/diagnostics12020298
15. Liu, K.: STBi-YOLO: a real-time object detection method for lung nodule recognition. IEEE Access **10**, 75385–75394 (2022). https://doi.org/10.1109/ACCESS.2022.3192034
16. Ma, L., Li, G., Feng, X., Fan, Q., Liu, L.: TiCNet: transformer in convolutional neural network for pulmonary nodule detection on CT images. J. Imaging Inform. Med. **37**, 196–208 (2024). https://doi.org/10.1007/s10278-023-00904-y
17. Mkindu, H., Wu, L., Zhao, Y.: Lung nodule detection in chest CT images based on vision transformer network with Bayesian optimization. Biomed. Sig. Process. Control **85**, 104866 (2023). https://doi.org/10.1016/j.bspc.2023.104866
18. Modak, S., Abdel-Raheem, E., Rueda, L.: Applications of deep learning in disease diagnosis of chest radiographs: a survey on materials and methods. Biomed. Eng. Adv. **5**, 100076 (2023). https://doi.org/10.1016/j.bea.2023.100076
19. Modak, S., Abdel-Raheem, E., Rueda, L.: Lung nodule segmentation on CT scan images using Patchwise Iterative Graph Clustering. In: 2023 IEEE International Symposium on Circuits and Systems (ISCAS), pp. 1–5. IEEE (2023). https://doi.org/10.1109/ISCAS46773.2023.10181811

20. Modak, S., Abdel-Raheem, E., Taha, L.Y.: A novel adaptive multilevel thresholding based algorithm for QRS detection. Biomed. Eng. Adv. **2**, 100016 (2021). https://doi.org/10.1016/j.bea.2021.100016
21. Nguyen, C.C., Tran, G.S., Burie, J.C., Nghiem, T.P., et al.: Pulmonary nodule detection based on faster R-CNN with adaptive anchor box. IEEE Access **9**, 154740–154751 (2021). https://doi.org/10.1109/ACCESS.2021.3128942
22. Redmon, J., Farhadi, A.: YOLOv3: an incremental improvement. arXiv preprint arXiv:1804.02767 (2018). https://doi.org/10.48550/arXiv.1804.02767
23. Usman, M., Rehman, A., Shahid, A., Latif, S., Shin, Y.G.: MEDS-Net: multi-encoder based self-distilled network with bidirectional maximum intensity projections fusion for lung nodule detection. Eng. Appl. Artif. Intell. **129**, 107597 (2024). https://doi.org/10.1016/j.engappai.2023.107597
24. Yagin, F.H., Alkhateeb, A., Colak, C., Azzeh, M., Yagin, B., Rueda, L.: A fecal-microbial-extracellular-vesicles-based metabolomics machine learning framework and biomarker discovery for predicting colorectal cancer patients. Metabolites **13**(5), 589 (2023). https://doi.org/10.3390/metabo13050589
25. Zhang, M., Liu, S., Zeng, B.: Hierarchical region proposal refinement network for weakly supervised object detection. In: 2021 IEEE International Conference on Image Processing (ICIP), pp. 669–673. IEEE (2021). https://doi.org/10.1109/ICIP42928.2021.9506087

Histology Image Artifact Restoration with Lightweight Transformer Based Diffusion Model

Chong Wang[1], Zhenqi He[2], Junjun He[3], Jin Ye[3], and Yiqing Shen[4(✉)]

[1] School of Biological Science and Medical Engineering, BeiHang University, Beijing, China
[2] Department of Mathematics, Faculty of Science, The University of Hong Kong, Kowloon, Hong Kong SAR
[3] Shanghai AI Laboratory, Shanghai, China
[4] Department of Computer Science, Johns Hopkins University, Baltimore, MD, USA
yshen92@jhu.edu

Abstract. Histology whole slide images (WSIs) are extensively used in tumor diagnosis and treatment planning. However, the presence of artifacts resulting from improper operations during WSI collection can impede both manual and deep learning-based analysis. To bridge this gap, we propose an innovative denoising diffusion model tailored for inpainting artifact-laden regions in histology WSIs. Our method focuses on preserving the intricate morphological structures, which are essential for accurate diagnosis. To ensure the preservation of morphological structures during regional artifact inpainting, we have developed a novel lightweight Transformer-based denoising network, that can capture the correlations between the regional artifact with the global morphological structures. In comparison to existing generative adversarial network (GAN) based solutions, our method minimizes changes in morphology while maximizing preservation of the stain style during the restoration of the artifact. By providing a more reliable and accurate restoration of artifact-affected areas, our model facilitates better analysis and interpretation of histological images, thereby potentially improving the accuracy of tumor diagnosis and treatment decisions. The code is available at https://github.com/zhenqi-he/artifact-restoration.

Keyword: Image Inpainting · Artifacts Restoration · Denoising Diffusion Model

1 Introduction

The analysis of histology whole slide images (WSIs) is crucial for various medical tasks, such as tumor diagnosis, nuclei localization, and treatment planning. However, the integrity of WSIs in histological analyses is frequently jeopardized by a multitude of variables, among which inaccuracies arising during the scanning or collection phases represent a predominant factor [12]. Such procedural deviations can induce significant morphological alterations in tissue structures, which are manifest as artifacts, including but not limited to tissue folds and bubbles [15]. The presence of such artifacts substantially complicates the accurate evaluation and interpretation of histological samples.

J. Finkelstein et al. (Eds.): AIME 2024, LNAI 14845, pp. 81–89, 2024.
https://doi.org/10.1007/978-3-031-66535-6_9

This complication elevates the likelihood of misdiagnosis among medical professionals and compromises the performance efficacy of Computer-Aided Diagnosis (CAD) systems [17]. Particularly concerning is the detrimental impact on deep learning models, which are increasingly deployed in histological analyses. These computational models are intrinsically vulnerable to the distorting influence of artifacts, thereby engendering potential misclassification or mislabeling of critical tissue structures. This, in turn, exacerbates the risk factors associated with both false positive and false negative diagnoses [19].

In clinical applications, although it is technically possible to rectify the issue of artifacts through re-scanning or re-preparation of the affected tissue slides, such a course of action imposes significant logistical and financial burdens, often requiring multiple iterations to achieve a satisfactory WSI [4]. As for CAD, the strategy of identifying and excluding artifact-laden regions does offer a solution; however, this tactic inevitably results in a forfeiture of valuable contextual information, thereby undermining the overall reliability of the diagnostic model [2]. Existing techniques for mitigating various types of artifacts, including but not limited to folds [11], ink markings [1], and out-of-focus [8] zones, have been introduced; yet, these methods remain inadequate in restoring occluded regions. More recently, approaches leveraging generative networks have been proposed for the remediation of artifact-afflicted areas [10]. These methods, however, suffer from inherent challenges related to mode collapse, complicating their training phases. Additionally, the propensity of generative-based approaches to induce alterations in structure or coloration detracts from the precision of the restored regions.

To ameliorate these challenges, we propose a novel Denoising Diffusion Probabilistic Model (DDPM) [7] with lightweight transformer as the denoising network, specifically engineered for the inpainting of artifact-laden regions within histological WSIs. Diverging significantly from the generative-based paradigms, the proposed methodology obviates the need for artifact-laden images during training, substantially facilitating the data acquisition process and streamlining the training pipeline. The key contributions of this study can be summarized as follows: (1) The introduction of an innovative, lightweight Transformer-based denoising network, purposed for the nuanced capturing of correlations between regional artifacts and global morphological structures. (2) The implementation of a conditional framework during the restoration process, selectively targeting artifact regions while scrupulously preserving the morphological and staining features of artifact-free zones.

2 Related Work

Artifact in Histology. The challenge of mitigating artifacts in medical image analysis is both enduring and complex, particularly for the field of histopathology. The presence of artifacts can significantly compromise the efficacy of computer-aided diagnosis systems, often leading deep learning algorithms to misclassify artifact-ridden areas as tumors. Artifacts are typically spawned by a variety of factors, such as equipment specifications, data acquisition protocols, and environmental conditions [12]. These intrusions not only degrade the overall quality of WSIs but also introduce errors in subsequent computational analyses. To address this, current practices often involve manually omitting images containing artifacts during the curation of histopathological datasets [6]. While this ensures data reliability, this procedure also poses a dilemma, namely the exclusion of these images leads to a loss of potentially valuable information, thereby generating a problem of dataset incompleteness.

Generative Model for Artifact Restoration. Generative models have shown their promise in ameliorating histological artifacts [16]. For example, CycleGAN has been adapted to tackle this specific problem [20]. However, existing generative techniques are not without their limitations. One significant drawback is the inadvertent alteration of staining styles, resulting in unnatural color variations across the histological images. Although recent advancements have mitigated some of these issues, a case in point being a refined CycleGAN model AR-CycleGAN that employs artifact and pseudo-color cycles [10], GAN-based approaches still face challenges. In particular, they can unintentionally introduce morphological changes in regions that are already artifact-free, putting at risk the integrity of critical tissue structure information.

3 Methods

The proposed method for denoising and restoring artifacts in medical images consists of two stages: training on artifact-free images and artifact restoration (Algorithm 1). During the training stage, the distribution of artifact-free regions is learned through a denoising diffusion model. In the inference stage, artifact regions are specifically targeted for restoration by substituting them with Gaussian noise and leveraging the contextual regions of artifact-free areas for recovery.

3.1 Diffusion Training Stage

In the training phase, local tissue structures are learned from artifact-free histological images for subsequent artifact restoration. Following the standard formulation of DDPM [7], it comprises a forward process and a reverse process. In the forward process, noises are gradually added until the structure of the image is fully destroyed, and the image at each timestep t can be determined by the states at the previous timestep $t-1$ through a Gaussian perturbation $q(\mathbf{x}_t|\mathbf{x}_{t-1}) = N\left(x_t; \sqrt{1-\beta_t}\mathbf{x}_{t-1}, \beta_t\mathbf{I}\right)$ parameterized by β_t. On the contrary, during the reverse phase, a denoising neural network is trained to approximate the conditional distribution $p(\mathbf{x}_{t-1}|\mathbf{x}_t)$, effectively reversing the effects of the noise-inducing operation $q(\mathbf{x}_t|\mathbf{x}_{t-1})$. The objective is to iteratively reconstruct the original data distribution from an initial condition of white noise ($\mathbf{x}_T$). Each reverse step, guided by the learned denoising function, refines the generated sample by removing noise and enhancing the essential features of the underlying data structure. Through this iterative process, the network essentially reverses the diffusion procedure, leading to a high-fidelity reconstruction of the original image or data point. Exemplary samples generated are demonstrated in Fig. 1.

Denoising diffusion methods predicated on UNet architecture are typically characterized by a substantial number of model parameters, resulting in protracted inference times. Inspired by SegFormer [18], to optimize efficiency for local deployment, a

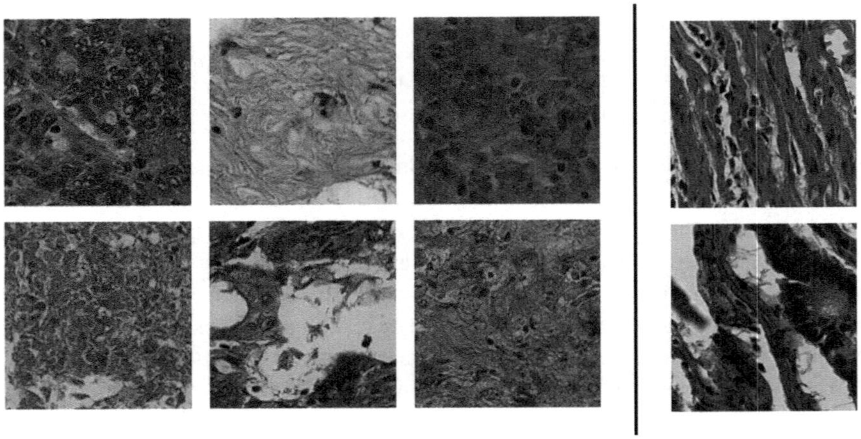

Fig. 1. The generated histological images by the diffusion model (left) and real-world histology data (right).

denoising network leveraging a lightweight Transformer architecture was engineered. Notably, relative to convolutional networks, the Transformer-based architecture demonstrates superior efficacy in capturing the intricate relationships between regional artifacts and global morphological structures (Fig. 2).

3.2 Artifact Restoration

In the artifact restoration stage, we utilize the diffusion model to restore the artifact region from an inpainting perspective (Fig. 3). Firstly, a threshold-based segmentation method is first used to detect the artifact regions in the input image $\mathbf{x}_0$. We then use this information to condition the denoising process only to the artifact regions, while maintaining the artifact-free regions to preserve the morphological structure and stain style. Formally, we write the artifact region as $\mathbf{x}_0 \odot \mathbf{m}$ and the artifact-free region as $\mathbf{x}_0 \odot (1 - \mathbf{m})$, where $\mathbf{m}$ is a boolean mask generated in the threshold method, and $\odot$ is the pixel-wise multiplication operator. It follows that the selective denoising process to inpaint the artifact region is formulated as $\mathbf{x}_t^{in} = \mathbf{x}_t^{sample} \odot (1 - \mathbf{m}) + \mathbf{x}_{t+1}^{out} \odot \mathbf{m}$, where $\mathbf{x}_t^{sample}$ follows the Gaussian perturbation in the forward process, and $\mathbf{x}_{t+1}^{out}$ is the output from the denoising network in the previous reverse step. We employ the Denoising Diffusion Implicit Model (DDIM) [13] to accelerate the restoration process.

Algorithm 1 Overall procedure for artifact restoration

Require: Original image x_0
Ensure: Denoised image x_T^{out}

```
                                        ▷ Non-Artifact Image Training Phase
Train Denoising Diffusion Model to learn p(x)
                                        ▷ Inference Phase
Identify artifact regions using threshold-based method
m ← Boolean mask of artifact regions
                                        ▷ Artifact Restoration Phase
for each timestep t = 0, 1, ..., T do
    if t is forward timestep then
        x_t ← ApplyNoise(x_{t−1}, β_t)
    else
        x_t ← Denoise(x_{t+1}, m)
    end if
    x_t^in ← x_t^sample ⊙ (1 − m) + x_{t+1}^out ⊙ m
end for
return x_T^out
```

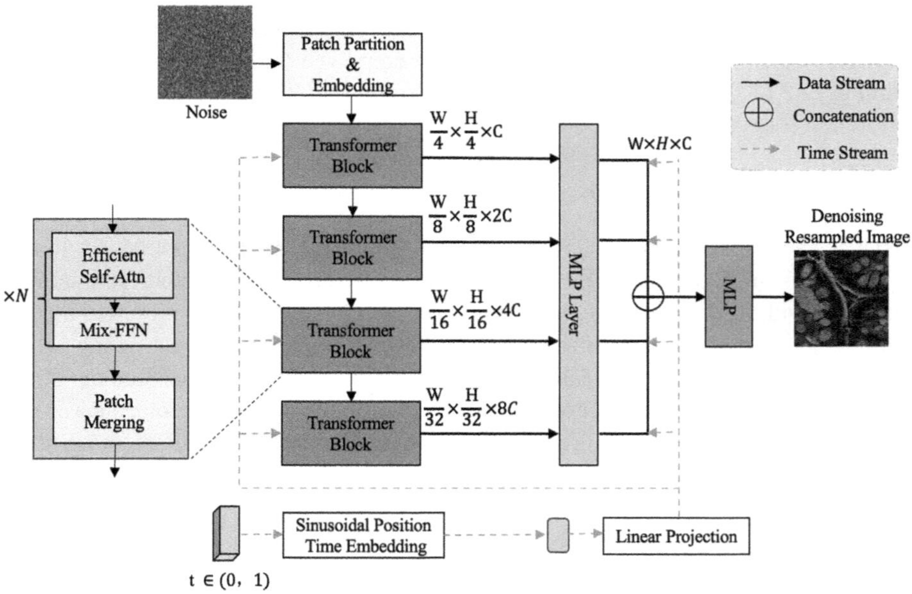

Fig. 2. The proposed lightweight Transformer-based denoising network. FFN, Feedforward Network; MLP, Multilayer Perceptron.

4 Experiments

4.1 Datasets and Settings

The efficacy of the proposed framework for artifact restoration has been rigorously evaluated in the context of both artifact removal and downstream tissue structure classification tasks. For the purpose of ensuring the reliability of the findings, the publicly accessible dataset, NCT-CRC-HE-100K [9], was employed for the training phase, while validation was conducted using the CRC-VAL-HE-7K dataset. We regard the images from the original dataset as clean and proceed to artificially synthesize Artifact images. Owing to the scarcity of comparable studies and open-source codes in the existing literature, CycleGAN [20] was chosen as the benchmark for comparison, and its performance was also evaluated on downstream classification tasks.

4.2 Implementation Details

We implemented the proposed method along with the associated downstream classification tasks using Python 3.8.10 and the PyTorch 1.10.0 library. To accelerate computations and facilitate parallelism, all experiments were run on a dual-GPU setup, specifically using two NVIDIA RTX A4000 graphics cards, each outfitted with 16 GiB of memory. Within the context of our denoising diffusion model, we adhered to a specific set of hyperparameters to optimize performance. We employed the Adam optimizer with a learning rate of the 1×10^4, and designated a total of 250 time steps for the artifact restoration for computational efficiency.

4.3 Evaluation of Artifact Restoration

In the evaluation of artifact restoration, one crucial aspect that must not be overlooked in the realm of medical imaging is the preservation of tissue structure and staining variations within artifact-free regions. As evidenced by the Fig. 4, the proposed method exhibits significant efficacy in retaining the structural integrity of the image. These results suggest that by strategically utilizing the distribution of artifact-free regions as a guide for the restoration process within artifact-afflicted areas, it becomes feasible to accurately replicate the tissue structure and staining characteristics inherent to artifact-free regions. This in turn culminates in a more authentic restoration of tissue images, thereby enhancing their clinical relevance.

4.4 Comparisons of Model Complexity

A noteworthy feature of our proposed method lies in its efficiency in terms of parameter count. Relative to other extant models, our approach operates with a substantially reduced number of parameters (Table 1). Specifically, when contrasted with UNet-based denoising models [7], our model manifests a 91.1% reduction in the number of parameters, diminishing from 1.081×10^8 to 9.62×10^6. This reduction in parameter complexity translates to accelerated inference times, which have been curtailed from 112.37 s to a mere 24.64 s. Such an improvement substantially augments the suitability of our model for real-world clinical deployment.

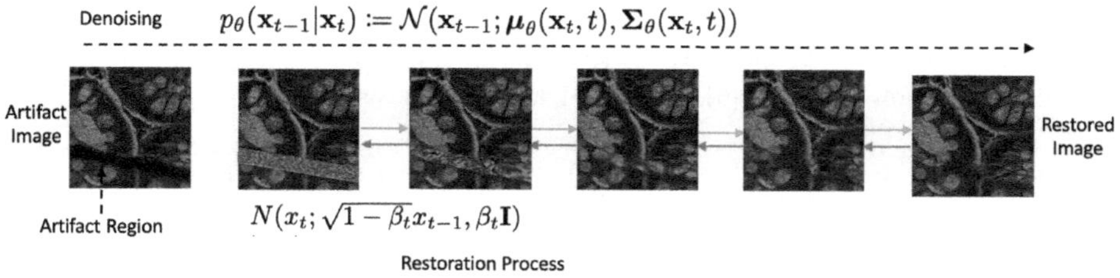

Fig. 3. The proposed artifact restoration pipeline.

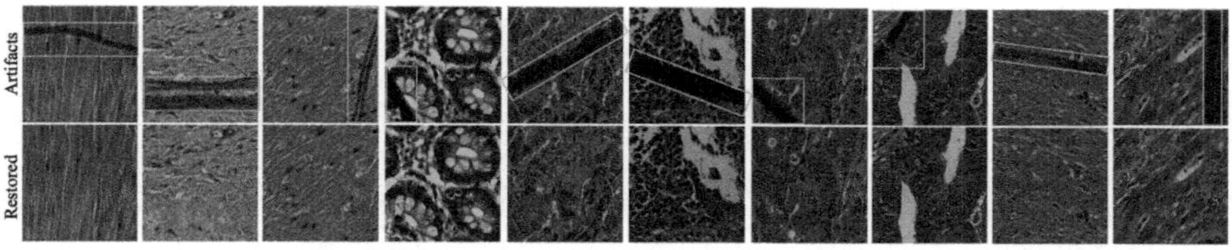

Fig. 4. Illustration of histology images with artifact, where the artifact region is highlighted by green rectangles (□); and restored images by our proposed denoising diffusion model.

4.5 Evaluation on Downstream Classification Tasks

To evaluate the effectiveness of the proposed artifact restoration framework, an assessment was conducted on downstream tissue structure classification tasks. Two baselines, denoted as 'Clean' and 'Artifacts', served as comparative standards. 'Clean' represents artifact-free images, while 'Artifacts' encapsulates images afflicted by artifacts. These baselines are critical for contextualizing the improvements furnished by our model. In Table 2, a comprehensive assessment against multiple classification benchmarks revealed that our method outperforms the CycleGAN approach across all metrics. Specifically, it was observed that the presence of artifacts led to a significant degradation in classification accuracy, with an average decline of 8.82%. However, the artifact removal augmented the classification accuracy of afflicted images. For instance, on the ResNet50 architecture, compared to the 'Artifacts' baseline, our method enhanced the accuracy by 7.13%, while CycleGAN improved it by 2.98%. Compared to the CycleGAN benchmark, our approach exhibited superior accuracy with an improvement of 3.54%, indicating enhanced capabilities for artifact restoration.

Table 1. Model complexity comparison in terms of the number of parameters and averaged inference time.

Methods	#Params ($\times 10^6$)	Time (s)
CycleGAN	28.28	1.065
DDPM (UNet)	108.41	112.37
Ours	9.62	24.64

Table 2. The effectiveness of the proposed artifact restoration framework on the downstream task-tissue classification task. We report the classification accuracy on the test set (%) with different network architectures of ResNet [5], RexNet [3], and EfficientNet [14].

Method	ResNet18	ResNet50	RexNet100	EfficientNetB0
Clean	95.529	94.833	95.487	95.808
Artifacts	80.302	85.012	90.446	90.626
CycleGAN	86.326	87.994	90.776	91.811
Ours	91.676	92.144	93.786	93.449

5 Conclusions

One salient finding was the marked degradation in classification accuracy in the presence of image artifacts—a challenge our method adeptly mitigates. Beyond merely enhancing accuracy rates, our framework sets itself apart through its exceptional efficiency in utilizing parameters. This efficiency not only showcases the robustness of our approach but also its practical viability in clinical environments. In such settings, the dual requirements of high accuracy and computational economy are critical, and our framework meets these demands admirably. Moreover, it is important to acknowledge that while our model does incur longer inference times, especially when compared to alternatives like CycleGAN, this outcome is a direct result of its intricate sequential diffusion steps. These steps, though time-consuming, are integral to the model's ability to accurately process and improve upon the input images. Future work will explore the application of distillation techniques to reduce the number of inference steps, thereby aiming to optimize the duration of the inference phase.

References

1. Ali, S., Alham, N.K., Verrill, C., Rittscher, J.: Ink removal from histopathology whole slide images by combining classification, detection and image generation models. In: 2019 IEEE 16th International Symposium on Biomedical Imaging (ISBI 2019), pp. 928–932. IEEE (2019)
2. Bautista, P.A., Yagi, Y.: Detection of tissue folds in whole slide images. In: 2009 Annual International Conference of the IEEE Engineering in Medicine and Biology Society, pp. 3669–3672. IEEE (2009)
3. Han, D., Yun, S., Heo, B., Yoo, Y.: RexNet: diminishing representational bottleneck on convolutional neural network (2020)
4. Han, L., Su, H., Yin, Z.: Phase contrast image restoration by formulating its imaging principle and reversing the formulation with deep neural networks. IEEE Trans. Med. Imaging **42**(4), 1068–1082 (2022)
5. He, K., Zhang, X., Ren, S., Sun, J.: Deep residual learning for image recognition. CoRR abs/1512.03385 (2015). http://arxiv.org/abs/1512.03385
6. He, Z., He, J., Ye, J., Shen, Y.: Artifact restoration in histology images with diffusion probabilistic models. arXiv preprint arXiv:2307.14262 (2023)

7. Ho, J., Jain, A., Abbeel, P.: Denoising diffusion probabilistic models. CoRR abs/2006.11239 (2020). https://arxiv.org/abs/2006.11239
8. Hosseini, M.S., Brawley-Hayes, J.A., Zhang, Y., Chan, L., Plataniotis, K.N., Damaskinos, S.: Focus quality assessment of high-throughput whole slide imaging in digital pathology. IEEE Trans. Med. Imaging **39**(1), 62–74 (2019)
9. Kather, J.N., et al.: Predicting survival from colorectal cancer histology slides using deep learning: a retrospective multicenter study. PLOS MED **16**(1), e1002730 (2019)
10. Ke, J., et al.: Artifact detection and restoration in histology images with stain-style and structural preservation. IEEE Trans. Med. Imaging **42**, 3487–3500 (2023)
11. Kothari, S., Phan, J.H., Wang, M.D.: Eliminating tissue-fold artifacts in histopathological whole-slide images for improved image-based prediction of cancer grade. J. Pathol. Inform. **4**(1), 22 (2013)
12. Seoane, J., Varela-Centelles, P., Ramírez, J., Cameselle-Teijeiro, J., Romero, M.: Artefacts in oral incisional biopsies in general dental practice: a pathology audit. Oral Dis. **10**(2), 113–117 (2004)
13. Song, J., Meng, C., Ermon, S.: Denoising diffusion implicit models. arXiv:2010.02502, October 2020. https://arxiv.org/abs/2010.02502
14. Tan, M., Le, Q.: EfficientNet: rethinking model scaling for convolutional neural networks (2020)
15. Taqi, S.A., Sami, S.A., Sami, L.B., Zaki, S.A.: A review of artifacts in histopathology. J. Oral Maxillofacial Pathol. JOMFP **22**(2), 279 (2018)
16. Venkatesh, B., Shaht, T., Chen, A., Ghafurian, S.: Restoration of marker occluded hematoxylin and eosin stained whole slide histology images using generative adversarial networks. In: 2020 IEEE 17th International Symposium on Biomedical Imaging (ISBI), pp. 591–595. IEEE (2020)
17. Wang, N.C., Kaplan, J., Lee, J., Hodgin, J., Udager, A., Rao, A.: Stress testing pathology models with generated artifacts. J. Pathol. Inform. **12**(1), 54 (2021)
18. Xie, E., Wang, W., Yu, Z., Anandkumar, A., Alvarez, J.M., Luo, P.: SegFormer: simple and efficient design for semantic segmentation with transformers. In: Neural Information Processing Systems (NeurIPS) (2021)
19. Zhang, Y., Sun, Y., Li, H., Zheng, S., Zhu, C., Yang, L.: Benchmarking the robustness of deep neural networks to common corruptions in digital pathology. In: Wang, L., Dou, Q., Fletcher, P.T., Speidel, S., Li, S. (eds.) International Conference on Medical Image Computing and Computer-Assisted Intervention, vol. 13432, pp. 242–252. Springer, Cham (2022). https://doi.org/10.1007/978-3-031-16434-7_24
20. Zhu, J.Y., Park, T., Isola, P., Efros, A.A.: Unpaired image-to-image translation using cycle-consistent adversarial networks. In: ICCV (2017). https://doi.org/10.1109/iccv.2017.244

Improved Glioma Grade Prediction with Mean Image Transformation

Shyam Sundhar Yathirajam and Sreedevi Gutta(✉)

California State University San Marcos, Computer Science and Information Systems, San Marcos, CA 92096, USA
sgutta@csusm.edu

Abstract. Accurate prediction of glioma grade is significant for treatment planning. Recently, several convolutional neural networks (CNN) were proposed that were trained on one or more combinations of T1, T2, T1 contrast enhanced (T1CE), and fluid attenuated inversion recovery (FLAIR) sequences. In this work, we explore the potential of training CNN by combining these four sequences to form a mean image that encapsulates core characteristics of tumor. The results indicate that the CNN trained with mean images has an F1-score of 94%, which is 7% higher than the CNN trained with stacked sequences. In addition, we explored the potential of machine learning models that were trained on features extracted from CNN. The results indicate that the ML models trained on features extracted from mean computed CNN has an F1-score of 98%, which is 8% superior compared to the CNN trained on stacked sequences. In addition, the proposed model is 2 times smaller with 5.7 million parameters compared to traditional method which has 11.4 million parameters. These findings highlight the potential of combining the sequences for enhancing the accuracy of glioma grading.

Keywords: glioma grading · mean computation · accuracy

1 Introduction

A brain tumor refers to an unusual growth of cells in the brain or its surrounding tissues within the central nervous system (CNS). The most prevalent malignant brain tumor among adults is Glioma [10]. Gliomas are categorized into grades I to IV. Grades I and II are termed as low-grade (least aggressive) and grades III and IV are termed as high-grade (most aggressive). Despite available treatment modalities such as chemotherapy, radiation therapy, individuals with high-grade gliomas experience poor survival rates [3].

The treatment approach for patients with glioma depends on tumor grade [6]. The standard way to determine the grade is through invasive methods such as biopsy, with pathologists examining tissue samples under a microscope. However, this process is expensive and time-consuming. Thus, there is a need for an accurate and non-invasive diagnostic method to predict tumor grade.

J. Finkelstein et al. (Eds.): AIME 2024, LNAI 14845, pp. 90–94, 2024.
https://doi.org/10.1007/978-3-031-66535-6_10

Previously, several studies have utilized radiomic features for non-invasive prediction of glioma grade [4, 7, 8]. These studies [4, 5, 7–9] first extracted radiomic features from Magnetic Resonance (MR) images and then these features were used by machine learning (ML) models for determination of glioma grade. Note that these radiomic features are hand-engineered containing limited information, thereby, restraining the performance of ML models. Therefore, it is required to build models that learn features directly from the data.

Recently, deep learning (DL) models, involving convolutional neural networks (CNNs) were proposed for glioma grade classification [1, 12]. The CNNs have the ability to learn features directly from MR images. Traditionally, one or more of the following MRI contrast sequences have been utilized to predict tumor grade: T1-weighted, T1-weighted contrast-enhanced (T1CE), T2-weighted, and T2 Fluid Attenuated Inversion Recovery (FLAIR). Each sequence offers unique information about the tumor [2]. Since these sequences are three-dimensional, they contribute to a four-dimensional patient data set. However, this substantial data volume necessitates the construction of huge models with a large number of parameters. This in turn requires extensive data for training and parameter optimization. In this paper, we propose a simple mean image transformation applied to all four sequences to reduce data size and alleviate the demand for extensive data requirement. We compare the proposed mean approach with the conventional method of stacking the sequences.

2 Imaging Data Description

The dataset used in this work were from the brain tumor segmentation 2018 (BraTS'18) challenge [2, 11]. The BraTS 2018 dataset consists of multi-institutional clinically-acquired pre-operative multimodal MRI scans of glioblastoma (GBM/HGG) and low-grade glioma (LGG) [2, 11]. The dataset consist of 210 patients with HGG and 75 patients with LGG, a total of 285 patients. For each patient, the following scans were collected: a) native (T1) and b) T1CE, c) T2-weighted (T2), and d) FLAIR volumes. The dataset were provided after co-registrating, skull-stripping, and manually segmenting the tumors by experienced neuro-radiologists [2, 11].

3 Methods

In this section, we first explain the pre-processing steps and then the state-of-the-art and proposed methods for glioma grading.

3.1 Pre-processing

The following pre-processing steps were applied in preparation of data for DL algorithms. First, patient-level train-test split was performed in the ratio of 75% and 25%. Next, for each patient, 2D images of size 240×240 were extracted for each sequence. These 2D images were resized from 240×240 to 190×190 to reduce the number of trainable parameters. The quality of the image was impacted with any further reduction in the image size. The resultant images were then cropped to 120 × 120 to eliminate the external excess black region around the skull part. Data augmentation was then performed on the training set by generating different variations of the image such as rotations, flips. This addresses the class imbalance problem and enhances the model's ability to generalize well. Normalization was then applied for faster training of CNN models. The resultant data was then used for model training.

3.2 CNN Trained on Stacked Sequences

The input to the CNN architecture employed to predict grade from stacked sequences is of shape (120,120,3), where 120 × 120 corresponds to the image dimensions and 3 corresponds to FLAIR, T2, and T1CE sequences. The architecture consists of a series of three convolutional layers with rectified linear unit (ReLU) activation function and max-pooling layers. The output is flattened and is given to two dense layers with ReLU activation function and dropout was used for regularization. The final output layer consists of a single neuron with a sigmoid activation function for binary classification. To predict the grade of a patient, majority voting technique was utilized.

3.3 Proposed Method: CNN Trained on Mean Computed Images

The input to the model are the mean images computed through pixel-wise calculations across all sequences. This comprehensive mean image encapsulates the tumor characteristics present in all the four modalities. The CNN architecture trained on mean computed images is illustrated in Fig. 1. The network comprises of convolutional, max-pooling, and dense layers with a ReLU activation function. To prevent overfitting, a dropout layer was added. The output layer consists of a single neuron with sigmoid activation function for glioma grade prediction.

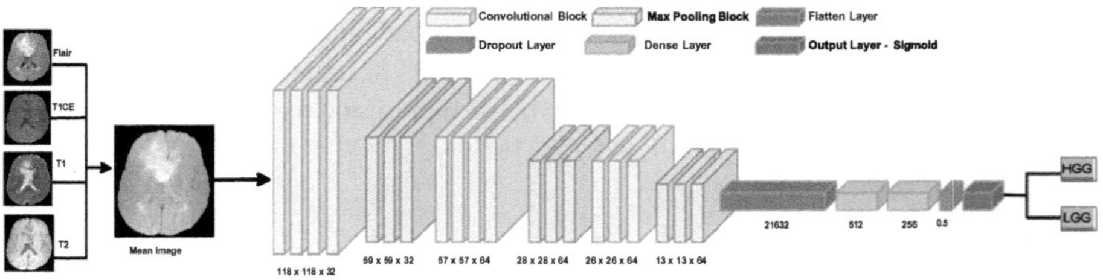

Fig. 1. The input to the network are the mean images computed from all the four sequences (FLAIR, T2, T1, T1CE). The network consists of convolutional and maxpooling layers for dimensionality reduction; and dense layers for feature extraction and classification.

4 Results and Discussion

The performance of CNN trained on mean images was compared with the conventional way of training on the stacked sequences. The results on the test set were presented in Table 1. The model trained on mean images obtained F1-score of 94%, which is 7% higher than the one trained on stacked sequences.

Table 1. Performance of CNN trained on stacked sequences and mean images. The network trained on mean images performed superior compared to the network trained on stacked sequences.

	Accuracy	Precision	Recall	F1-Score
Stacked CNN	89	90	88	87
Proposed Mean Method	**95**	**96**	**93**	**94**

To further improve the performance of grade prediction, we built ML models, support vector machine (SVM), decision tree, random forests, and Xtreme Gradient Boosting (XGBoost), trained on features learned by the CNN models. The results were presented in Table 2. As can be seen, the ML models trained on features learned by mean computed CNN performed better than the one trained on stacked sequences. Also, the performance obtained by SVM and XGBoost trained on features learned by proposed mean method was 4% higher than the results presented in Table 1. Note that all the hyperparameters in the ML models were tuned using five-fold cross-validation.

Table 2. Performance of ML models, SVM, decision tree, random forest, and XGBoost, trained on features extracted from Stacked CNN and proposed mean method. The SVM and XGBoost models trained on features learned from the proposed method had an F1-score of 98%, which is at least 7% higher than features learned from Stacked CNN.

Model	SVM				Decision Tree				Random Forest				XGBoost			
	A	P	R	F1	A	P	R	F1	A	P	R	F1	A	P	R	F1
Stacked CNN	87	92	90	91	81	94	84	88	93	92	91	91	89	95	90	90
Proposed	**98**	**98**	**98**	**98**	97	**98**	**98**	95	95	97	97	97	**98**	**98**	**98**	**98**

Abbreviations: A (Accuracy), P (Precision), R (Recall), F1 (F1 Score).

Additionally, we performed a paired t-test to determine the p-value for the results obtained by the proposed method. We obtained a p-value of 0.004, indicating our results to be statistically significant.

The training time for the proposed mean method is 897 s, whereas for the stacked CNN it is 2390 s. All the experiments were run on an Intel(R) Xeon(R) Silver 4214R CPU @ 2.40 GHz server with a total RAM of 62 GB. In addition, the number of parameters of the proposed method reduced by half from 11.4 million to 5.7 million when compared to the traditional stacked CNN method. This improvement makes the proposed method not only more computationally feasible but also more robust against overfitting, a common challenge in deep learning models. These results highlight the effectiveness of the proposed approach of training on mean computed images in enhancing the performance of glioma grade prediction, especially when paired with ML algorithms.

5 Conclusion

In this work, we proposed training CNN with mean images formed by combining T1, T1CE, T2, and FLAIR sequences. The proposed approach is compared with the traditional way of training CNN with stacked sequences. The proposed approach yielded impressive results with an F1-score of 94% compared to 87% obtained by stacking sequences. The performance further improved by 4% by extracting features from CNN and training ML algorithms. These insights are valuable for improving the performance of glioma grade prediction.

References

1. Anaraki, A.K., Ayati, M., Kazemi, F.: Magnetic resonance imaging-based brain tumor grades classification and grading via convolutional neural networks and genetic algorithms. Biocybernet. Biomed. Eng. **39**(1), 63–74 (2019)
2. Bakas, S., Reyes, M., Jakab, M., et al.: Identifying the best machine learning algorithms for brain tumor segmentation, progression assessment, and overall survival prediction in the brats challenge. arXiv preprint arXiv:1811.02629 (2018)
3. Behin, A., Hoang-Xuan, K., Carpentier, A.F., Delattre, J.Y.: Primary brain tumours in adults. The Lancet **361**(9354), 323–331 (2003)
4. Kumar, R., Gupta, A., et al.: CGHF: a computational decision support system for glioma classification using hybrid radiomics-and stationary wavelet-based features. IEEE Access **8**, 79440–79458 (2020)
5. Niu, J., Tan, Q., Zou, X., Jin, S.: Accurate prediction of glioma grades from radiomics using a multi-filter and multi-objective-based method. Math. Biosci. Eng. **20**(2), 2890–2907 (2023)
6. Ostrom, Q.T., Gittleman, H., Truitt, G., Boscia, A., Kruchko, C., Barnholtz-Sloan, J.S.: Cbtrus statistical report: primary brain and other central nervous system tumors diagnosed in the united states in 2011–2015. Neuro-oncology **20**(suppl 4), iv1–iv86 (2018)
7. Skogen, K., Schulz, A., Dormagen, J.B., Ganeshan, B., Helseth, E., Server, A.: Diagnostic performance of texture analysis on MRI in grading cerebral gliomas. Eur. J. Radiol. **85**(4), 824–829 (2016)
8. Tian, Q., Yan, L.F., Zhang, Y.Z., et al.: Radiomics strategy for glioma grading using texture features from multiparametric MRI. J. Magn. Reson. Imaging **48**(6), 1518–1528 (2018)
9. Ullah, F., et al.: Evolutionary model for brain cancer-grading and classification. IEEE Access (2023)
10. Wen, P.Y., Kesari, S.: Malignant gliomas in adults. N. Engl. J. Med. **359**(5), 492–507 (2008)
11. Weninger, L., Rippel, O., Koppers, S., Merhof, D.: Segmentation of brain tumors and patient survival prediction: methods for the brats 2018 challenge. In: Brainlesion: Glioma, Multiple Sclerosis, Stroke and Traumatic Brain Injuries: 4th International Workshop, BrainLes 2018, Held in Conjunction with MICCAI 2018, Granada, 16 September 2018, Revised Selected Papers, Part II 4, pp. 3–12. Springer (2019)
12. Yang, Y., Yan, L.F., Zhang, D.L., et al.: Glioma grading on conventional MR images: a deep learning study with transfer learning. Front. Neurosci. **12**, 804 (2018)

Learning to Predict the Optimal Template in Stain Normalization for Histology Image Analysis

Shiling Luo[1], Junxin Feng[1], Yiqing Shen[2], and Qiongxiong Ma[1(✉)]

[1] Guangdong Provincial Key Laboratory of Nanophotonic Functional Materials and Devices, School of Information and Optoelectronic Science and Engineering, South China Normal University, Guangzhou, China
{20213232093,20213231036,maqx}@m.scnu.edu.cn

[2] Department of Computer Science, Johns Hopkins University, Baltimore, MD, USA
yshen92@jhu.edu

Abstract. Histology images are the gold standard for tumor diagnosis, and numerous studies have investigated deep learning methods for their analysis. However, stain heterogeneity, resulting from variations in staining protocols and batches across laboratories, impairs the generalization ability of these methods. Although stain normalization can mitigate this issue, current approaches rely on the manual selection of template images, which is time-consuming, requires domain expertise, and is susceptible to subjective biases. To address this challenge, we propose Learnable Automatic Normalization (LAN), an innovative two-stage method that automates the stain normalization process. In the first stage, a deep network in LAN learns to predict the optimal stain normalization template for each histology image, eliminating the need for manual template selection. Then, LAN generates virtually normalized histology samples using the predicted template. These normalized samples are then used in the second stage for a downstream task, with the supervision signal from this stage updating the network in the first stage. We evaluated LAN on the NCT-CRC-HE-100K-NORMAL dataset and demonstrated its superior performance in terms of accuracy compared to existing stain normalization methods. By automating and optimizing the stain normalization process, LAN has the potential to streamline the workflow in histology image analysis, ultimately contributing to more efficient and accurate pathological diagnoses. The code is available at https://github.com/Christyshiling/LAN.

Keywords: Histology Image · Stain Normalization · Deep Learning

1 Introduction

Histology images are the gold standard for tumor diagnosis, but they often exhibit significant variations in stain due to discrepancies in staining protocols

J. Finkelstein et al. (Eds.): AIME 2024, LNAI 14845, pp. 95–103, 2024.
https://doi.org/10.1007/978-3-031-66535-6_11

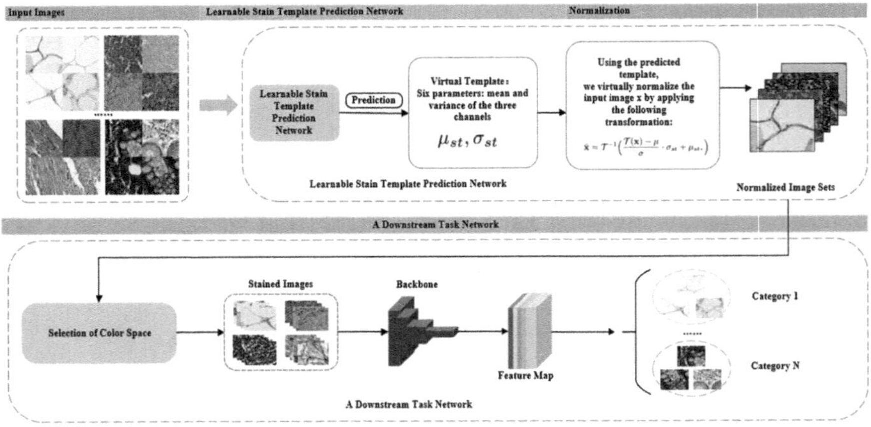

Fig. 1. The overall framework for the proposed LAN.

across different laboratories and even within a single laboratory during different staining batches [1,3,4,8,15]. Such stain variations can impact the performance of deep learning methods in downstream tasks [22], thereby limiting both the accuracy and robustness [7] of computer aided diagnostic systems. Stain normalization methods have been proposed to address this issue. However, automated processing and analysis of histology images still require substantial manual efforts on selecting the stain template for normalization [10], which is time-consuming and may introduce potential inconsistencies arising from subjective judgments. To narrow this gap, we propose Learnable Automatic Normalization (LAN) [16,21], a novel learnable method to reduce the manual cost of template selection. Specifically, LAN enables the model to automatically adjust the normalization template [13,22] without relying on manual annotation. The major contribution of this work is two-fold, summarized as follows. First, we innovatively formulate the end-to-end [18] stain normalization processs as an optimal stain template prediction task. Secondly, based upon our formulation, we introduce a two stage deep learning framework LAN, which learns the optimal stain template for normalization to automatize the process.

2 Methodology

Method Overview. LAN consists of two main components: a learnable stain template prediction network and a downstream task network. The stain template prediction network takes a histology image as input and predicts a set of six parameters that define the optimal stain normalization template for that image. These parameters correspond to the mean and standard deviation of each RGB color channel in the normalized image space. By learning to predict these parameters directly from the data, LAN eliminates the need for manual template selection and parameter tuning, which are time-consuming and prone to subjective biases. Once the optimal stain normalization template is predicted, it is used to virtually normalize the input histology image. This normalized image is then passed to the downstream task network, which can be any task-specific architecture designed for applications such as classification, segmentation, or detection. During the training phase, the stain template prediction network and the downstream task network are optimized jointly in an end-to-end manner. The supervision signal from the downstream task is propagated back to the stain template prediction network, allowing it to learn to predict normalization templates that are optimal for the specific downstream task. This end-to-end optimization ensures that the learned stain normalization templates are tailored to the characteristics of the input data and the requirements of the target application. The overall framework of LAN is depicted in Fig. 1.

Stain Template Prediction Network. The stain template prediction network (f_{st}) is a crucial component of the LAN framework, responsible for predicting the optimal stain normalization template for each input histology image. The network is trained using the supervision signal from the downstream task network, enabling it to learn task-specific normalization templates that enhance the performance of the overall system. The architecture of f_{st} consists of a series of convolutional layers that act as feature extractors. These layers are optimized to identify and encode key color patterns and morphological characteristics in the input images, which are subsequently projected into a stain template. The learned features are then used to generate a six-dimensional staining template, which captures the optimal color distribution for the specific downstream task. Formally, given an input histology image $\mathbf{x} = (x_{ijk}) \in \mathbb{R}^{H \times W \times 3}$, the stain template prediction network is defined as $[\mu_{st}, \sigma_{st}] = f_{st}(\mathbf{x})$, where $\mu_{st}, \sigma_{st} \in \mathbb{R}^3$ represent the predicted channel-wise mean and standard deviation of in the optimal stain normalization template, respectively. Using the predicted template, we virtually normalize the input image $\mathbf{x}$ by applying the following transformation [19]:

$$\tilde{\mathbf{x}} = \mathcal{T}^{-1}\left(\frac{\mathcal{T}(\mathbf{x}) - \mu}{\sigma} \cdot \sigma_{st} + \mu_{st},\right) \quad (1)$$

where $\mathcal{T}$ denotes the transformation from the RGB color space to a more suitable color space for normalization (e.g., HSV), and $\mathcal{T}^{-1}$ is its inverse transformation. The variables $\mu \in \mathbb{R}^3$ and $\sigma \in \mathbb{R}^3$ represent the channel-wise mean and standard deviation computed from the input image $\mathbf{x}$, respectively.

Downstream Task Network and Overall Training. The downstream task network (f_{dt}) is the second main component of the LAN framework, designed to perform the specific analysis task on the virtually normalized histology images. The choice of the downstream network architecture depends on the nature of the task, such as classification, segmentation, or detection. For instance, in a classification setting, f_{dt} can be a convolutional neural network (CNN) with a softmax output layer, trained to predict the class labels of the input images. The downstream task network takes the normalized image $\tilde{\mathbf{x}}$ as input and produces the corresponding task-specific output $\mathbf{y} = f_{dt}(\tilde{\mathbf{x}})$. The training objective for f_{dt} is to minimize the downstream task loss function $\mathcal{L}_{dt}$ associated with the specific task, such as cross-entropy loss for classification or dice loss for segmentation. During the training phase, the stain template prediction network f_{st} and the downstream task network f_{dt} are optimized jointly in an end-to-end manner. The overall training objective is to minimize the combined loss function:

$$\mathcal{L} = \mathcal{L}_{dt}(\mathbf{y}, \mathbf{y}^*) + \lambda \cdot \mathcal{R}(\mu st, \sigma_{st}) \tag{2}$$

where $\mathbf{y}^*$ denotes the ground truth labels for the downstream task, $\mathcal{R}$ is a regularization term imposed on the predicted stain normalization parameters μ_{st} and σ_{st}, and λ is a hyperparameter controlling the balance between the two loss terms. The joint optimization of f_{st} and f_{dt} allows the stain template prediction network to learn normalization templates that are tailored to the specific downstream task. The supervision signal from the downstream task is propagated back to f_{st} through gradient backpropagation, guiding it to predict templates that enhance the discriminative power of the normalized images for the given task.

Inference Stage. During inference, a test image is first passed through the stain template prediction network to obtain the optimal normalization parameters. The image is then virtually normalized using these parameters and fed into the downstream task network for the final prediction. This end-to-end inference process ensures that the stain normalization step is seamlessly integrated with the downstream analysis, enabling the LAN framework to handle variations in staining and improve the overall usability of the entire system.

3 Experiments

Dataset Details. We evaluate our proposed LAN model using two primary datasets for training: NCT-CRC-HE-100K-NORM ('NORM'), and NCT-CRC-HE-100K-NONORM ('NONORM'). The NCT-CRC-HE-100K and NCT-CRC-HE-100K-NONORM dataset each consist of 100,000 high-resolution histology images of colorectal cancer tissue, each annotated with detailed class labels. We externally validate it using the CRC-VAL-HE-7K dataset [12], which contains 7,180 images of pathological slices with heterogeneous staining characteristics sourced from multiple data centers.

Implementation Details. Our implementation is based on Python 3.8 and PyTorch 1.10, and the experiments are conducted on an NVIDIA GeForce RTX 3090 Ti GPU and an Intel Core i9-12900 CPU. All input images are resized to a uniform dimension of 224 × 224 pixels before being fed into the network. To investigate the impact of different downstream classification networks on the performance of our LAN model, we consider several popular architectures, including ResNet18 [9], ResNet50 [9], MobileNet [11], ViT [6], and Swin Transformer [17]. These architectures represent a diverse range of network designs, from lightweight models to more complex and powerful ones, allowing for a comprehensive evaluation of our proposed method. The hyperparameter settings are kept consistent across the two training phases. We use a learning rate of 0.0001, a batch size of 64, and train each phase for 20 epochs. In the first phase, we employ the Adam optimizer to train the stain template prediction network. For the second phase, which involves joint optimization of the entire LAN framework, we switch to the AdamW optimizer [14] and use the cross-entropy loss function [5] to guide the learning process. To quantify the model's performance, we use the top-1 classification accuracy [2] as our primary evaluation metric.

Table 1. Comparison of test accuracy on the tissue type classification task using different methods, datasets, and network architectures between the proposed LAN method and the baseline (Reinhard's method [19]). LAN is implemented for three color spaces (RGB, LAB, HSV) and also with ImageNet pretrained weights.

Methods	Dataset	ResNet18	ResNet50	MobileNet	ViT	Swin Transformer
Baseline	NONORM	82.86	79.59	81.53	90.22	89.41
	NORM	93.59	94.57	95.57	96.23	96.78
LAN (RGB)	NONORM	88.40	88.53	81.63	80.10	90.36
	NORM	94.90	92.29	91.05	92.18	94.01
LAN (LAB)	NONORM	54.91	55.10	53.22	60.00	62.25
	NORM	87.42	88.31	89.50	90.84	91.71
LAN (HSV)	NONORM	50.23	52.71	53.05	58.79	61.30
	NORM	90.70	92.14	93.57	94.08	94.92
LAB (ImageNet)	NONORM	52.30	53.10	91.64	89.00	94.20
	NORM	34.50	39.10	95.74	96.97	97.50

Results. Table 1 presents the test accuracy comparison between the proposed LAN method and the baseline i.e., Reinhard's method [19], on the tissue type classification task using different datasets and network architectures. The results demonstrate that LAN consistently outperforms the baseline method across various color spaces (RGB, LAB, HSV) [20]and network architectures on the NONORM dataset, which contains images with significant stain variations. On the NONORM dataset, LAN in RGB space achieves the highest accuracy of 90.36% using the Swin Transformer architecture, surpassing the baseline method by 0.95%. LAN in RGB space also shows improved performance compared to the baseline across other network architectures, with an increase of 5.54%, 8.94%, and 0.10% in accuracy for ResNet18, ResNet50, and MobileNet, respectively. However, LAN in RGB space performs slightly worse than the baseline with the ViT architecture, with a decrease of 10.12% in accuracy. The results also highlight the impact of different color spaces on the performance of LAN. While LAN

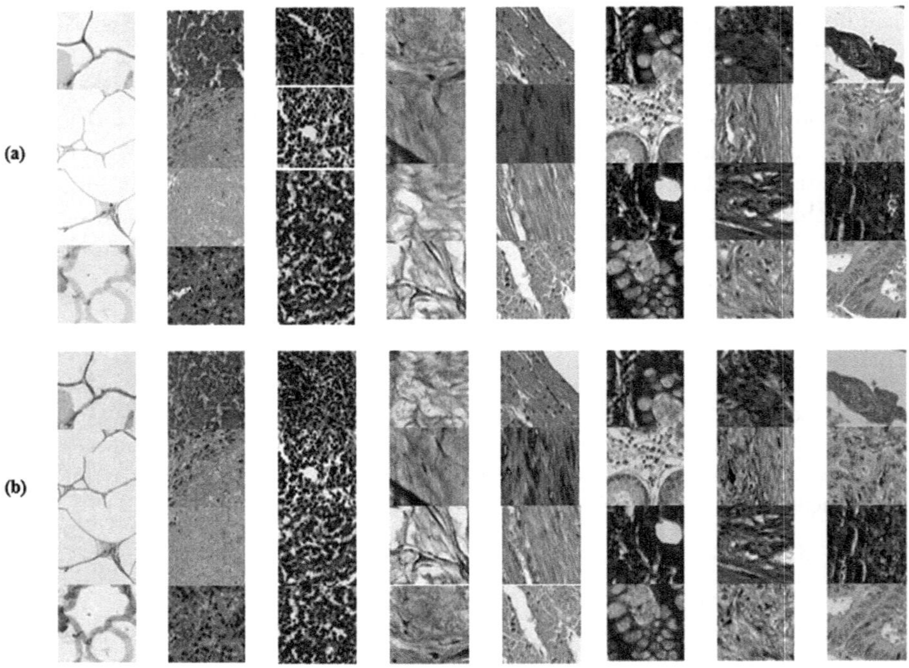

Fig. 2. Visual comparison of histology images before and after stain normalization using the proposed LAN method. (a) Original histology images with stain variations. (b) Corresponding stain normalized images generated by LAN, demonstrating improved color consistency and enhanced visual clarity.

in RGB space generally outperforms the baseline, LAN in LAB space and LAN in HSV space show a significant drop in accuracy on the NONORM dataset across all network architectures. This suggests that the choice of color space plays a crucial role in the effectiveness of stain normalization using LAN. Interestingly, when using ImageNet pretrained weights, LAN achieves the highest accuracy of 94.20% with the Swin Transformer on the NONORM dataset, surpassing all other methods. This indicates that leveraging pretrained weights from large-scale datasets can further enhance the performance of LAN, particularly in the LAB color space. On the NORM dataset, which contains images with consistent staining, the baseline method generally outperforms LAN across all color spaces and network architectures. This is expected, as stain normalization techniques may not provide significant benefits when the staining is already consistent. We provide a visualization in Fig. 2. In summary, the proposed LAN method demonstrates improved performance compared to the baseline on the NONORM dataset, particularly in the RGB color space. The choice of color space and the use of pretrained weights significantly impact the effectiveness of LAN in handling stain variations. These findings highlight the potential of

LAN as a promising stain normalization technique for improving the accuracy of tissue type classification in histopathological image analysis.

4 Conclusion

To addresses the challenge of selecting stain normalization template automatically, we introduce Learnable Automatic Normalization (LAN), a novel two-stage method that automates the stain normalization process in histology image analysis. The proposed framework consists of a stain template prediction network that learns to predict the optimal normalization template for each histology image, eliminating the need for manual template selection. The predicted template is then used to virtually normalize the input image, which is subsequently passed to a downstream task network for further analysis. We evaluated LAN on the NCT-CRC-HE-100K-NORMAL dataset and demonstrated its superior performance compared to existing stain normalization methods. The joint optimization of the stain template prediction network and the downstream task network enables LAN to learn task-specific normalization templates, enhancing the overall performance of the system. The experimental results, conducted using various popular downstream classification architectures, such as ResNet, MobileNet, ViT, and Swin Transformer, validate the effectiveness of LAN in improving classification accuracy. By automating and optimizing the stain normalization process, LAN streamlines the workflow in histology image analysis and contributes to more efficient and accurate pathological diagnoses. Future work will focus on extending the LAN framework to other medical image analysis tasks and exploring its potential in clinical settings. The integration of LAN with existing computer-aided diagnostic systems could improve their usability.

Acknowledgements. This work was partly supported by the Guangdong Basic and Applied Basic Research Foundation (2023A1515012966, 2023A1515011641), the Special Funds for the Cultivation of Guangdong College Students' Scientific and Technological Innovation ("Climbing Program" Special Funds) (No. pdjh 2023b0142), and the Key field research projects in Foshan City (Grant No. 2120001009232).

References

1. Abe, T., Yamaguchi, M., Murakami, Y., Ohyama, N., Yagi, Y.: Color correction of pathological images for different staining-condition slides. In: Proceedings. 6th International Workshop on Enterprise Networking and Computing in Healthcare Industry - Healthcom 2004 (IEEE Cat. No.04EX842), pp. 218–223 (2004). https://doi.org/10.1109/HEALTH.2004.1324526
2. Boyd, S., Mohri, M., Radovanovic, A.: Accuracy at the top. Adv. Neural. Inf. Process. Syst. **2**, 1–30 (2012)
3. Chang, J.R., et al.: Stain mix-up: unsupervised domain generalization for histopathology images. In: de Bruijne, M., et al. (eds.) Medical Image Computing and Computer Assisted Intervention - MICCAI 2021, pp. 117–126. Springer International Publishing, Cham (2021)

4. Ciompi, F., et al.: The importance of stain normalization in colorectal tissue classification with convolutional networks. In: 2017 IEEE 14th International Symposium on Biomedical Imaging (ISBI 2017), pp. 160–163 (2017). https://doi.org/10.1109/ISBI.2017.7950492
5. Das, R., Chaudhuri, S.: On the separability of classes with the cross-entropy loss function (2019)
6. Dosovitskiy, A., et al.: An image is worth 16x16 words: Transformers for image recognition at scale. CoRR **abs/2010.11929** (2020). https://arxiv.org/abs/2010.11929
7. Gupta, V., Singh, A., Sharma, K., Bhavsar, A.: Automated classification for breast cancer histopathology images: Is stain normalization important? In: Cardoso, M.J., et al. (eds.) Computer Assisted and Robotic Endoscopy and Clinical Image-Based Procedures, pp. 160–169. Springer International Publishing, Cham (2017)
8. Gurcan, M.N., Boucheron, L.E., Can, A., Madabhushi, A., Rajpoot, N.M., Yener, B.: Histopathological image analysis: a review. IEEE Rev. Biomed. Eng. **2**, 147–171 (2009). https://doi.org/10.1109/RBME.2009.2034865
9. He, K., Zhang, X., Ren, S., Sun, J.: Deep residual learning for image recognition. In: 2016 IEEE Conference on Computer Vision and Pattern Recognition (CVPR), pp. 770–778 (2016). https://doi.org/10.1109/CVPR.2016.90
10. Hoque, M.Z., Keskinarkaus, A., Nyberg, P., Seppänen, T.: Stain normalization methods for histopathology image analysis: A comprehensive review and experimental comparison. Inform. Fusion **102**, 101997 (2024). https://doi.org/10.1016/j.inffus.2023.101997, https://www.sciencedirect.com/science/article/pii/S1566253523003135
11. Howard, A., et al.: Searching for mobilenetv3. In: 2019 IEEE/CVF International Conference on Computer Vision (ICCV), pp. 1314–1324 (2019). https://doi.org/10.1109/ICCV.2019.00140
12. Kather, J.N., et al.: Predicting survival from colorectal cancer histology slides using deep learning: a retrospective multicenter study. PLoS Med. **16**(1), e1002730 (2019)
13. Ke, J., Shen, Y., Jiang, X., Guo, Y., Chen, Y., Liang, X.: Multiple-datasets and multiple-label based color normalization in histopathology with cgan. In: Medical Imaging (2021). https://api.semanticscholar.org/CorpusID:232282100
14. Kingma, D., Ba, J.: Adam: A method for stochastic optimization. Computer Science (2014)
15. Komura, D., Ishikawa, S.: Machine learning methods for histopathological image analysis. Comput. Struct. Biotechnol. J. **16**, 34–42 (2018)
16. Li, Y., Zhou, H., Liu, N., Shen, Y.: Stain normalization and augmentation in frequency space for histology analysis. In: 2023 IEEE International Conference on Bioinformatics and Biomedicine (BIBM), pp. 2031–2035 (2023). https://doi.org/10.1109/BIBM58861.2023.10385658
17. Liu, Z., et al.: Swin transformer: Hierarchical vision transformer using shifted windows. In: 2021 IEEE/CVF International Conference on Computer Vision (ICCV), pp. 9992–10002 (2021). https://api.semanticscholar.org/CorpusID:232352874
18. Monga, I., et al.: Software-defined network for end-to-end networked science at the exascale. Futur. Gener. Comput. Syst. **110**, 181–201 (2020). https://doi.org/10.1016/j.future.2020.04.018, https://www.sciencedirect.com/science/article/pii/S0167739X19305618
19. Reinhard, E., Adhikhmin, M., Gooch, B., Shirley, P.: Color transfer between images. IEEE Comput. Graphics Appl. **21**(5), 34–41 (2001). https://doi.org/10.1109/38.946629

20. Schwarz, M.W., Cowan, W.B., Beatty, J.C.: An experimental comparison of RGB, YIG, lab, HSV, and opponent colour models. ACM Trans. Graph. **6**(2), 123–158 (1987)
21. Shen, Y., Ke, J.: Staindiff: transfer stain styles of histology images with denoising diffusion probabilistic models and self-ensemble. In: Greenspan, H., Madabhushi, A., Mousavi, P., Salcudean, S., Duncan, J., Syeda-Mahmood, T., Taylor, R. (eds.) Medical Image Computing and Computer Assisted Intervention - MICCAI 2023, pp. 549–559. Springer Nature Switzerland, Cham (2023)
22. Shen, Y., Luo, Y., Shen, D., Ke, J.: Randstainna: learning stain-agnostic features from histology slides by bridging stain augmentation and normalization. In: Wang, L., Dou, Q., Fletcher, P.T., Speidel, S., Li, S. (eds.) Medical Image Computing and Computer Assisted Intervention - MICCAI 2022, pp. 212–221. Springer Nature Switzerland, Cham (2022)

MRI Brain Cancer Image Detection: Application of an Integrated U-Net and ResNet50 Architecture

Mahshid Benchari(✉) and Michael W. Totaro

University of Louisiana at Lafayette, Lafayette, LA 70503, USA
{mahshid.benchari1,michael.totaro}@louisiana.edu

Abstract. Brain cancer results in the deaths of many people each year. Magnetic Resonance Imaging (MRI) is used to segment different regions of brain tumors, such as edema and tumor cores, which is challenging due to differences in location, size, shape, and intensity. In this study, we apply an integrated CNN model that combines aspects of both ResNet50 and U-Net, leveraging the strengths of each. Specifically, we replaced the last 5 layers of ResNet50 with 10 additional layers, thereby serving as the encoder in the U-Net architecture. Our results with ResNet50 offer precision and IoU scores of 0.98 and 0.97, respectively. Our approach appears to show great promise in terms of how well the model detects and pinpoints brain cancer.

Keywords: Magnetic Resonance Imaging (MRI) · Brain Cancer · Convolutional Neural Networks (CNNs) · U-Net · ResNet50

1 Introduction

As with many forms of cancer, early detection of brain tumors increases the effectiveness of medical treatments. The application of Deep Learning (DL) can enhance MRI-based cancer detection. Recent studies show that: (a) ResNet50 achieved 95% accuracy [11], (b) Singular Value Decomposition (SVD) resulted in 96.66% accuracy [5], and (c) ResNet50 modifications attained 97.2% accuracy, thereby suggesting its effectiveness for use in such computer-aided systems [2].

Our research question in this paper is as follows: *How might we detect brain tumors from MRI images using artificial neural networks, attaining a high accuracy rate while also minimizing the number of epochs during model training?* We propose using ResNet50 in a Convolutional Neural Network (CNN). Pre-training the U-Net model's encoder with ResNet50 and adding 10 layers should enhance performance. Our dataset includes brain MRI images with manual FLAIR abnormality segmentation masks. We train the data using our proposed deep learning model, since successful results may be achieved in large databases. The remainder of this paper includes our experimental design (Methodology), results, and their implications. We conclude by summarizing our findings, as well as future work.

J. Finkelstein et al. (Eds.): AIME 2024, LNAI 14845, pp. 104–108, 2024.
https://doi.org/10.1007/978-3-031-66535-6_12

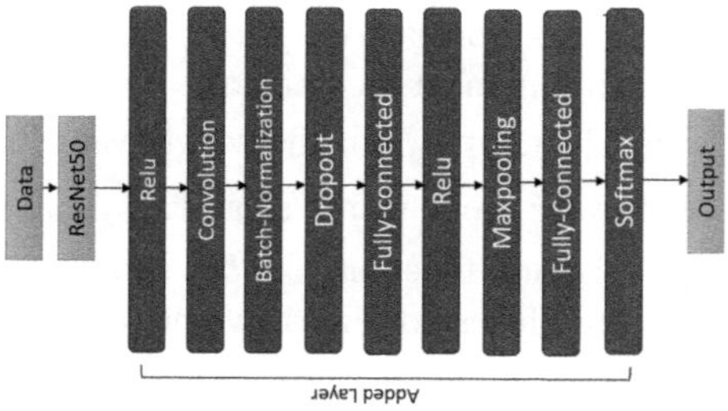

Fig. 1. The structure of the enhanced model.

2 Methodology

2.1 Dataset

The dataset for this research, obtained from Kaggle, comprised of brain MRI images and manual FLAIR abnormality segmentation masks, was collected from TCGA and TCIA [1, 3], for 110 patients with low-grade malignant nervous system tumors. We developed our application using Python. During the training of the CNN models and the newly created model, we use key parameters such as: 50 epochs, an initial learning rate of 0.0001, a batch size of 8, SoftMax activation function, verbosity set to 1, shuffle strategy of every-epoch, and a total of 4 classes.

2.2 ResNet50 and Adjusted Framework

In our approach, we use ResNet50 [7], a CNN designed to identify objects in extensive datasets, featuring convolutional, pooling, and fully connected layers, and is particularly useful in detecting brain tumors by identifying crucial features. We employed the encoder of a pre-trained ResNet50-based U-Net model for image segmentation [9], which consists of encoder and decoder paths. Additionally, we utilized the ResNet50 model due to: (a) its prior training knowledge, and (b) its history of success with biomedical data, thereby making it a suitable choice for our study.

As shown in Fig. 1, we replaced the last 5 layers of ResNet50 with 10 new layers. Our model processes input data of size 256x256x3, which is then normalized for stability and faster training. Batch normalization is applied to standardize the output from convolution or fully connected layers before activation, thereby accelerating training and enhancing overall network performance [10]. This is shown in $Y_i = \frac{X_i - \mu_B}{\sqrt{\sigma_\beta^2 + \epsilon}}$, where Y_i is the normalization process, and μ_B, the average value of the stack, is $\frac{1}{M}\sum_{i=1}^{M} X_i$. The standard deviation of the stack μ_B is equal to $\frac{1}{M}\sum_{i=1}^{M}(X_i - \mu_B)^2$, where M is the number of inputs.

Next, we apply a 2D convolution operation with 64 filters and a specified kernel size, followed by normalizing the output. To prevent overfitting, we include a dropout layer, randomly deactivating nodes and connections during training. Neurons in the fully connected layer process information from all components of the next layer, treating it as a one-dimensional matrix [4]. In such a layer, $u_i{}^l = \sum_j w_{ji}^{l-1} y_j{}^{l-1}$ is employed during the feedforward process. In this context and based on $y_i{}^l = f(u_i{}^l) + b^{(l)}$, l is the layer number, while i and j are specific neuron numbers. y_i^l Indicates the output value produced by a particular neuron in a layer. The term w_{ji}^{l-1} is the weight value between neurons in the hidden layer, and y_j^{l-1} denotes the input neuron's value. u_i^l represents the output layer's value before applying the activation function, and $b^{(l)}$ signifies the deviation value for that layer. ReLU activation introduces non-linearity and max pooling to reduce spatial dimensions. We also include convolutional layers, batch normalization, activations, and skip connections.

Fig. 2. Training and Validation IoU.

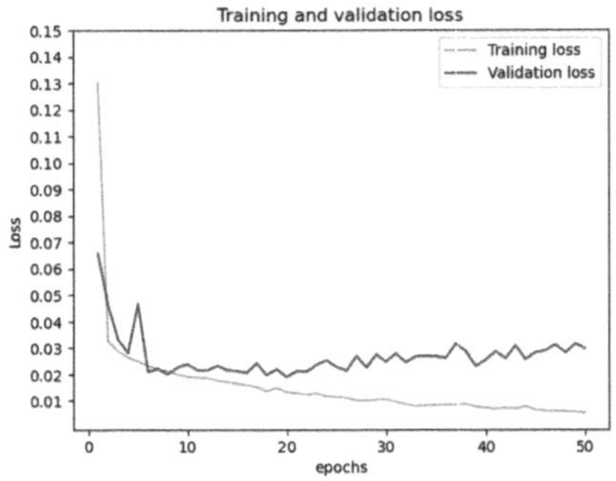

Fig. 3. Training and Validation Loss.

3 Results and Discussion

Combining both ResNet50 and U-Net offers a robust means by which to detect and segment brain tumors, thereby leading to an overall improvement in accuracy. The U-Net model enhances ResNet50's brain tumor detection accuracy by generating pixel-wise probability maps. We also calculate Intersection over Union (IoU) values for various classes and provide visual representations of images with their masks for interpretation. IoU measures the accuracy of an object detector on a specific dataset. It compares true positives (*Tp*) and false positives (*Fp*) to dataset classification true positives (*Tp*) and false negatives (*Fn*). Equation (1) shows both the IoU and accuracy by comparing correctly predicted images to the total tested.

$$IoU = \frac{T_p}{T_p + F_p + F_n}, Accuracy = \frac{T_p + T_n}{T_p + F_p + T_n + F_n} \tag{1}$$

Figures 2 and 3 illustrate our model's improved accuracy and minimal loss. Our model demonstrates a significant decrease in loss and achieves impressive IoU and prediction accuracy scores, underscoring its effectiveness in consistently delivering accurate predictions. Figure 4 displays a test image, its ground truth, and the segmentation mask predicted by the trained model. This method offers valuable clinical insights for diagnosis and treatment planning.

Table 1 compares our study's results with other related studies. Integrating U-Net with ResNet50 achieved an accuracy of 0.98 and an IoU of 0.97, surpassing other models

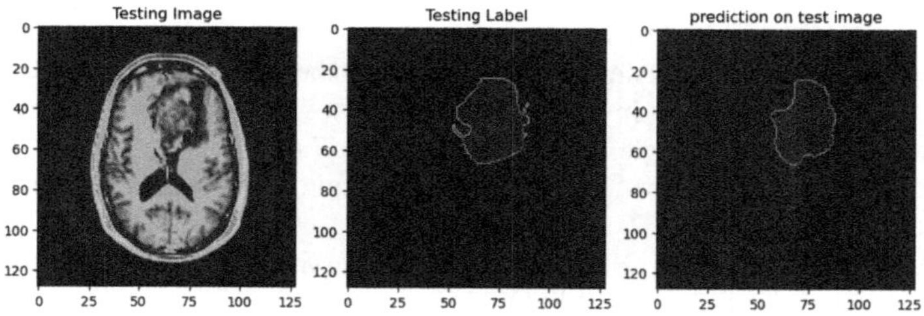

Fig. 4. Testing Image, Testing Label, and Prediction on test image.

in precision. The loss function also decreased notably, suggesting areas for performance enhancement. We refined our model by reducing epochs to 50 and the learning rate to 0.0001, employing techniques like data augmentation and dropout regularization to prevent overfitting.

Comparative studies show promising results: VGG-19 and ResNet-50 out-perform training "from scratch" with a recall of 92.03% [8]. VGG-16 achieved 94.42% accuracy and 83.86% recall [12], while ResNet50 reached 96.2% accuracy in MRI segmentation tasks [6]. These findings suggest improved healthcare outcomes and medical imaging.

Table 1. Comparison with various models.

Method	Model	Dataset	Results
[8]	Lake-Net	Chest X-ray images	Prediction 0.9836, Recall 0.9203
[12]	VGC-16	253 MRI images	Prediction 0.9442, Recall 0.8386
[6]	ResNet50	5856 chest X-rays	Prediction 0.962
Our model	ResNet50	MRI brain images	Prediction 0.98, IoU 0.97

4 Conclusion

Recall our research question: *How can we detect brain tumors from MRI images using artificial neural networks, achieving high accuracy with fewer training epochs?* We integrated a CNN model with ResNet50 and U-Net, replacing the last 5 layers of ResNet50 with 10 additional layers, serving as the encoder for U-Net. We reduced the number of epochs and learning rates during training to 50 and 0.0001, respectively. Our results yielded a precision and IoU of 0.98 and 0.97, respectively, along with a notable reduction in the model's loss function. Future work will explore various architectures and new methods to boost performance.

References

1. The Cancer Genome Atlas. http://cancergenome.nih.gov/abouttcga

2. Çinar, A., Yildirim, M.: Detection of tumors on brain mri images using the hybrid convolutional neural network architecture. Med. Hypotheses **139**, 109684 (2020)
3. Clark, K., et al.: The cancer imaging archive (TCIA): maintaining and operating a public information repository. J. Digit. Imaging **26**, 1045–1057 (2013)
4. Dan, C., Ueli, M., Jonathan, M., Jürgen, S.: Multi-column deep neural network for traffic sign classification. Neural Netw. **32**(1), 333–338 (2012)
5. El Abbadi, N.K., Kadhim, N.E.: Brain tumor classification based on singular value decomposition. Brain **5**(8), 10–17148 (2016)
6. Hashem, M., Yousof, N.A.A.-K.A., Saeed, B., Radman, M., Al-qershi, D.A., Ba, A.E.: Detection of pneumonia by using nine pre-trained transfer learning models based on deep learning techniques. Iraqi J. Comput. Inform. **47**, 1 (2021)
7. He, K., Zhang, X., Ren, S., Sun, J.: Deep residual learning for image recognition. In: Proceedings of the IEEE Conference on Computer Vision and Pattern Recognition, pp. 770–778 (2016)
8. Ikechukwu, A.V., Murali, S., Deepu, R., Shivamurthy, R.: Resnet-50 vs vgg-19 vs training from scratch: a comparative analysis of the segmentation and classification of pneumonia from chest x-ray images. Glob. Trans. Proc. **2**(2), 375–381 (2021)
9. Ronneberger, O., Fischer, P., Brox, T. U-net: convolutional networks for biomedical image segmentation. In: Medical Image Computing and Computer-Assisted Intervention–MICCAI 2015: 18th International Conference, Munich, Germany, 5–9 October 2015, Proceedings, Part III 18, pp. 234–241. Springer (2015)
10. Salimans, T., Kingma, D.P.: Weight normalization: a simple reparameterization to accelerate training of deep neural networks. Adv. Neural Inf. Process. Syst. **29** (2016)
11. Saxena, P., Maheshwari, A., Maheshwari, S.: Predictive modeling of brain tumor: a deep learning approach. In: Sharma, M.K., Dhaka, V.S., Perumal, T., Dey, N., João Manuel Tavares, R.S. (eds.) Innovations in Computational Intelligence and Computer Vision. AISC, vol. 1189, pp. 275–285. Springer, Singapore (2021). https://doi.org/10.1007/978-981-15-6067-5_30
12. Sevli, O.: Performance comparison of different pre-trained deep learning models in classifying brain MRI images. Acta Infologica **5**(1), 141–154 (2021)

MRI Scan Synthesis Methods Based on Clustering and Pix2Pix

Giulia Baldini[1(✉)], Melanie Schmidt[1(✉)], Charlotte Zäske[2], and Liliana L. Caldeira[3]

[1] Department of Computer Science, Heinrich-Heine-University Düsseldorf, Universitätsstraße 1, 40225 Düsseldorf, Germany
{gbaldini,mschmidt}@hhu.de

[2] Department of Diagnostic and Interventional Radiology, University Hospital Aachen, Pauwelsstraße 30, 52074 Aachen, Germany
charlotte.zaeske@rwth-aachen.de

[3] Department of Radiology, University Hospital of Cologne, Kerpener Street 62, 50937 Cologne, Germany
liliana.caldeira@uk-koeln.de

Abstract. We consider a missing data problem in the context of automatic segmentation methods for Magnetic Resonance Imaging (MRI) brain scans. Usually, automated MRI scan segmentation is based on multiple scans (e.g., T1-weighted, T2-weighted, T1CE, FLAIR). However, quite often a scan is blurry, missing or otherwise unusable. We investigate the question whether a missing scan can be synthesized. We exemplify that this is in principle possible by synthesizing a T2-weighted scan from a given T1-weighted scan.

Our first aim is to compute a picture that resembles the missing scan closely, measured by *average mean squared error* (MSE). We develop/use several methods for this, including a random baseline approach, a clustering based method and pixel-to-pixel translation method by Isola et al. [15] (Pix2Pix) which is based on conditional GANs. The lowest MSE is achieved by our clustering-based method.

Our second aim is to compare the methods with respect to the effect that using the synthesized scan has on the segmentation process. For this, we use a DeepMedic model trained with the four input scan modalities named above. We replace the T2-weighted scan by the synthesized picture and evaluate the segmentations with respect to the tumor identification, using Dice scores as numerical evaluation. The evaluation shows that the segmentation works well with synthesized scans (in particular, with Pix2Pix methods) in many cases.

Keywords: MRI Scan Synthesis Methods · Clustering · Pix2Pix

1 Introduction

Automated brain tumor segmentation is an active and highly relevant research field [26,30,38] with its own annual data challenge called BraTS (*Brain Tumor*

L. L. Caldeira and M. Schmidt—have jointly supervised this project.

J. Finkelstein et al. (Eds.): AIME 2024, LNAI 14845, pp. 109–125, 2024.
https://doi.org/10.1007/978-3-031-66535-6_13

Segmentation Challenge). Given a set of MRI scans, the objective of automated tumor segmentation is to find and segment a tumor in the depicted brain automatically. Segmentation software (e.g. DeepMedic [17], nnU-Net [14]) is not only able to find the exact tumor location, but also to distinguish between different subregions of the tumor, such as edema and necrosis. These methods [14,17] use a set of four modalities to achieve this purpose, namely T1W (T1-weighted), T2W (T2-weighted), T1CE (T1-weighted contrast enhanced) and FLAIR. However, in the clinical routine, it is uncommon that all modalities are available for a single examination. While it is plausible to create separate networks for each subset of available scans, a more convenient approach is to address the missing data directly and replace the missing scan. Indeed, a recent BraTS challenge [19] states that *the ability to substitute missing modalities and gain segmentation performance is highly desirable and necessary for the broader adoption of these algorithms in the clinical routine.*[1] We showcase that creating synthesized scans is possible to a certain extent by investigating the following question:

How can we produce a reasonable synthesized T2W image from a T1W scan of the same patient?

Figure 1 shows a slice of a T1W scan and the corresponding slice of the T2W scan of the same patient (these scans are 3D, but we only depict one slice). Since T2W looks complementary to T1W, a first idea might be to inverse every voxel, i.e., to transform the voxel intensity p_{T1W} of T1W to $p_{\mathrm{T2W}} = \min(\mathrm{T1W}) + \max(\mathrm{T1W}) - p_{\mathrm{T1W}}$ where $\min(\mathrm{T1W})$ and $\max(\mathrm{T1W})$ are the minimum and maximum values of all voxels in T1W. This transformation is what created the third picture in Fig. 1. Our goal is to substantially beat this baseline approach.

We develop and evaluate two methods. The first one is based on clustering. It uses a training data set to understand the relationship between T1W and T2W. For a set of given T1W and T2W scans, the method learns the intensity spectrum of each tissue in T1W and how it is translated into the corresponding spectrum that the tissue has in T2W. This is done by identifying areas of similar color in both pictures via clustering and then mapping compatible areas to each other in order to compute a color transformation table. While clustering methods have been applied in medical imaging, the primary focus has been on aiding brain tissue segmentation [3,10,20,21,24].

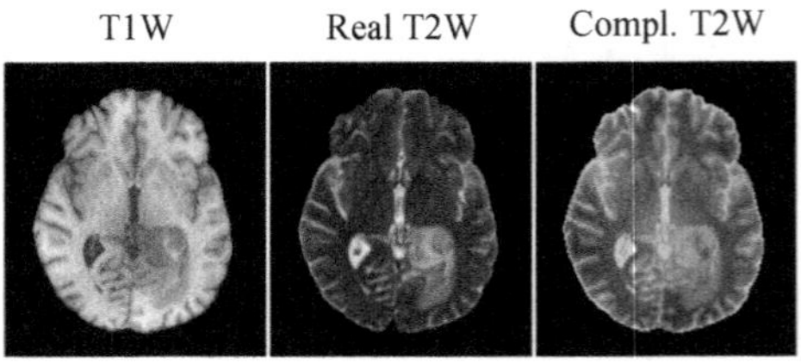

Fig. 1. A slice of a T1W and a T2W scan, and of the T1W complement (Patient `BraTS19 CBICA BLI 1`).

[1] Note, however, that our work was mostly done before that challenge was posed, so in particular, we are using the BraTS 2019 data set for our experiments.

Our BrainClustering approach is conceptually different from other works in the area and might lead to further work on related problems. We describe our clustering method in detail in Sect. 2 and in the full version of the paper [6]. The second method is an application of a known tool named Pix2Pix [15,35] that has already been used in a variety of medical applications for image-to-image translation [1,7,12,39]. We use it with slight modifications of the software that we describe in Sect. 3 and in the full version of the paper [6]. In follow-up work, Raut et al. [31] uses this same modified version of Pix2Pix to generate and evaluate different mappings (e.g., T2W to FLAIR).

Our evaluation is based on a data set based on 460 patients from BraTS 19 [4,5,22]. We both evaluate how close the synthesized picture is to the true scan with respect to minimum mean squared error (MSE) as well as how using it affects the segmentation process. For 335 patients, the BraTS 19 data is supplemented with a ground truth segmentation. A radiologist segmented the scans for 15 additional patients to increase the number of cases with ground truth. However, 110 cases remain that have no ground truth.

2 BrainClustering

We name our clustering algorithm BrainClustering. It computes a synthesized T2W image by applying a chain of several clustering and mapping steps. The clusterings are used to build mapping tables, and these tables are in turn used to generate a synthesized T2W. We refer to the process of building these mapping tables as *training*, even though it does not involve a neural network model. Figure 2 gives an overview of the process for a single patient, i.e., for one pair of T1W and T2W. The training steps are applied to pairs of given T1W and T2W which we process to understand how to construct a T2W image from a T1W image. The method is described in detail in the full version of the paper [6].

3 Pix2Pix

Pix2Pix [15,35] is a general solution to the pixel-to-pixel translation problem. Given an input image and an output image, this neural network framework is able to learn a relationship between the two. At inference time, it applies the learned function to create a new, previously unseen image. Pix2Pix is based on *conditional GANs* (cGAN) [23]. A cGAN is a neural network framework that simultaneously trains two models, a *generator* G and a *discriminator* D, which engage in an adversarial process. During training, the generator starts with random noise as input, aiming to produce data resembling the training set. The discriminator evaluates the generated data, providing feedback to adjust the generator's parameters. Simultaneously, the discriminator learns to better distinguish between real and generated data, refining its parameters to maximize correct classifications.

The Pix2Pix objective function is a fusion between the traditional cGAN loss [23] and the L1 loss. We build our objective function by further expanding this

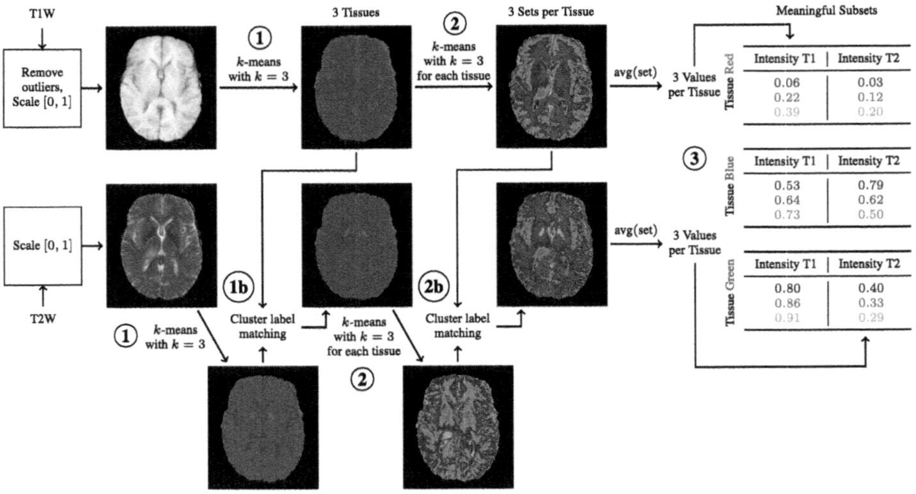

Fig. 2. [Best viewed in color] Schematic representation of the BrainClustering process: First, we use k-means1d to obtain a macro clustering of the scan according to the desired number of macro clusters. Second, each macro cluster is clustered into micro clusters. Finally, the points belonging to these micro clusters are averaged and stored in one table for each tissue, where there is an entry for each corresponding value of T1W and T2W.

concept. Let G and D be any non-linear mapping functions and let x, y represent our input and our desired output respectively. The generator is asked to learn the distribution of the data y from noise z such that it represents a mapping from a given input x, i.e., $G : \{x, z\} \to y$. The discriminator with mapping $D(y)$ outputs the probability that y comes from the training data. In fact, we train D to maximize the probability that the real data y is identified as real, while we train G to maximize the probability that the data generated from noise $G(z)$ is detected as real by the discriminator. The latter objective corresponds to minimizing $1 - D(G(z))$. The final form of Eq. (1) below comes from formalizing these high-level objectives with the log-loss principle of multi-class classification problems. Additionally, Isola et al. [15] add the term (2), which means that the objective of the discriminator is unchanged, whereas the generator is asked to trick the discriminator while maintaining a small loss. Since our objective is to have a low mean squared error in the tumor area, such that tumors can be easily identified, we added (3) to the loss function. T is a matrix that is zero where there is no tumor and one otherwise, $\odot$ is the Hadamard product and $\text{sum}(T)$ is the grand sum of the matrix T, i.e., the number of tumor voxels. Thus, our final loss function is

$$= \mathbb{E}_{x,y}\left[\log D(x,y)\right] + \mathbb{E}_{x,z}\left[\log(1 - D(x, G(x,z)))\right] \quad (1)$$
$$+\lambda \cdot \mathbb{E}_{x,y,z}\left[\|y - G(x,z)\|_1\right] \quad (2)$$
$$+\gamma \cdot \mathbb{E}_{x,y,z}\left[\frac{1}{\mathrm{sum}(T)} \cdot \|T \odot y - G(T \odot x, T \odot z)\|_2^2\right]. \quad (3)$$

As it is usual in machine learning, the best values for λ and γ can be found with a hyperparameter optimization. Further modifications of Pix2Pix are described in the full version of the paper [6].

The default generator architecture for Pix2Pix is a UNet [32], more specifically a UNet-256, which requires an input with side length multiple of 256. Given the 3D nature of the scans and the increased memory usage, we opt for a UNet-128 model with seven downsampling and seven upsampling blocks. During training, the images are preprocessed to have side length 128 by performing random crops, while during inference they are padded to have side length 256. Additional augmentation is done by randomly flipping the images. We also test whether ResNets [13], which are used by Pix2PixHD [35], are a better choice for our model. We apply the Adam solver [18] with a learning rate of 0.0002, the momentum parameters $\beta_1 = 0.5, \beta_2 = 0.999$ and a batch size of 1. All other network parameters are the default ones given by Pix2Pix. We train 16 models per generator architecture (UNet or ResNet) with the values $\lambda, \gamma \in \{0, 100, 500, 1000\}$ for a total of 32 different models.

4 Experimental Evaluation

We test BrainClustering and Pix2Pix. As baseline, we use the complementary approach described in Sect. 1 and a simple method that we name Random which consists of filling all non-background voxels by uniformly distributed random values.

4.1 Data

Our collective contains a total of 460 patients from BraTS 19 [4,5,22]. Each patient has four three-dimensional MRI brain scans (T1W, T2W, T1CE, FLAIR) of size (240, 240, 155) in NIfTI (*Neuroimaging Informatics Technology Initiative*) format. All scans have been preprocessed to have the same voxel size (1mm^3), are skull-stripped and registered to the same anatomical template. A ground truth segmentation that shows the exact position of the tumor is present for 350 of the patients. Most of the tumor ground truth comes from BraTS, and a portion of the scans has been manually segmented by a radiologist from the University Hospital of Cologne (C.Z.) and the segmentation has later been converted to the BraTS labels. A detailed representation of the BraTS 19 tumor labels is given by Bakas et al. [4]. Each voxel is identified as either background (no tumor), non-enhancing part of the tumor and necrotic core, excess of fluid associated with the tumor (edema), or contrast enhanced part of the tumor.

We divide the data into a training set, a validation set, and three test as described in Table 1. The training set is used to train both approaches, the validation set to optimize the hyperparameters, and the three test sets to evaluate the approaches. We evaluate `testBraTS` and `testOUR` separately due to the different sources of ground truth.

4.2 Code and System

BrainClustering is written in C++17 and ported to Python with Boost.Python [2], while our Pix2Pix implementation is written in Python3.8 and uses the machine learning framework PyTorch [27]. Our implementations are available on GitHub at https://github.com/giuliabaldini/brainclustering and https://github.com/giuliabaldini/Pix2PixNIfTI. The BrainClustering scans are produced using a computer cluster with 36 nodes and two Intel Xeon CPU E5-2690 v2 CPUs per node with 10 cores each and 128 GB of RAM. We compute the Pix2Pix scans, the segmentation and the evaluation results on a machine with eight 32 GB NVIDIA Tesla V100-SXM2.

Table 1. Splits for the BraTS 19 data.

Data Set	N. Patients	Tumor Ground Truth
`training`	270	BraTS 19
`validation`	30	BraTS 19
`testBraTS`	35	BraTS 19
`testOUR`	15	anonymous
`no-ground-truth`	110	/

4.3 Evaluation Methods

We evaluate:

a) Visual Similarity to Real Scans, evaluated by computing *Average Mean Squared Errors* (MSEs) to the original scans. We compute the MSE between the real T2W and each synthesized T2W in the brain mask voxels and, if the tumor ground truth is available, we additionally compute the MSE restricted to the tumor area. Finally, we normalize all values to $[0, 1]$.

b) Usefulness for Segmentation: We use a DeepMedic [17] model to evaluate the usefulness of the synthesized images for segmentation. The model was computed on the `training` data set with all four MRI scans and the corresponding tumor ground truth as input. The hyperparameters were chosen according to a clinical evaluation on DeepMedic [29]. The model is trained to predict three tumor classes: necrotic/non-enhancing tumor, edema and contrast enhanced tumor.

For the evaluation, we compute the segmentation with the DeepMedic model on the test data sets using the synthesized T2W (instead of the real T2W scan) and the real T1W, T1CE and FLAIR. We use Dice scores [8] and undirected 95th Hausdorff distance [33] (HD95) as evaluation metrics. For the case of `testBraTS` and `testOUR`, we compute the Dice score and the HD95 between the segmentation produced with the synthesized images and the ground truth segmentation.

For `no-ground-truth`, these values are computed in comparison to the segmentation produced with the real T2W (instead of the ground truth segmentation). Following the BraTS evaluation [4,5,22], we assess the tumor classes based on three categories: whole tumor (necrotic/non-enhancing tumor, edema and contrast enhanced tumor), tumor core (necrotic/non-enhancing tumor and contrast enhanced tumor), and active tumor (contrast-enhanced tumor) and then compute the average of the three cases.

5 Results

Figures and diagrams could not be included into this short version due to space restrictions. The full version [6] contains a visual impression of the resulting synthesized images as well as the numerical evaluation that is summarized in the following.

5.1 Run Time

BrainClustering Train&Test takes 2.2 h on average to train on 270 patients, and testing takes around 9 s per scan. By comparison, Search 5 and Search 10 take respectively 2.5 min and 4.8 min per patient. The Pix2Pix ResNet training takes around 9 h, while the Pix2Pix UNet training takes around 3 h. The testing process takes around 7 seconds per scan. The baseline methods can be computed in negligible time, with Complementary requiring only one pass over T1W and Random not needing any input data.

5.2 Mean Squared Error Evaluation

Random a much higher MSE than all other methods, averaging at 0.37 on the brain mask and 0.11 on the brain tumor area The second baseline, Complementary evaluates to reasonable MSEs, e.g., it has 0.04 on average in the brain mask and 0.03 in the tumor area on `testBraTS`. All our methods improve upon these values. The scans generated by Pix2Pix have a better MSE than BrainClustering in the brain mask area. In the tumor area, the average MSE of both methods is comparable, and it is overall better, although the median MSE is still below 0.02. The best results are produced by Pix2Pix ResNet with $\lambda = 500$ and $\gamma = 500$, but Pix2Pix UNet with $\lambda = 100$ and $\gamma = 0$ also produces good results. Notably, BrainClustering with Search with 10 patients and 5 clusters achieves the best median MSE in the tumor area among all methods we tested. We conclude that it is possible to compute synthesized scans that closely resemble the true scan.

5.3 Segmentation Evaluation

With respect to the segmentation, all our methods perform better than Complementary, and Pix2Pix performs better than BrainClustering. For Pix2Pix, the ResNet models are in most cases better than the UNet ones. The segmentations with the original T2W obtain an average Dice score of 0.833 on testOUR and of 0.748 on testBraTS, and an average of 0.774 for both data sets together. The overall best performing Pix2Pix approach is with $\lambda = 100$ and $\gamma = 0$, with an average Dice of 0.807 on testOUR and of 0.744 on testBraTS, and an average of 0.763 for both data sets together. For BrainClustering, the best method is Search with 10 scans, 5 macro clusters and 100 micro clusters, with an average Dice of 0.741 on testOUR and of 0.739 on testBraTS, and an average of 0.739 for both data sets together. This method achieves an average HD95 of 12.047 on testOUR, of 8.021 on testBraTS and of 9.229 for both data sets together. On no-ground-truth, the best Pix2Pix model achieved a Dice score of 0.858 and an HD95 of 4.706 and the best BrainClustering model achieved a Dice score of 0.794 and an HD95 of 6.213.

So note that for the data set with ground truth (testOUR and testBraTS), the best performing Pix2Pix method leads to an average Dice score of 0.763 compared to an average score of 0.774 when using the real T2-weighted, so it is very close (BrainClustering performs a bit worse, achieving an average score of 0.739). As an additional indicator, we computed the difference in the metrics when replacing the real T2W by a synthesized T2W scan. The Pix2Pix ResNet approach with $\lambda = 100$ and $\gamma = 0$ decreases the Dice score by less than 0.01 on testBraTS, and by less than 0.03 on testOUR. The average Dice score of the original segmentations on testBraTS and testOUR are 0.748 and 0.833, so the average decreases in Dice score are below 1% and 4% of the original score.

An interesting byproduct of our work is that we observe that substituting the T2W image with random noise also produced acceptable results. This is a common practice that gets some justification from our evaluation. However, the best tested methods outperform this workaround.

Acknowledgement. This work was partly funded by the German Research Foundation (DFG), project numbers 416767905 and 456558332.

Ethics Approval. Ethical approval was not required for this study, as it exclusively utilizes data made available during the BraTS segmentation challenge.

Appendix

A Detailed Description of BrainClustering

Figure 2 gives an overview of the BrainClustering process, which can be divided into multiple steps.

Training Step 1: Both images are segmented into *macro clusters*. The segmentation is done by solving a one-dimensional k-means problem optimally by dynamic programming with the implementation k-means1d [11,34,37]. In the example, the number of macro clusters k is 3. Since the macro clusters are supposed to correspond to the different tissues, they should intuitively be equal to the number of different tissues. However, the process works better if we allow a little more macro clusters to account for varieties of tissues. Empirically, a good range is between three and six clusters.

Training Step 1b: As an intermediate step after finding the macro clusters, we need to identify which cluster represents what tissue. The tissues in T1W can be identified by ordering the clusters according to the average intensity value of their pixels. We could do a similar approach to finding the tissues in T2W, but we instead find them by matching them to the found clusters in T1W as described below in Step 2b (cluster label matching).

Training Step 2: Next, we compute a *micro clustering* for each tissue which captures the shadings of the tissues and the relation between the shading in T1W and T2W. We again implement this step by optimally solving a one-dimensional k-means problem. The number of micro clusters corresponds to the number of different shades that we allow. In Fig. 2, we use 3 micro clusters per tissue for visualization purposes, while in practice, we use at least 100 micro clusters.

Training Step 2b: Now, there are micro clusters for every tissue, both in T1W and in T2W. They are small patches of the same shading, and we aim to identify how such a patch in T1W is mapped to T2W. For this, we need to match the clusters in T1W and T2W. We call this process *cluster label matching.* The matching is done by first excluding voxels that are only present in one of the scans (since the scans are registered to the same template, these are only a few voxels), and then finding the label that maximizes the number of voxels that the matched clusters have in common. The problem reduces to maximum weighted matching problem, which we solve with the Hungarian method [16].

After this step, we have a macro clustering and every macro clustering is subdivided into small patches, and every small patch in T1W has its specified counter part in T2W.

Training Step 3: The next step is to capture the relationship between the shadings in T1W and T2W. The idea behind this is that we assume that it is possible to map the shading of a specific tissue in T1W to the shading of the same tissue in T2W. We model this by using a function f_t for every tissue which is supposed to translate intensity values from T1W to intensity values in T2W. We later want to use the function $f_t(i_1) = i_2$ to predict the T2W intensities of a patient whose T2W scan is missing.

For every tissue type t, we have computed a micro clustering and matched the micro clusters between the two scans. Now we compute the average of the points in each micro cluster of T1W, which we call a_1. Then, we compute the average for each corresponding micro cluster of T2W, which we call a_2. Finally, we add a new entry (a_1, a_2) to f_t, which is the map corresponding to the current macro cluster/tissue type. It may happen that there already is an entry with key a_1 stored in the map. In this case, we compute the average with the moving average formula:

$$f_t(a_1) \leftarrow f_t(a_1) + \frac{a_2 - f_t(a_1)}{\#_t(a_1)}, \tag{4}$$

where $\#_t(a_1)$ is the number of times we tried to add an entry with key a_1 to the map f_t. Note that consequently we have to keep track of $\#_t(a_1)$, i.e., we have to annotate each entry of the map with the corresponding cardinality.

Training Step 1-3: are done for all T1W/T2W scan pairs available in the training data set, and the values found in Step 3 are always inserted into the same tables. So we get as many tables as we have macro clusters, and up to as many rows as we have micro clusters for *all* training images. Since we process many scans, the full tables get too large, and in reality, we only store a meaningful subset of the rows. When querying for a T2W, we can compute missing values by interpolation.

What is left to describe is how we now synthesize images based on our mapping tables. We implemented two options (the second one is faster).

Synthesizing T2W **:** For a patient whose T2W is missing, we preprocess the T1W scan and then cluster it with k-means1d to obtain a macro clustering. We load all tables f_t into the memory which were computed with the same number of macro clusters. Let t be one of the macro clusters. We consider each point p belonging to cluster t. We find the two rows of f_t with the closest intensity values to p and interpolate $f_t(p)$ from these two rows. The resulting intensities form a synthesized scan. Once it is computed, it is postprocessed using a 3×3 median filter [9] to remove *salt-and-pepper* noise which was present in the synthesized images.

Synthesizing T2W **in Search Mode:** We noticed that having large tables computed by training on many images (e.g. 200) produces noise in the results. Thus, we propose an additional method that does not precompute a large model before querying, but instead computes a small model every time we want to answer a query, which we call *Search*. Given a T1W query image, we first search in the training data set for the w patients whose T1W has the smallest mean squared error to the query T1W (where w is a small constant, e.g. $w = 5$). Then, we create a small model by performing the training process only on these w patients, and produce the synthesized T2W with this model. This approach has the downside that we have to create a new model for every single query, thus significantly increasing the query time. However, since we choose small values for w, a synthesized T2W image for one input can be computed in around ten minutes, without the need of computing a model beforehand.

We compare the classic Train&Test mode and Search mode with $w \in \{5, 10\}$ and using $3, 4, 5$ and 6 macro clusters, for a total of 12 different models.

B Additional Modifications to Pix2Pix

We implement some modifications to Pix2Pix:

1. We modify the loading functionality to accept three-dimensional NIfTI images.
2. We extend the data augmentation process with an additional library, TorchIO [28], which implements useful preprocessing and data augmentation routines for medical imaging.
3. We add the mixed precision training functionality from NVIDIA APEX (available at https://github.com/NVIDIA/apex). Operations like matrix to matrix multiplication and convolution are then performed in half-precision floating-point format, which results in a speed-up at training time.
4. Since a big portion of our scans is background, the L1 loss is only computed on the actual brain voxels.
5. Recent studies have shown that the transpose convolution operation during upsampling creates checkerboard artifacts [25,36]. We implement the solution of Wojna et al. [36]: *linear additive upsampling*. This method has been successfully used for the generator of Pix2PixHD [35] and applied on CT scans [12]. We substitute the transpose convolution operation with a two-factor upsampling followed by a 4-factor reduction of the number of channels, and finally by a 3×3 convolution with stride 1. We use bilinear upsampling if the scans are two-dimensional, and trilinear upsampling for the three-dimensional case.

C Additional Figures, Diagrams and Tables

(See Figs. 3, 4, 5, 6, 7 and Table 2).

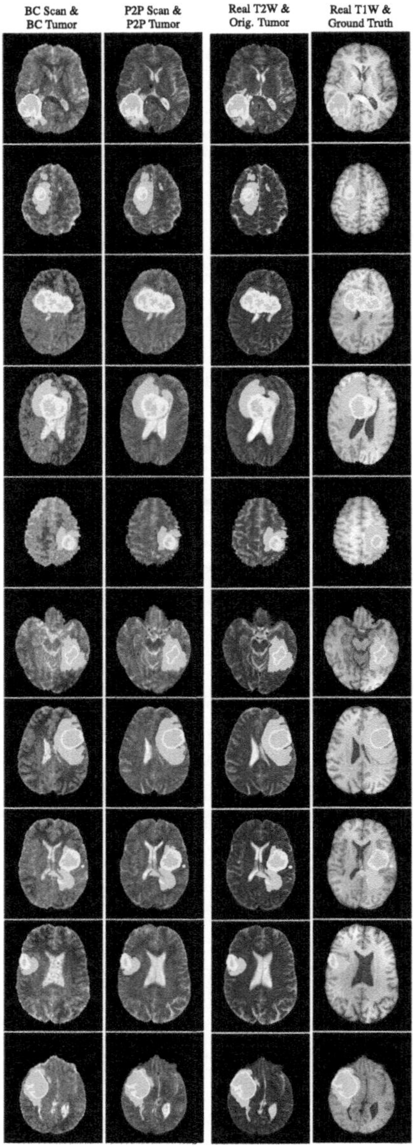

Fig. 3. BrainClustering (BC), Pix2Pix (P2P) and original T2W scans and corresponding superimposed segmentation generated with DeepMedic (together with the real T1W, T1CE and FLAIR), shown for all cases from the `testOUR` data set. In the last column, the ground truth segmentation is superimposed on the T1W scan.

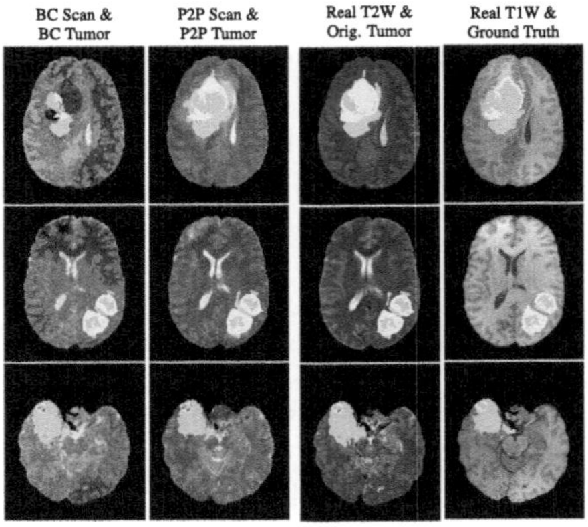

Fig. 4. Continuation of Fig. 3.

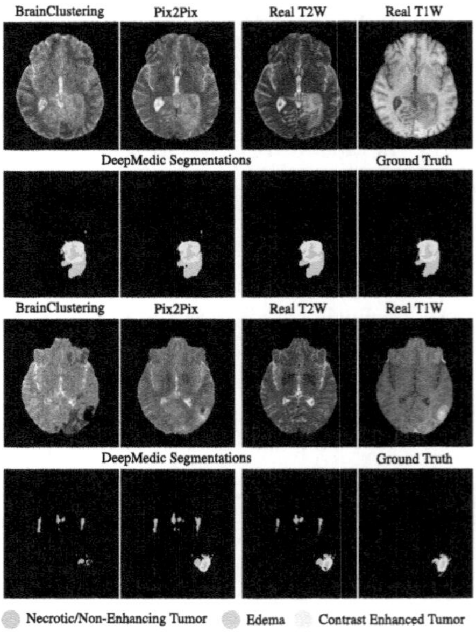

Fig. 5. The segmentations in the second and in the fourth row have been created using the respective T2W and the three original T1W, T1CE, FLAIR. The rightmost image in the "Real T1W" column shows the real T1W image and the ground truth segmentation.

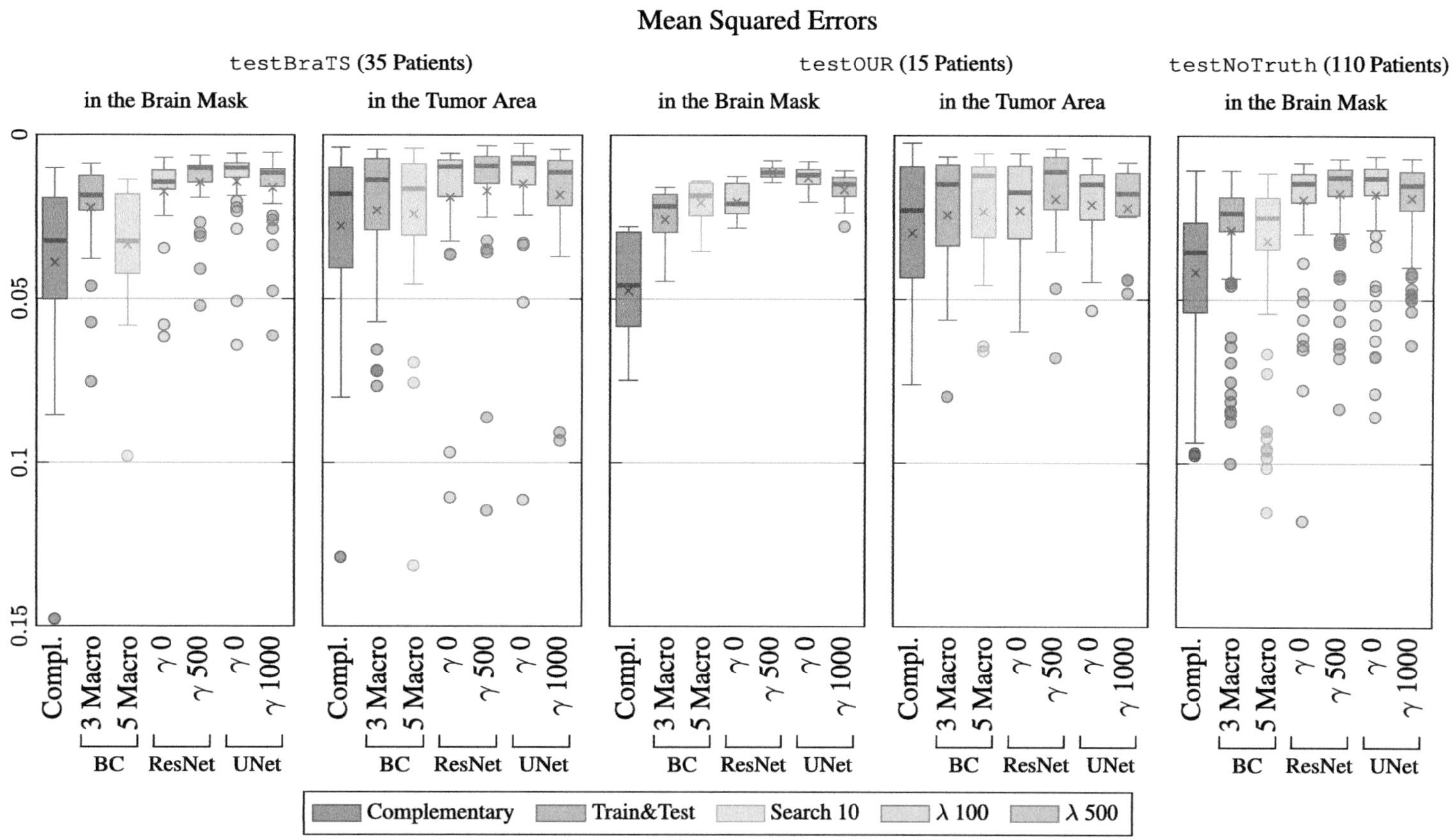

Fig. 6. [Best viewed in color] Comparison of different approaches with respect to MSE in the brain mask and (if available) in the tumor areas. The scores represent the distribution of the patients. A lower value indicates a better score. The crosses represent the averages, the thick bars are the medians and the dots are the outliers.

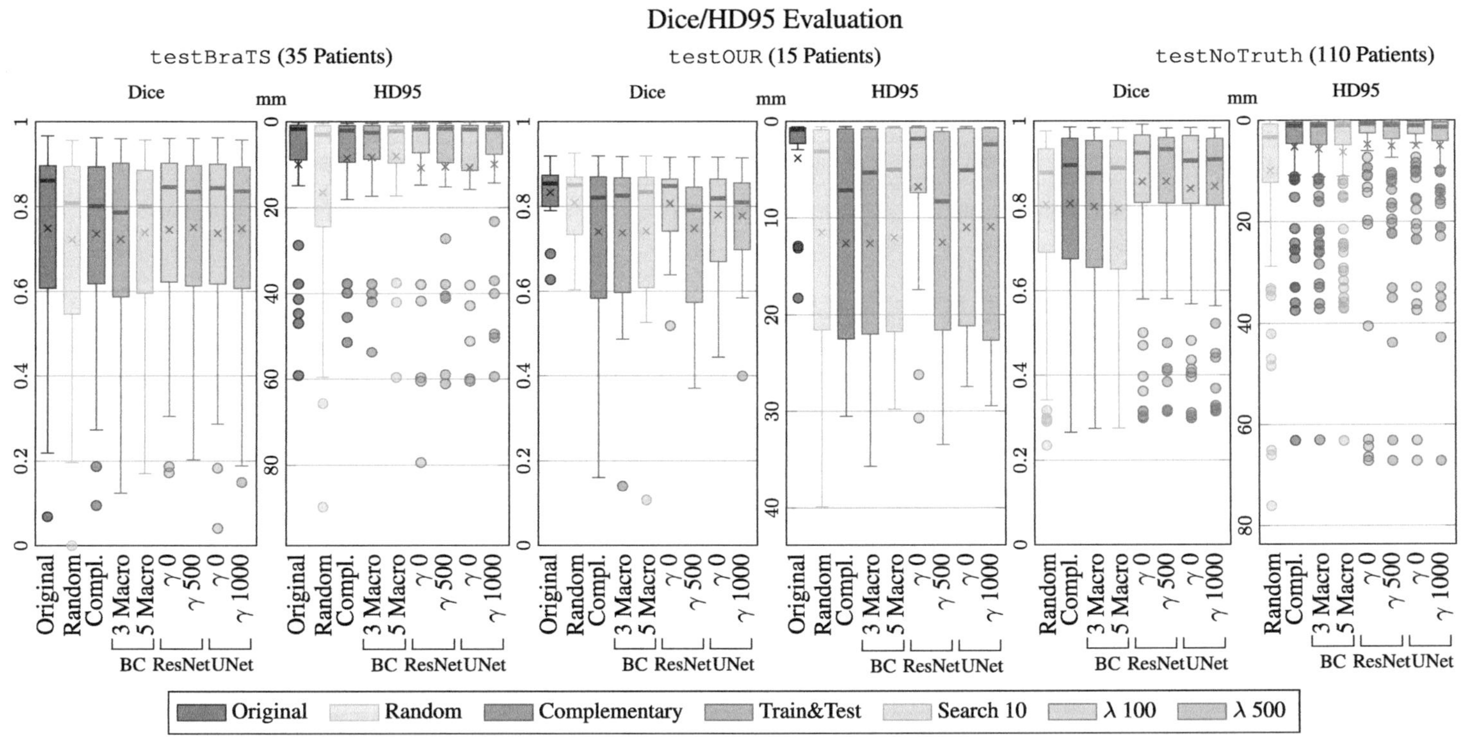

Fig. 7. [Best viewed in color] Comparison with respect to the true tumor segmentation (for `testBraTS` and `testOUR`) and the original tumor segmentation (for `testNoTruth`) using Dice and undirected 95th Hausdorff distance (HD95). For `testBraTS` and `testOUR`, we also include the segmentation which is produced by using the real T2W as *Original* to give a positive baseline as to what scores are achievable if T2W was reproduced perfectly. We compute the Dice score and the undirected 95% percent Hausdorff distance on three different regions and average the scores. The box plots represent the distribution of the patients. For the Dice score a higher value indicates a better value whereas for the Hausdorff distance the opposite holds. The crosses represent the averages, whereas the thick bars are the means and the dots are the outliers.

Table 2. Mean and standard deviation of (Method-Dice - Original-Dice) and (Original-HD95 - Method-HD95). Negative values indicate a decrease in quality.

Approach	`testBraTS`	`testOUR`
Average Change in Dice Score		
Random	-0.03 ± 0.08	-0.02 ± 0.07
Complementary	-0.01 ± 0.09	-0.09 ± 0.17
BC Train/3 Macro	-0.03 ± 0.07	-0.10 ± 0.18
BC-S 10/5 Macro	-0.01 ± 0.10	-0.09 ± 0.19
ResNet 100/0	-0.00 ± 0.06	-0.03 ± 0.08
ResNet 500/500	$+0.00 \pm 0.06$	-0.09 ± 0.13
UNet 100/0	-0.01 ± 0.06	-0.05 ± 0.09
UNet 500/1000	-0.00 ± 0.05	-0.06 ± 0.10
Average Change in 95% Hausdorff Distance		
Random	-6.55 ± 12.76	-7.72 ± 10.94
Complementary	$+1.46 \pm 11.12$	-8.84 ± 10.35
BC Train/3 Macro	$+1.63 \pm 11.50$	-8.85 ± 11.02
BC-S 10/5 Macro	$+1.90 \pm 10.60$	-8.26 ± 10.15
ResNet 100/0	-0.78 ± 7.41	-2.96 ± 7.94
ResNet 500/500	-0.57 ± 5.79	-8.75 ± 9.73
UNet 100/0	-0.77 ± 5.46	-7.20 ± 9.19
UNet 500/1000	$+0.03 \pm 4.49$	-7.14 ± 10.54

References

1. Abdelmotaal, H., Abdou, A.A., Omar, A.F., El-Sebaity, D.M., Abdelazeem, K.: Pix2pix conditional generative adversarial networks for scheimpflug camera color-coded corneal tomography image generation. Transl. Vis. Sci. Technol. **10**(7), 21 (2021)
2. Abrahams, D., Seefeld, S.: Boost.Python. https://www.boost.org/doc/libs/1_75_0/libs/python/doc/html/index.html. Accessed 6 Aug 2023
3. Al-Dmour, H., Al-Ani, A.: MR brain image segmentation based on unsupervised and semi-supervised fuzzy clustering methods. In: Proc. IEEE DICTA, pp. 1–7. Gold Coast, QLD, Australia (Nov 2016). https://doi.org/10.1109/DICTA.2016.7797066
4. Bakas, S., Akbari, H., Sotiras, A., Bilello, M., Rozycki, M., et al.: Advancing the cancer genome atlas glioma MRI collections with expert segmentation labels and radiomic features. Sci. Data **4**(170117) (Sep 2017)
5. Bakas, S., Reyes, M., Jakab, A., Bauer, S., Rempfler, M., et al.: Identifying the best machine learning algorithms for brain tumor segmentation, progression assessment, and overall survival prediction in the BraTS challenge (2018). https://doi.org/10.48550/arXiv.1811.02629
6. Baldini, G., Schmidt, M., Zäske, C., Caldeira, L.L.: MRI scan synthesis methods based on clustering and Pix2Pix (2023). https://doi.org/10.48550/arXiv.2312.05176

7. Bazangani, F., Richard, F.J.P., Ghattas, B., Guedj, E.: Alzheimer's Disease Neuroimaging Initiative: FDG-PET to T1 weighted MRI translation with 3D elicit generative adversarial network (E-GAN). Sensors (Basel) **22**(12), 4640 (2022)
8. Bertels, J., Eelbode, T., Berman, M., Vandermeulen, D., Maes, F., et al.: Optimizing the dice score and jaccard index for medical image segmentation: Theory & practice (2019). https://doi.org/10.48550/arXiv.1911.01685
9. Bradski, G.: The OpenCV Library (2000). https://opencv.org/, Dr. Dobb's Journal of Software Tools
10. Caldeira, L., Almeida, P., Seabra, J.: MR brain tumor segmentation using clustering. In: Proc. ESMRMB Congress, pp. 48–49. Antalya, Turkey (Sep 2009). https://doi.org/10.1007/s10334-009-0175-1
11. Grønlund, A., Larsen, K.G., Mathiasen, A., Nielsen, J.S., Schneider, S., et al.: Fast exact k-means, k-medians and bregman divergence clustering in 1D (2018). https://doi.org/10.48550/arXiv.1701.07204
12. Haubold, J., Hosch, R., Umutlu, L., Wetter, A., Haubold, P., et al.: Contrast agent dose reduction in computed tomography with deep learning using a conditional generative adversarial network. Eur. Radiol. (2021). https://doi.org/10.1007/s00330-021-07714-2
13. He, K., Zhang, X., Ren, S., Sun, J.: Deep residual learning for image recognition (2015). https://doi.org/10.48550/arXiv.1512.03385
14. Isensee, F., Jaeger, P.F., Kohl, S.A.A., Petersen, J., Maier-Hein, K.H.: nnU-Net: a self-configuring method for deep learning-based biomedical image segmentation. Nat. Methods **18**, 203–211 (2020). https://doi.org/10.1038/s41592-020-01008-z
15. Isola, P., Zhu, J.Y., Zhou, T., Efros, A.A.: Image-to-image translation with conditional adversarial networks. In: Proc. IEEE CVPR, pp. 5967–5976. Honolulu, HI, USA (Jul 2017). https://doi.org/10.1109/CVPR.2017.632
16. Jonker, R., Volgenant, A.: A shortest augmenting path algorithm for dense and sparse linear assignment problems. Computing **38**, 325–340 (1987). https://doi.org/10.1007/BF02278710
17. Kamnitsas, K., Ledig, C., Newcombe, V.F.J., Simpson, J.P., Kane, A.D., et al.: Efficient multi-scale 3D CNN with fully connected CRF for accurate brain lesion segmentation. Med. Image Anal. **36**, 61–78 (2016). https://doi.org/10.1016/j.media.2016.10.004
18. Kingma, D.P., Ba, J.: Adam: A method for stochastic optimization (2017). https://doi.org/10.48550/arXiv.1412.6980
19. Li, H.B., Conte, G.M., Anwar, S.M., Kofler, F., Leemput, K.V., et al.: The brain tumor segmentation (BraTS) challenge 2023: Brain MR image synthesis for tumor segmentation (BraSyn) (2023). https://doi.org/10.48550/arXiv.2305.09011
20. Li, M., Zhou, J., Wang, D., Peng, P., Yu, Y.: Application of clustering-based analysis in MRI brain tissue segmentation. Comput. Math. Methods Med. **2022**, 7401184 (2022)
21. Malathi, M., Sinthia, P.: MRI brain Tumour segmentation using hybrid clustering and classification by back propagation algorithm. Asian Pacific J. Cancer Prevent. **19**(11), 3257–3263 (Nov 2018). https://doi.org/10.31557/APJCP.2018.19.11.3257
22. Menze, B.H., Jakab, A., Bauer, S., Kalpathy-Cramer, J., Farahani, K., et al.: The multimodal brain tumor image segmentation benchmark (BRATS). IEEE Trans. Med. Imaging **34**(10), 1993–2024 (2015). https://doi.org/10.1109/TMI.2014.2377694
23. Mirza, M., Osindero, S.: Conditional generative adversarial nets (2014). https://doi.org/10.48550/arXiv.1411.1784

24. Mirzaei, G., Adeli, H.: Segmentation and clustering in brain MRI imaging. Rev. Neurosci. **30**(1), 31–44 (2018). https://doi.org/10.1515/revneuro-2018-0050
25. Odena, A., Dumoulin, V., Olah, C.: Deconvolution and checkerboard artifacts. Distill (2016). https://doi.org/10.23915/distill.00003
26. Padmapriya, T., Sriramakrishnan, P., Kalaiselvi, T., Somasundaram, K.: Advancements of MRI-based brain tumor segmentation from traditional to recent trends: a review. Curr. Med. Imaging Rev. **18**(12), 1261–1275 (2022)
27. Paszke, A., Gross, S., Massa, F., Lerer, A., Bradbury, J., et al.: Pytorch: An imperative style, high-performance deep learning library (2019). https://doi.org/10.48550/arXiv.1912.01703
28. Pérez-García, F., Sparks, R., Ourselin, S.: TorchIO: a Python library for efficient loading, preprocessing, augmentation and patch-based sampling of medical images in deep learning (2020). https://doi.org/10.48550/arXiv.2003.04696
29. Perkuhn, M., Stavrinou, P., Thiele, F., Shakirin, G., Mohan, M., et al.: Clinical evaluation of a multiparametric deep learning model for glioblastoma segmentation using heterogeneous magnetic resonance imaging data from clinical routine. Invest. Radiol. **53**(11), 647–654 (2018). https://doi.org/10.1097/RLI.0000000000000484
30. Ranjbarzadeh, R., Caputo, A., Tirkolaee, E.B., Jafarzadeh Ghoushchi, S., Bendechache, M.: Brain tumor segmentation of MRI images: a comprehensive review on the application of artificial intelligence tools. Comput. Biol. Med. **152**(106405), 106405 (2023)
31. Raut, P., Baldini, G., Schöneck, M., Caldeira, L.: Using a generative adversarial network to generate synthetic MRI images for multi-class automatic segmentation of brain tumors. Front. Radiol. **3** (2024). https://doi.org/10.3389/fradi.2023.1336902
32. Ronneberger, O., Fischer, P., Brox, T.: U-net: Convolutional networks for biomedical image segmentation (2015). https://doi.org/10.48550/arXiv.1505.04597
33. Sharp, G., Wu, Z., Peroni, M., Lee, J., Li, R., et al.: Plastimatch (2011). http://plastimatch.org/
34. Steinberg, D.: kmeans1d (2020). https://github.com/dstein64/kmeans1d
35. Wang, T.C., Liu, M.Y., Zhu, J.Y., Tao, A., Kautz, J., et al.: High-resolution image synthesis and semantic manipulation with conditional gans. In: Proc. IEEE CVPR, pp. 8798–8807. Salt Lake City, UT, USA (Jun 2018). https://doi.org/10.1109/CVPR.2018.00917
36. Wojna, Z., Ferrari, V., Guadarrama, S., Silberman, N., Chen, L.C., et al.: The devil is in the decoder: Classification, regression and GANs (2019). https://doi.org/10.48550/arXiv.1707.05847
37. Wu, X.: Optimal quantization by matrix searching. J. Algorithms **12**, 663–673 (1991). https://doi.org/10.1016/0196-6774(91)90039-2
38. Yan, B., Cao, M., Gong, W., Wei, B.: Multi-scale brain tumor segmentation combined with deep supervision. Int. J. Comput. Assist. Radiol. Surg. **17**(3), 561–568 (2022)
39. Yang, Q., Li, N., Zhao, Z., Fan, X., Chang, E.I.C., et al.: MRI cross-modality image-to-image translation. Sci. Rep. **10**(1), 3753 (2020)

Supervised Pectoral Muscle Removal in Mammography Images

Parvaneh Aliniya(✉), Mircea Nicolescu, Monica Nicolescu, and George Bebis

University of Nevada, Reno, NV 89557, USA
paliniya@unr.edu

Abstract. In this paper, we provide the segmentation masks of the pectoral muscle for INbreast, MIAS, and a CBIS-DDSM subset datasets, which will enable the development of supervised methods and the utilization of deep learning for pectoral muscle removal from mammography images. We trained AU-Net separately on the INbreast and CBIS-DDSM subset for the segmentation of the pectoral muscle. We used cross-dataset testing to evaluate the performance of the models on an unseen dataset. The experimental results show that cross-dataset testing achieves a comparable performance to the same-dataset experiments. In addition, the models were tested on the entire MIAS dataset, and they outperformed previous methods. The segmentation masks are available at https://github.com/Parvaneh-Aliniya/pectoral_muscle_groundtruth_segmentation.

Keywords: Breast Cancer · Pectoral Muscle · Mammography

1 Introduction

Breast cancer is one of the main cancer types in the female population, with a high mortality rate. Mammograms are images taken from two different views of breast. These views are Craniocaudal (CC) and mediolateral oblique (MLO). CC is the view in which the breast is compressed horizontally, and in the MLO view, the compression is diagonal. Mammography images are the most commonly used tool for breast cancer screening due to their availability and lower cost. Therefore, the development of automated cancer detection methods for these images is of high importance due to the benefits they bring to the patients by increasing survival chances when detecting the abnormalities accurately in early stages [1].

In this paper, we target the task of removing the pectoral muscle in mammography images, which is of high importance for automated density estimation. Due to the fact that normally the segmentation mask of the pectoral muscle is not available in the publicly available datasets or in the examinations for breast cancer screening in current practice, most of the pectoral removal approaches use traditional machine learning approaches. Since the location, shape, density, and position of the muscle vary between the images [2], the performance of these methods is limited. Ideally, utilizing deep learning-based approaches would help mitigate these hurdles to a great extent if annotations for the pectoral muscle were available.

J. Finkelstein et al. (Eds.): AIME 2024, LNAI 14845, pp. 126–130, 2024.
https://doi.org/10.1007/978-3-031-66535-6_14

We argue that providing annotations even for several datasets will enable researchers to train supervised pectoral muscle removal and utilize them for new unseen datasets with high performance. Hence, we provide pectoral muscle segmentation for three benchmark datasets, INbreast [3], MIAS [4], and a widely used subset of CBIS-DDSM [5]. We separately trained AU-Net [6], which is a widely used segmentation method for mammography images, on the datasets. To validate the proposed method, same-dataset and cross-dataset tests were used. In same-dataset tests, train and test sets belong to one of the datasets. In the cross-dataset experiments, train and test sets are from two different datasets. The models achieve high accuracy (99.24 ± 0.16 on average) for both same-dataset and cross-dataset experiments.

The contributions of this paper are three-fold: (1) Generating the segmentation masks of pectoral muscle for INbreast, MIAS, and CBIS-DDSM subset datasets. (2) Training pectoral muscle removal models using the AU-Net architectures separately for INbreast and CBIS-DDSM datasets. (3) Evaluating the models by same-dataset and cross-dataset testing to measure the generalizability of the supervisely trained models on the same and new datasets.

2 Related Work

The segmentation of the pectoral muscle is not available in mammography datasets; hence, most of the proposed methods are traditional machine learning-based methods. These methods, in general, aim to use the appearance of the muscle and its location to detect and eliminate it. According to a recent study [1], thresholding [7] and region growing [8] are among the widely used approaches.

The general idea for thresholding is to use the observation that the brightness of the pectoral muscle is generally higher than the neighboring regions; therefore, by eliminating pixels lower than a certain threshold, the region for the pectoral muscle will the extracted. This idea, coupled with the utilization of the orientation of the breast (whether the breast is on the left or right side of the image) and the generic shape of the muscle, has been used in the literature [9].

In general, region-growing-based methods start with initial seeds, and then, according to certain similarity metrics, they continue adding a new neighboring pixel to a region until a termination criterion is met [9, 10]. Aside from the previous methods that aim to use region growing and thresholding, graph-cut [11], Hough Transform [12], line estimation, polynomial fitting and curve estimation [13], k-means [14], active contours [15] and contour growing [16] are also used in several methods.

3 Proposed Method

3.1 Ground Truth Generation for Pectoral Muscle

For the segmentation of pectoral muscle, LabelMe [17] was used, in which polygons were fitted to the pectoral muscle for the MLO images in INbreast, a subset of CBIS-DDSM and MIAS datasets. For images with higher density or lower visibility of the boundaries of the pectoral muscle, the portion of the muscle that was clearly distinguishable from the breast tissues was selected. The segmentation masks are generated in JSON and image formats with two classes, the pectoral muscle, and background (the rest of breast tissues and image background).

3.2 Pectoral Muscle Segmentation

The main motivation for this study is to provide the segmentation of the pectoral muscle for several datasets, which enables the training of pectoral muscle removal that could also be applied to new unseen datasets. The main use-case of these segmentation masks will be in the removal of pectoral muscle from images in the preprocessing step for tasks such as the classification of images (for instance, benign/malignant), segmentation of the masses and other abnormalities, and density estimation. To use the segmentation masks for a new dataset; first, a segmentation model should be trained using the provided muscle segmentation masks, and then the model could be used for the segmentation of the pectoral muscles in the new dataset.

Given the segmentation masks of the pectoral muscles, we propose to use deep learning-based methods for pectoral muscle segmentation. To this end, we selected AU-Net which is a state-of-the-art method for segmentation in mammography images. AU-Net [6] is an improved version of the U-Net [18] in which the encoding and decoding paths are not symmetrical. ResUnit and the basic decoder proposed in the AU-Net have been used for the encoder and the decoder. The details of the novel idea of AU-Net, the Attention guided Up-sampling Block (AU Block), are presented in the AU-Net approach [6]. The binary cross-entropy loss function was used in the proposed method.

After training on the INbreast and CBIS-DDSM subsets separately, the models have been used for the cross-dataset tests to evaluate their performance on unseen datasets.

4 Experimental Results

4.1 Datasets and Preprocessing

The pectoral muscle masks for all of the MLO view images in the INbreast dataset have been provided in this study and used for the experiments. A 5-fold cross-validation is used for the INbreast dataset. CBIS-DDSM is an enhanced subset of the DDSM dataset. We provided the pectoral muscle segmentation masks for all the 447 MLO view images in a commonly used subset of CBIS-DDSM. In the experiments, the original split for testing and training is used. MIAS dataset contains only MLO view images. We also included the segmentation masks for all the images in the MIAS dataset (unless the muscle was not visible in the images). We used MIAS for cross-dataset testing using models trained on INbreast, CBIS-DDSM subset, and a combination of datasets. For all of the datasets, cropping, padding, resizing, and artifact removal have been performed as needed. The models were trained with a learning rate of 0.0001 for 100 epochs.

4.2 Results for INbreast, CBIS-DDSM Subset, and MIAS Datasets

The experimental results and comparison with state-of-the-art methods are presented in Table 1. The Dice Similarity Coefficient (DSC), Sensitivity, and Accuracy have been used as evaluation metrics in the experiments. The following format is used for the names of the experiments: "train dataset name - test dataset name". We used 'CBIS' instead of 'CBIS-DDSM subset' in the table for convenience. As shown, the proposed method has accurate prediction for same and cross-dataset tests and also outperforms the previous methods on the MIAS dataset. It should be noted that the entire MIAS dataset was used only in test time for all the experiments with the MIAS dataset.

Table 1. Results for pectoral muscle segmentation for same and cross-dataset experiments. In the name of the experiments, the first term is the training dataset, and the second is the test dataset.

Train-Test Pair	DSC↑	Sensitivity↑	Accuracy↑
[16]	–	–	98.1
[13]	–	–	96.81
[7]	–	–	98
[10]	–	–	97.8
CBIS-CBIS	96.61	98.07	99.50
INbreast-INbreast	95.09	95.54	99.55
INbreast-CBIS	91.87	97.09	99.00
CBIS-INbreast	86.89	78.87	98.93
CBIS-MIAS	95.11	95.04	99.53
INbreast-MIAS	90.64	95.44	99.03
Combined-MIAS	95.73	95.55	99.59

5 Conclusion

In this study, we provide the segmentation masks for the INbreast, MIAS, and CBIS-DDSM subset datasets. We also trained AU-Net separately on the INbreast, CBIS-DDSM subset, and a combination of both datasets. We tested both models on the entire MIAS dataset, resulting in accuracies of 99.03%, 99.53%, and 99.59% for models trained on the INbreast and CBIS-DDSM subset and the combination of both datasets, respectively.

References

1. Moghbel, M., Ooi, C.Y., Ismail, N., Hau, Y.W., Memari, N.: A review of breast boundary and pectoral muscle segmentation methods in computer-aided detection/diagnosis of breast mammography. Artif. Intell. Rev. **53**, 1873–1918 (2020)
2. Haris, U., Kabeer, V.: Automated breast pectoral muscle segmentation in digital mammograms using holistic regularization method. J. Clin. Otorhinolaryngol. Head Neck Surg.
3. Moreira, I.C., Amaral, I., Domingues, I., Cardoso, A., Cardoso, M.J., Cardoso, J.S.: Inbreast: toward a full-field digital mammographic database. Acad. Radiol. **19**, 236–248 (2012)

4. Suckling, J., Parker, J., Dance, D., et al.: Mammographic Image Analysis Society (MIAS) Database v1.21. Apollo - University of Cambridge Repository (2015)
5. Lee, R.S., Gimenez, F., Hoogi, A., Miyake, K.K., Gorovoy, M., Rubin, D.L.: A curated mammography data set for use in computer-aided detection and diagnosis research. Sci. Data (2017)
6. Sun, H., et al.: AUNet: attention-guided dense-upsampling networks for breast mass segmentation in whole mammograms. Phys. Med. Biol. **65**(5), 055005 (2020)
7. Czaplicka, K., Włodarczyk, J.: Automatic breast-line and pectoral muscle segmentation. Schedae Informaticae **20** (2011)
8. Ergin, S., Esener, İ.I., Yüksel, T.: A genuine GLCM-based feature extraction for breast tissue classification on mammograms. Int. J. Intell. Syst. Appl. Eng. **4**, 124–129 (2016)
9. Subashini, T.S., Ramalingam, V., Palanivel, S.: Pectoral muscle removal and detection of masses in digital mammogram using CCL. Int. J. Comput. Appl. 71–76 (2010)
10. Chen, Z., Zwiggelaar, R.: A combined method for automatic identification of the breast boundary in mammograms. In: 2012 5th International Conference on BioMedical Engineering and Informatics (2012)
11. Camilus, K.S., Govindan, V.K., Sathidevi, P.S.: Computer-aided identification of the pectoral muscle in digitized mammograms. J. Digit. Imaging **23**, 562–580 (2010)
12. Xu, W., Li, L., Liu, W.: A novel pectoral muscle segmentation algorithm based on polyline fitting and elastic thread approaching. In: 2007 1st International Conference on Bioinformatics and Biomedical Engineering, pp. 837–840. IEEE (2007)
13. Shen, R., Yan, K., Xiao, F., Chang, J., Jiang, C., Zhou, K.: Automatic pectoral muscle region segmentation in mammograms using genetic algorithm and morphological selection. J. Digit. Imaging **31**, 680–691 (2018)
14. Slavković, I.M., Gavrovska, A., Milivojević, M., Reljin, I., Reljin, B.: Breast region segmentation and pectoral muscle removal in mammograms. Telfor J. **8**(1), 50–55 (2016)
15. Yin, K., Yan, S., Song, C., Zheng, B.: A robust method for segmenting pectoral muscle in mediolateral oblique (MLO) mammograms. Int. J. Comput. Assist. Radiol. Surg. **14**, 237–248 (2019)
16. Rampun, A., Morrow, P.J., Scotney, B.W., Winder, J.: Fully automated breast boundary and pectoral muscle segmentation in mammograms. Artif. Intell. Med. **79**, 28–41 (2017)
17. LabelMe. https://github.com/labelmeai/labelme
18. Ronneberger, O., Fischer, P., Brox, T.: U-net: convolutional networks for biomedical image segmentation.In: Medical Image Computing and Computer-Assisted Intervention–MICCAI 2015, pp. 234–241. Springer (2015)

TinySAM-Med3D: A Lightweight Segment Anything Model for Volumetric Medical Imaging with Mixture of Experts

Tianyuan Song[1], Guixia Kang[1(✉)], and Yiqing Shen[2]

[1] Key Laboratory of Universal Wireless Communications, Beijing University of Posts and Telecommunications, Beijing, China
{songty,gxkang}@bupt.edu.cn

[2] Department of Compter Science, Johns Hopkins University, Baltimore, MD, USA
yshen92@jhu.edu

Abstract. Segment anything models (SAMs) demonstrate exceptional zero-shot segmentation capabilities on natural images when provided appropriate prompts. However, directly applying SAMs to medical images presents challenges due to the complexity and diversity of medical data compared to natural images. Some research aims to integrate medical knowledge into SAM by fine-tuning the model. However, the encoders are based on the ViT architecture, which incurs high computational costs when applied directly to 3D medical data, limiting real-time performance on resource-constrained devices. To address these limitations, we introduce TinySAM-Med3D, an efficient SAM tailored for 3D medical image segmentation. TinySAM-Med3D builds on SAM-Med3D by distilling the encoder to a lightweight TinyViT and substituting the multilayer perceptron with a Mixture of Experts (MoEs) to preserve performance while significantly reducing computational and memory costs. This enables real-time segmentation on resource-constrained devices without sacrificing performance. Evaluations on the abdominal CT dataset of TotalSegmentator reveal that TinySAM-Med3D attains a 0.8440 Dice score while using only 33.87% of SAM-Med3D parameters. It also accelerates inference by 3.36× over SAM-Med3D. Thus, TinySAM-Med3D facilitates deploying SAMs for fast 3D medical image segmentation. Our code and models are available at https://github.com/songty21110133/TinySAM-Med3D.

Keywords: Medical Image Segmentation · Foundation Model · Knowledge Distillation · Mixture of Experts · Resource Efficiency

1 Introduction

Segment Anything Models (SAMs) [8] are vision foundation models demonstrating exceptional zero-shot segmentation capabilities when provided with

This work was supported by Fundamental Research Funds for the Central Universities (2020XD-A06-1), the State Key Program of the National Natural Science Foundation of China (82030037).

J. Finkelstein et al. (Eds.): AIME 2024, LNAI 14845, pp. 131–139, 2024.
https://doi.org/10.1007/978-3-031-66535-6_15

appropriate prompts. SAMs contain a vision transformer (ViT) encoder to extract spatial-semantic representations, a lightweight decoder to produce segmentation masks, and a prompt encoder to guide the model [2]. Trained with over 1 billion masks, SAM exhibits outstanding zero-shot generalization ability to unseen data [3,9]. Although trained on over 1 billion masks from natural images, directly applying SAMs to medical images impacts performance due to the unique properties of medical data such as variable imaging protocols, small anatomical structures, and intricate shape variations.

Consequently, research has adapted SAMs for medical image analysis via additional training on medical data [17]. For example, MedSAM [10] fine-tunes the image encoder while preserving all the architectures frozen. SAM-Med2D [1] fine-tunes the model with adapters. However, these approaches process 3D volumetric scans slice-by-slice, losing valuable 3D spatial context. SAM-Med3D [18] addresses this limitation by transforming SAM into a fully 3D architecture trained from scratch on over 131K expert-annotated 3D segmentation masks across 247 anatomical categories sampled from various modalities. This captures rich 3D spatial context to improve segmentation accuracy. However, its ViT encoder contains over more than 90 million parameters, incurring heavy computational costs when applied to high-resolution 3D medical images. This hinders real-time analysis on resource-constrained devices prevalent in clinical settings.

In natural image domains, knowledge distillation transfers latent knowledge from large, complex teacher models to smaller, faster student models [5,14,15]. Although recent works have explored knowledge distillation to compress SAMs [16,20,22], directly applying these approaches to medical images risks unacceptable performance decreases as they exclusively target natural images. However, applying these methods directly to the field of medical imaging can result in a performance decrease as they are targeting natural images. To address this challenge, we propose TinySAM-Med3D which distill the complex teacher encoder to a Mixtures-of-Experts (MoEs) based efficient student encoder [7,21] to retain high performance while maximizing efficiency Specifically, we distill the computationally heavy ViT encoder into a lightweight 3D TinyViT model and augment the network capacity via soft-gated MoEs [11] without increasing inference latency. This allows efficiently scaling TinySAM-Med3D to large 3D medical datasets.

Our major contributions are two-fold, summarized as follows. Firstly, we propose TinySAM-Med3D, an efficient medical image segmentation framework that combines knowledge distillation and soft-gated Mixtures-of-Experts (MoEs). Specifically, we distill a large vision transformer encoder into a lightweight 3D TinyViT model and enhance representation power via soft MoEs without increasing inference latency. This allows efficiently scaling to high resolution 3D medical images while reducing parameters by 36.11 s and FLOPs by 154.48 GFLOPS compared to prior state-of-the-art SAM-Med3D. Secondly, we demonstrate the practical utility of TinySAM-Med3D for 3D medical image analysis on resource-constrained devices. Evaluations on the abdominal CT segmentation benchmark

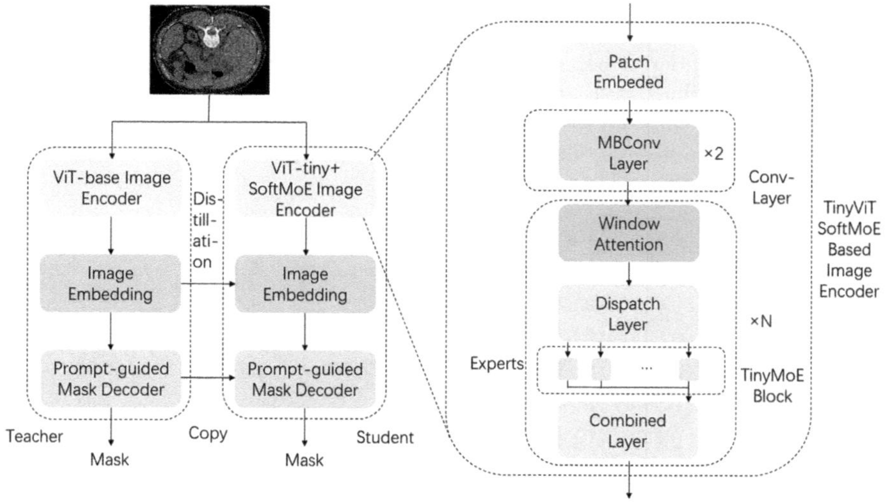

Fig. 1. The overall architecture of our TinySAM-Med3D. **Left**: Distillation from the ViT image encoder in SAM-Med3D to a TinyViT encoder. **Right**: The architecture of our encoder with MoE. counterparts.

TotalSegmentator reveal TinySAM-Med3D achieves a competitive Dice score of 84% while using just a third of SAM-Med3D's parameters.

2 Method

Method Overview. We selected SAM-Med3D as the base model for our framework due to its state-of-the-art performance in 3D medical image segmentation from capturing rich 3D spatial contexts. However, SAM-Med3D relies on a computationally ViT encoder, presenting challenges for real-time analysis on constrained devices. As shown in Fig. 1, we first distill the cumbersome ViT encoder into a lightweight 3D TinyViT model to significantly reduce computational overhead while preserving representation power, following the design of MobileSAM [22] for 2D natural images Specifically, TinyViT matches SAM-Med3D's feature dimensions, allowing us to replace the teacher ViT with the student TinyViT. Additionally, we boost TinyViT's capacity via soft MoEs which route tokens to expert sub-networks based on computed gates. Replacing TinyViT's MLP with a soft MoE preserves high performance without increasing inference latency.

SAM Backbone. To effectively capture 3D spatial contexts critical for medical image analysis, we transform TinyViT into a 3D encoder-decoder architecture. The encoder contains four stages to hierarchically extract features at different resolutions. The first stage uses efficient mobile inverted bottleneck convolutional (MBConv) blocks [6] to learn low-level 3D representations and detailed shape

Table 1. Comparison of 3D medical image segmentation models on the abdominal CT dataset from TotalSegmentator. We report intersection-over-union (IoU), Dice score, FLOPs, MACs, number of parameters (Params), and inference time. **Bold** indicates best score and underlined is second best. For fair comparison, all methods utilize a consistent decoder with embeddings projected to equivalent dimensions. TinySAM-Med3D achieves competitive accuracy using substantially fewer parameters and computations versus state-of-the-art SAM-Med3D.

Encoder	IoU	Dice	FLOPs (G)	MACs (G)	Params (M)	InferTime (s)
UNet [12]	0.6353	0.7805	169.83	84.66	**5.84**	15.83
SwinUNetr [4]	0.6974	0.8253	74.06	36.21	12.33	84.52
TinyViT [19]	0.6931	0.8013	30.24	15.01	9.35	**14.41**
MedNext [13]	0.6840	0.8109	119.67	58.67	11.04	34.53
SAM-Med3D [18]	**0.7947**	**0.8693**	180.39	89.41	92.92	51.41
TinySAM-Med3D	0.7506	0.8440	**25.91**	**12.47**	31.48	15.30

cues. The subsequent stages employ Transformer blocks with windowed self-attention for long-range context aggregation while focusing computations over local 3D regions to minimize cost [19]. Specifically, the self-attention computes correlations between features within a 3D attention window rather than the full volume. Window sizes gradually increase across stages to widen receptive fields while maintaining efficiency. We adapt all components to 3D, including adjusting feature dimensions in the early stages and attention window sizes to accurately model relationships in 3D space. The decoder contains upsampling blocks to recover spatial resolution and concatenate encoder features for precise localization. This hierarchical design captures multi-scale 3D semantics to effectively segment medical images.

SoftMoE. To further enhance TinyViT's representation power, we incorporate a Soft Mixture of Experts (SoftMoE) [11] layer in place of the MLP in each Transformer block. SoftMoE combines multiple expert neural networks to improve overall model capacity and performance. The key idea is to softly route input tokens to each expert based on relevance, then convexly combine expert outputs to obtain enriched feature representations. Specifically, SoftMoE first computes routing weights between input tokens and each expert using a gating module. These gates reflect similarity between token embeddings and expert specializations. Input tokens are then weightedly fed into experts for processing. This soft allocation automatically matches tokens with appropriate experts rather than hard assignment. Finally, expert outputs are blended using normalized gates as coefficients. This allows dynamically adjusting contributions based on current input-expert relevance. By leveraging diverse experts within a single model, SoftMoE enhances TinyViT's flexibility to handle complex medical images. The gating mechanism facilitates specialization among experts while convex combination integrates their complementary strengths. Thus, SoftMoE

Table 2. Ablation study evaluating SoftMoE configurations in TinySAM-Med3D on the abdominal CT segmentation task. From the table, it can be observed that the model achieves a good balance between performance and parameter count when number of experts is set to 64.

Experts	IoU	Dice	FLOPs (G)	MACs (G)	Params (M)	InferTime (s)
8	0.6850	0.7986	25.24	12.47	9.37	14.69
16	0.7321	0.8280	25.33	12.47	12.53	14.79
32	0.7442	0.8368	25.52	12.47	18.85	15.05
64	0.7506	0.8440	25.91	12.47	31.48	15.30
128	0.7549	0.8447	26.67	12.47	56.75	18.79

effectively boosts TinyViT's representation power, translating to significant segmentation accuracy gains while maintaining high computational efficiency.

3 Experiments

Implementations. Our work was implemented in PyTorch and trained on two NVIDIA RTX3090 GPUs, each with 24 GB of memory. The overall model training process consists of two parts: model distillation and fine-tuning on the dataset. During the distillation process, we used the AdamW optimizer with an initial learning rate of 5e−4, and the model was trained for a total of 50 epochs. For the fine-tuning process, we followed the fine-tuning hyperparameters of SAM-Med3D. During training, our work also built upon SAM-Med3D, and we resized all images to $128 \times 128 \times 128$ by cropping. The loss function used was DiceCELoss. For data augmentation, we applied RandomFlip and ZNormalization on each image data.

Table 3. Ablation on expert slot capacity in TinySAM-Med3D. All settings use 64 experts and the distilled TinyViT encoder. Increasing slot capacity marginally impacts accuracy while growing latency and parameters. Thus, 1 slot per expert provides the best efficiency-accuracy trade-off.

Slots/Expert	IoU	Dice	FLOPs (G)	MACs (G)	Params (M)	InferTime (s)
1	0.7506	0.8440	25.91	12.47	31.48	15.30
2	0.7532	0.8443	26.67	12.47	31.60	18.91

Overall Performance. We demonstrate TinySAM-Med3D's effectiveness on the abdominal CT segmentation task from TotalSegmentator. Table 1 presents the performance metrics of UNet [12], SwinUnetr [4], MedNext [13], SAM-Med3D, and our proposed TinySAM-Med3D on the test set. It is important to

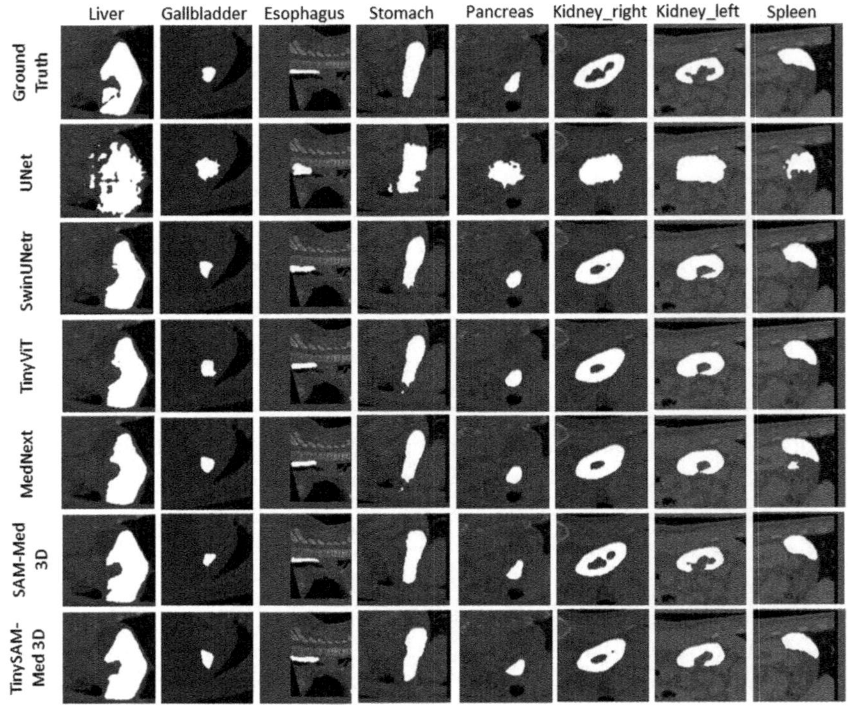

Fig. 2. Qualitative 3D segmentation results comparing SAM-Med3D and our proposed TinySAM-Med3D on axial slices from abdominal CT scans. We visualize outputs across anatomical structures using varying prompt lengths. White depicts predicted segmentation masks. TinySAM-Med3D achieves accurate sparse and dense region detection despite the compact encoder and lower computations. This demonstrates the practical utility of our efficient approach without sacrificing quality.

note that, for a fair comparison, we only utilized the encoders of the other models for the control experiments and maintained consistent feature dimensions using the neck layer. Results demonstrate our approach achieves a Dice score of 84.40%, on par with top-performing SAM-Med3D which attained 86.93% Dice. However, TinySAM-Med3D reduces the number of parameters to 33.87% from 92.92M in SAM-Med3D down to 31.48M. This substantial parameter reduction stems from replacing the large ViT encoder with our knowledge distilled TinyViT model. Additionally, TinySAM-Med3D accelerate the inference by 3.36× compared to SAM-Med3D, enabling real-time performance. Furthermore, it is observed that irrespective of the anatomical and structural disparities among various organs, the SAM architecture demonstrates its capability to deliver effective segmentation outcomes under the direction of prompts.

Ablation Study. We systematically evaluate how key hyperparameters of the SoftMoE module, including the number of experts and token slots per expert, impact the accuracy and efficiency trade-offs of TinySAM-Med3D. As shown in Table 2, we ablate the number of experts while fixing other settings like the distilled TinyViT encoder. More experts significantly expands model capacity to capture greater feature diversity, evident by the steady Dice score improvements from 0.7986 with 8 experts up to 0.8440 with 64 experts. However, as experts grow, so does model size and computations. We find 64 experts strikes the best balance, achieving 0.8440 Dice score while adding having 31.48M parameters compared to the 9.37M baseline TinyViT, which is crucial for constrained devices. Similarly in Table 3, additional token slots per expert marginally improves Dice to 0.8443 but introduces more calculations and latency. Since slots primarily enhance training stability rather than representation power, we select 1 slot to maximize efficiency without quality degradation during inference.

Visualization. Figure 2 provides qualitative results contrasting 3D segmentations from SAM-Med3D versus our distilled TinySAM-Med3D. We sample axial slices following the protocol of SAM-Med3D for consistent visualization. As shown, TinySAM-Med3D achieves precise contour detection rivaling SAM-Med3D across diverse anatomical structures and segmentation prompts. The comparable quality underscores the effectiveness of our knowledge transfer which preserves SAM-Med3D's learned feature representations within the lightweight TinyViT encoder. Moreover, the addition of SoftMoE further enhances these learned embeddings to improve localization and region differentiation capabilities. Thus through efficient distillation and expertise integration, TinySAM-Med3D retains competitive performance while enabling real-time analysis.

4 Conclusion

In this work, we distil SAM-Med3D to a MoE-based tiny model, namely TinySAM-Med3D. Building upon the advanced SAM-Med3D framework, our approach transfers knowledge from the heavy vision transformer encoder to a lightweight 3D TinyViT model augmented with soft-gated dynamic routing expert layers. This preserves SAM-Med3D's state-of-the-art segmentation quality while accelerate the inference and lower the model size. Experimental results demonstrate that our approach achieves substantial weight compression (92.92M to 31.48M) and improves inference speed (51.40 s to 15.30 s). Since our work preserves all the pipelines of the original SAM-Med3D and only replaces the image encoder, it allows for easy plug-and-play deployment for clinical usage.

References

1. Cheng, J., et al.: SAM-Med2D. arXiv preprint arXiv:2308.16184 (2023)
2. Dosovitskiy, A., et al.: An image is worth 16 × 16 words: transformers for image recognition at scale. arXiv preprint arXiv:2010.11929 (2020)
3. Gao, H., Li, Y., Long, K., Yang, M., Shen, Y.: A survey for foundation models in autonomous driving. arXiv preprint arXiv:2402.01105 (2024)
4. Hatamizadeh, A., Nath, V., Tang, Y., Yang, D., Roth, H.R., Xu, D.: Swin UNETR: Swin transformers for semantic segmentation of brain tumors in MRI images. In: Crimi, A., Bakas, S. (eds.) International MICCAI Brainlesion Workshop. LNCS, vol. 12962, pp. 272–284. Springer, Cham (2021). https://doi.org/10.1007/978-3-031-08999-2_22
5. Hinton, G., Vinyals, O., Dean, J.: Distilling the knowledge in a neural network. arXiv preprint arXiv:1503.02531 (2015)
6. Howard, A., Zhmoginov, A., Chen, L.C., Sandler, M., Zhu, M.: Inverted residuals and linear bottlenecks: mobile networks for classification, detection and segmentation (2018)
7. Jacobs, R.A., Jordan, M.I., Nowlan, S.J., Hinton, G.E.: Adaptive mixtures of local experts. Neural Comput. **3**(1), 79–87 (1991)
8. Kirillov, A., et al.: Segment anything. arXiv preprint arXiv:2304.02643 (2023)
9. Liu, X., et al.: Towards the unification of generative and discriminative visual foundation model: a survey. arXiv preprint arXiv:2312.10163 (2023)
10. Ma, J., He, Y., Li, F., Han, L., You, C., Wang, B.: Segment anything in medical images. Nat. Commun. **15**(1), 654 (2024)
11. Puigcerver, J., Riquelme, C., Mustafa, B., Houlsby, N.: From sparse to soft mixtures of experts. arXiv preprint arXiv:2308.00951 (2023)
12. Ronneberger, O., Fischer, P., Brox, T.: U-Net: convolutional networks for biomedical image segmentation. In: Navab, N., Hornegger, J., Wells, W., Frangi, A. (eds.) Medical Image Computing and Computer-Assisted Intervention–MICCAI 2015: 18th International Conference, Munich, Germany, 5–9 October 2015, Proceedings, Part III 18, pp. 234–241. Springer, Cham (2015). https://doi.org/10.1007/978-3-319-24574-4_28
13. Roy, S., et al.: MedNeXt: transformer-driven scaling of convnets for medical image segmentation. In: Greenspan, H., et al. (eds.) International Conference on Medical Image Computing and Computer-Assisted Intervention, pp. 405–415. Springer, Cham (2023). https://doi.org/10.1007/978-3-031-43901-8_39
14. Shen, Y., Xu, L., Yang, Y., Li, Y., Guo, Y.: Self-distillation from the last mini-batch for consistency regularization. In: Proceedings of the IEEE/CVF Conference on Computer Vision and Pattern Recognition, pp. 11943–11952 (2022)
15. Shen, Y., Xu, L., Yang, Y., Li, Y., Guo, Y.: Mixed sample augmentation for online distillation. In: ICASSP 2023-2023 IEEE International Conference on Acoustics, Speech and Signal Processing (ICASSP), pp. 1–5. IEEE (2023)
16. Shu, H., et al.: TinySAM: pushing the envelope for efficient segment anything model. arXiv preprint arXiv:2312.13789 (2023)
17. Wang, C., et al.: SAM-Med: a medical image annotation framework based on large vision model. arXiv preprint arXiv:2307.05617 (2023)
18. Wang, H., et al.: SAM-Med3D. arXiv preprint arXiv:2310.15161 (2023)
19. Wu, K., et al.: TinyVit: fast pretraining distillation for small vision transformers. In: Avidan, S., Brostow, G., Cissé, M., Farinella, G.M., Hassner, T. (eds.) European Conference on Computer Vision, pp. 68–85. Springer, Cham (2022). https://doi.org/10.1007/978-3-031-19803-8_5

20. Xiong, Y., et al.: EfficientSAM: leveraged masked image pretraining for efficient segment anything. arXiv preprint arXiv:2312.00863 (2023)
21. Xue, F., Shi, Z., Wei, F., Lou, Y., Liu, Y., You, Y.: Go wider instead of deeper. In: Proceedings of the AAAI Conference on Artificial Intelligence, vol. 36, pp. 8779–8787 (2022)
22. Zhang, C., et al.: Faster segment anything: towards lightweight SAM for mobile applications. arXiv preprint arXiv:2306.14289 (2023)

Towards a Formal Description of Artificial Intelligence Models and Datasets in Radiology

Charles E. Kahn Jr.[1(✉)], Abhinav Suri[2], Safwan Halabi[3], and Hari Trivedi[4]

[1] University of Pennsylvania, Philadelphia, PA, USA
ckahn@upenn.edu
[2] National Institutes of Health Clinical Center, Bethesda, MD, USA
[3] Ann and Robert H. Lurie Children's Hospital, Chicago, IL, USA
[4] Emory University, Atlanta, GA, USA

Abstract. The Radiology Model and Dataset Ontology (RMDO) has been created to provide formal descriptions for the growing number of artificial intelligence (AI) models and datasets in diagnostic radiology. RMDO builds upon generalized "model cards" and "datasheets for datasets" by highlighting features specific to radiology and by referencing concepts from related external ontologies and coding schemes, such as RadLex, the LOINC/RSNA Radiology Playbook, and radiology common data elements. RMDO also incorporates the tasks and methods hierarchies from the Papers With Code web resource. In accordance with the FAIR guiding principles, application of the ontology will allow AI resources to be more readily discoverable and reusable. Its application also is expected to improve the ability to match AI models with relevant datasets and to facilitate detection of potential biases in released AI models.

Keywords: Ontology · Radiology · Model Cards · Datasheets · Metadata · FAIR Principles

1 Introduction

Artificial intelligence (AI) models have become increasingly important in diagnostic radiology. To assure that machine learning models are used in the settings for which they are well suited, models ideally should be accompanied by documentation that details their performance, intended use, and potential limitations. "Model cards" have been proposed as brief documents to accompany trained AI models to provide benchmarked evaluation for a variety of conditions, such as demographic, cultural, or phenotypic subgroups that are relevant to the model's intended application [1]. Similarly, "datasheets for datasets" have been proposed to increase transparency and accountability, to mitigate unwanted biases in AI models, and to enable great reproducibility of machine learning results [2]. An AI model is intimately bound to the datasets used to train and test it. Hence, one can view model and dataset descriptors as two sides of the same coin.

J. Finkelstein et al. (Eds.): AIME 2024, LNAI 14845, pp. 140–144, 2024.
https://doi.org/10.1007/978-3-031-66535-6_16

1.1 Goals

The authors sought to formalize the description of radiology AI models and datasets to achieve several goals:

(1) Promote the reproducibility of the performance of trained AI models,
(2) Identify and mitigate potential biases in the development and clinical application of AI models,
(3) Support FAIR (Findable, Accessible, Interoperable, Reusable) guiding principles to assure that AI models and associated data are fully discoverable and interoperable [3], and
(4) Enable users to quickly assess utility of AI tools and datasets for utilization in practice or future projects.

1.2 Ontology Basics

An ontology is a human-readable and machine-computable description of a domain of discourse [4, 5]. The Radiology Model and Dataset Ontology (RMDO) defines the formal attributes (descriptors) and their allowable values for AI models and datasets that have been published or are under development in radiology.

RMDO's terminology is extended by reference to terms from external ontologies and vocabularies, including RadLex (version 4.1; *radlex.org*), an ontology of more than 45,000 radiology concepts, including radiologically relevant anatomy, diseases, imaging modalities, image acquisition characteristics, and imaging findings [6]; LOINC/RSNA Radiology Playbook (version 2.77; *loinc.org*), a systematic and standardized nomenclature of imaging procedures, linked to RadLex concepts, that has been incorporated into Logical Observation Identifiers Names and Codes (LOINC) terminology standard [7]; a standardized data dictionary of radiology common data elements (*radelement.org*) [8]; and the Papers With Code (PWC) classification of machine learning methods and tasks [9]. PWC offers an open, web-based resource that indexes publications, code, and datasets; as of 1 April 2024, PWC cataloged 2,218 distinct machine learning methods.

2 Ontology

2.1 Overall Structure

The primary entity of the Radiology Model and Dataset Ontology (RMDO) is a *Project*. RMDO incorporates metadata about a Project to describe its individual and institutional authors, versioning, availability, and licensing. Each *Project* consists of zero or more *Models* and zero of more *Datasets*. In that way, one can use the ontology to describe an isolated AI model, a dataset, or one or more models and their associated datasets. The attributes that describe AI models and datasets are described below.

2.2 Model Description

The *Model* entity describes a specific, trained machine learning model. The details about the model's architecture, intended uses, metrics, and ethical considerations are meant to provide information to those who would seek to use or reproduce the model about the problem the model aims to solve, the approach the authors took, the evaluation metrics used, links to code, and other information about how to run the model. For those readers of a scientific article about the model, the description can provide critical details that may be difficult to identify. The descriptions also can encourage authors to actively consider ethical issues when planning experiments and writing their manuscript (Table 1).

Table 1. A partial list of RMDO model descriptors.

Attribute	Description	Examples
Model details	Version, type, licensing, contact information, publications	
Uses	Intended, out-of-scope, and excluded uses	Risk assessment, image reconstruction, triage, diagnosis
Users	Intended, out-of-scope, and excluded users	Radiologists, radiology technologists
Task type	Papers With Code task label	Keypoint detection (task/keypoint-detection)
Methods	Papers With Code method label	Mask R-CNN (paper/mask-r-cnn)
Input data	LOINC/RSNA Radiology Playbook procedure names	XR Spine (RPID2571)
Output data	Radiology common data elements	Cobb angle (RDE264)
Metrics	Model performance measures, decision thresholds	Dice coefficient, area under the receiver-operating characteristic curve
Ethical considerations	Performance analyzed across disadvantaged groups	

2.3 Dataset Description

A critical aspect of a trained AI model is the dataset that is used to train it. Projects such as open challenges are associated with datasets by design [10]. Additionally, publications of AI models may incorporate one or more datasets as part of an evaluation benchmarking paper. Accordingly, a *Dataset* description section to the RMDO is essential to capture information about datasets used to train AI models part of a project. Each project has one or more datasets and each model may refer to one or more datasets as part of its training data (Table 2).

Table 2. A partial list of RMDO dataset descriptors.

Attribute	Description	Examples
Imaging modality	RadLex descriptor	Computed tomography (RID10321)
Imaging procedure	LOINC/RSNA Radiology Playbook entity	CT Abdomen and Pelvis without IV Contrast (RPID144)
Counts	Patient count, study count, image count	14120 patients
Image file format	Datafile type	DICOM, JPEG, NiFTI
Output information	Radiology common data elements	Spleen volume (mL) (RDE1221)
Availability	Data use agreement, licensing scheme	CC BY-NC 4.0
Image characteristics	Resolution, pre-processing, burned-in PHI	Rescaled to 224x224
Patients	Inclusion and exclusion criteria; demographic characteristics	
Data partitions	Partitioning scheme; demographic characteristics	
Annotation	Methodology, output	Patient label, study label, 2D image label, bounding box
Annotator characteristics	Scope of annotation, qualifications, experience	
References	Relevant publication, data repository	Digital Object Identifier (DOI), GitHub reference

3 Application

RMDO is under active development by the Radiology AI Data Standards Committee of the Radiological Society of North America (RSNA), a 54,000-member professional organization. RMDO incorporates terminology specific to radiology AI models and datasets, and will be integrated with existing ontologies, such as RadLex (Fig. 1). The JavaScript Object Notation (JSON) language will be used to represent RMDO documents based on a JSON Schema document.

Example model cards have been developed for a variety of trained AI models, and datasheets have been created for datasets created and published as part of the RSNA's AI competitions. These resources have helped guide the elaboration of RMDO's descriptors and their allowable values. Further experimentation is being done to establish an online web-interface that can aid project, model, and dataset creators to fill out fields of the RMDO to create a standardized model card.

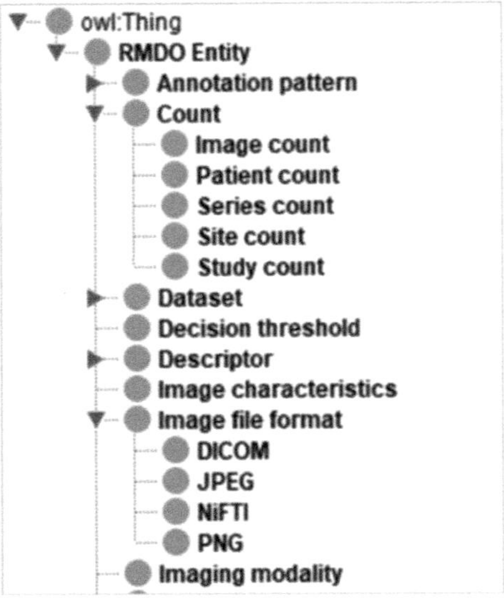

Fig. 1. Part of RMDO's concept hierarchy, as shown in Protégé.

Disclosure of Interests. The authors have no competing interests to declare that are relevant to the content of this article.

References

1. Mitchell, M., Wu, S., Zaldivar, A., et al.: Model cards for model reporting. In: Proceedings of the Conference on Fairness, Accountability, and Transparency, pp. 220–229 (2019)
2. Gebru, T., Morgenstern, J., Vecchione, B., et al.: Datasheets for datasets. Commun. ACM **64**, 86–92 (2021)
3. Wilkinson, M.D., Dumontier, M., Aalbersberg, I.J., et al.: The FAIR guiding principles for scientific data management and stewardship. Sci. Data **3**, 160018 (2016)
4. Bodenreider, O.: Biomedical ontologies in action: role in knowledge management, data integration and decision support. Yearb. Med. Inform. **17**(1), 67–79 (2008)
5. Smith, B., Arabandi, S., Brochhausen, M., et al.: Biomedical imaging ontologies: a survey and proposal for future work. J. Pathol. Inform. **6**, 37 (2015)
6. RadLex. http://www.radlex.org/. Accessed 31 Mar 2024
7. Vreeman, D.J., Abhyankar, S., Wang, K.C., et al.: The LOINC RSNA radiology playbook - a unified terminology for radiology procedures. J. Am. Med. Inform. Assoc. **25**, 885–893 (2018)
8. Kohli, M., Alkasab, T., Wang, K., et al.: Bending the artificial intelligence curve for radiology: informatics tools from ACR and RSNA. J. Am. Coll. Radiol. **16**, 1464–1470 (2019)
9. Papers with Code., https://paperswithcode.com/. Accessed 02 Apr 2024
10. Kitamura, F.C., Prevedello, L.M., Colak, E., et al.: Lessons learned in building expertly annotated multi-institution datasets and hosting the RSNA AI challenges. Radiol. Artif. Intell. **6**, e230227 (2024)

Towards Aleatoric and Epistemic Uncertainty in Medical Image Classification

Timo Löhr[1,2,3(✉)], Michael Ingrisch[2,3], and Eyke Hüllermeier[1,2]

[1] Department of Informatics, LMU Munich, Munich, Germany
[2] Munich Center for Machine Learning (MCML), Munich, Germany
[3] Department of Radiology, University Hospital, LMU Munich, Munich, Germany
t.loehr@lmu.de

Abstract. Medical domain applications require a detailed understanding of the decision making process, in particular when data-driven modeling via machine learning is involved, and quantifying uncertainty in the process adds trust and interpretability to predictive models. However, current uncertainty measures in medical imaging are mostly monolithic and do not distinguish between different sources and types of uncertainty. In this paper, we advocate the distinction between so-called aleatoric and epistemic uncertainty in the medical domain and illustrate its potential in clinical decision making for the case of PET/CT image classification.

Keywords: Medical Image Analysis · Machine Learning · Uncertainty Quantification · PET/CT Data

1 Introduction

Over the last decade, machine learning (ML) has showcased remarkable potential in analyzing medical data [23] but also introduced critical risks. Incorrect predictions could result in misdiagnoses leading to improper treatments [10]. Consequently, medical applications demand a deeper comprehension of the underlying ML models and a reliable decision-making framework. A pathway towards trust and interpretability involves quantifying uncertainty in the image analysis process, which can originate from multiple sources [11]. A common differentiation is drawn between so-called aleatoric uncertainty, which stems from the inherent randomness within the data, and epistemic uncertainty, which is caused by the lack of knowledge about the ideal predictor for a specific task [13, 21].

Identifying and measuring these uncertainties allow for the detection of issues in the learning and prediction process, especially in clinical settings, where critical decisions such as interpreting PET/CT scans are made by physicians. Moreover, it helps to determine the most appropriate way of action to improve performance and reliability, e.g., if additional training data should be supplied by physicians or the model should be adjusted by the data scientist. For example, information about aleatoric uncertainty empowers physicians to assess whether a dependable decision can be reached based

J. Finkelstein et al. (Eds.): AIME 2024, LNAI 14845, pp. 145–155, 2024.
https://doi.org/10.1007/978-3-031-66535-6_17

on the patient information at hand, or if further examinations are necessary. Likewise, understanding epistemic uncertainty provides valuable insights into whether a given predictor is sufficiently informed to make reliable decisions or requires additional training examples.

Due to these reasons, we are convinced that the distinction between aleatoric and epistemic uncertainty is crucial in the medical domain and has a high potential to improve clinical decision making. This paper is meant to elaborate on this potential and illustrate it for the case of PET/CT image classification.

2 Related Work

Uncertainty quantification has received increasing interest in machine learning research in the recent past [13]. Methods developed in this field have been used, for instance, to generate resilient predictions, implement active-learning techniques, or enhance the interpretability of neural networks [15, 24, 27].

Lambert et al. [18] recently highlighted the considerable interest in UQ for medical images. They indicate that a notable portion of research focuses on quantifying total predictive uncertainty across diverse medical applications. Many researchers employ techniques like Monte Carlo dropout or deep ensembles to quantify uncertainty, aiming at selective and high-confidence classifications, for example, in diabetic retinopathy or histopathology images [4, 8, 14, 19]. However, most approaches concentrate on quantifying only the overall predictive uncertainty [1, 22], with only a few notable exceptions focusing either on aleatoric or epistemic uncertainty. For instance, Wang et al. [26] use test time augmentation (TTA) to quantify aleatoric uncertainty for tumor segmentation in MRI images, while Ayhan et al. [2] assess the same method for fundus image classification.

Even less approaches quantify *both* aleatoric and epistemic uncertainty in the medical domain, for instance, by combining multiple uncertainty measures [22]. Combalia et al. [5] attempt to merge Monte Carlo dropout and TTA to estimate both uncertainties in skin lesion classification, reporting successful identification of difficult and out-of-distribution samples. This underscores the importance of a distinction between uncertainties in medical image classification tasks.

To the best of our knowledge, a comprehensive analysis and theoretical differentiation between aleatoric and epistemic uncertainty in real-world medical image data, alongside a practical application of these concepts, are still lacking. Therefore, this paper aims to clarify these concepts and investigate their usage in the medical scenario of tumor classification in PET/CT images.

3 Machine Learning and Uncertainty Quantification

In this work, we focus on the supervised learning paradigm, where a learner is provided with training data

$$\mathcal{D} = \{(\mathbf{x}_1, y_1), ..., (\mathbf{x}_N, y_N)\} \subset \mathcal{X} \times \mathcal{Y} .$$

Here, $\mathcal{X}$ represents the instance space and $\mathcal{Y}$ a (finite) set of possible categories. In a medical context like tumor classification within PET/CT images, each $\mathbf{x}_i$ might be a PET/CT image paired with a tumor label y_i (e.g., lung cancer, lymphoma, melanoma, negative). A (probabilistic) predictor is a mapping $h : \mathcal{X} \rightarrow \Delta_{\mathcal{Y}}$ that assigns a probability distribution $(p_1, \ldots, p_C) = h(\mathbf{x}) \in \Delta_{\mathcal{Y}}$ to each instance $\mathbf{x}$, with $C = |\mathcal{Y}|$ the number of categories and $p_j \in [0, 1]$ the (estimated) probability that $\mathbf{x}$ belongs to the jth category. Given a class $\mathcal{H}$ of predictors—the *hypothesis space* in ML jargon—to choose from and a loss function $\ell : \Delta_{\mathcal{Y}} \times \mathcal{Y} \rightarrow \mathbb{R}$ that penalizes predictions in light of observed outcomes, the learner aims at finding the optimal (Bayes) predictor, namely, the predictor that generalizes best in the sense of having the lowest expected loss:

$$h^* = \operatorname*{argmin}_{h \in \mathcal{H}} \mathbb{E}_{(\mathbf{x},y)\sim P}\big[\ell(h(\mathbf{x}), y)\big]$$

Assuming that the training data $\mathcal{D}$ as well as future (query) data are generated (independently) by the same distribution P on $\mathcal{X} \times \mathcal{Y}$, and hence that the former is representative of the data-generating process, the learner leverages $\mathcal{D}$ to produce a predictor h (e.g., through empirical risk minimization, i.e., minimizing the loss on $\mathcal{D}$). Obviously, since the training data is finite, h can only be an approximation of the truly best predictor h^*. Consequently, uncertainty persists regarding h^* as well as the approximation quality of h, and this uncertainty translates into uncertainty about the quality of predictions $h(\mathbf{x})$ [13].

This uncertainty can be broken down into different parts to gain a more detailed understanding of its sources. Aleatoric uncertainty (data uncertainty) refers to the inherent, irreducible uncertainty stemming from the non-deterministic nature of the input-output dependence governed by P. Thus, even if the learner had perfect knowledge about P, it could not uniquely determine a single outcome $y \in \mathcal{Y}$. Instead, the latter is only characterized by the conditional probability

$$P(y \,|\, \mathbf{x}) = \frac{P(\mathbf{x}, y)}{P(\mathbf{x})} .$$

In the medical context, for example, even highly experienced physicians cannot make a definitive class assignment for some images, e.g., due to the image resolution, artifacts, or noise. Increasing the dataset size or intensifying expert training cannot overcome this issue, as the relevant information cannot be extracted from these specific input samples. As long as the instance representation, i.e., the space $\mathcal{X}$, is not expanded through additional information (e.g., another medical test), this uncertainty remains irreducible.

Epistemic uncertainty stems from the lack of knowledge regarding the ground-truth measure P and hence the optimal predictor h^*, which is caused by the learner's limited empirical experience. This uncertainty is hence reducible by gathering additional experience, e.g., by adding new samples to the training data. This is somehow comparable to a physician, who improves her image classification skills by studying more cases over time. To represent epistemic uncertainty, a learner must go beyond the induction of a single predictor h, which only captures aleatoric uncertainty. More specifically, it needs a way to represent its uncertainty about h. . In the Bayesian approach, this is accomplished by a (second-order) probability distribution Q on the hypothesis space $\mathcal{H}$, and learning essentially consists of replacing a prior on this space by a posterior distribution. Broadly speaking, the more concentrated this distribution becomes, the less (epistemically) uncertain the learner is.

The goal of *uncertainty quantification* is to measure the amount of uncertainty in a prediction the learner makes for a query instance $\mathbf{x}$, i.e., to provide a numerical measure of (total) uncertainty as well as a decomposition of this uncertainty into an aleatoric and an epistemic part. Various measures of that kind have been proposed in the literature. The most common approach in (Bayesian) machine learning relies on a result from information theory, according to which (Shannon) entropy can be expressed as the sum of conditional entropy and mutual information [7]: The total uncertainty is the entropy

$$\mathrm{TU}(\mathbf{x}) = H\left[\,\bar{p}(\cdot \mid \mathbf{x})\,\right] = -\sum_{y \in \mathcal{Y}} \bar{p}(y \mid \mathbf{x}) \log_2 \bar{p}(y \mid \mathbf{x}) \, ,$$

where the posterior predictive distribution $\bar{p}$ is obtained through (Bayesian) model averaging, i.e., averaging the probability predictions $p(y \mid h, \mathbf{x})$ made by the individual models $h \in \mathcal{H}$, weighted by their posterior probability:

$$\bar{p}(y \mid \mathbf{x}) = \int_{\mathcal{H}} p(y \mid h, \mathbf{x}) \, d\, Q(h) \, .$$

As fixing a model h means committing to a single predictive distribution $p(y \mid h, \mathbf{x})$ and hence removing all epistemic uncertainty, the entropy of such a distribution is a suitable measure of aleatoric uncertainty. The conditional entropy is the expectation of this measure with regard to Q:

$$\mathrm{AU}(\mathbf{x}) = \int_{\mathcal{H}} H\big[p(\cdot \mid h, \mathbf{x})\big] \, d\, Q(h) \, .$$

Epistemic uncertainty is then computed as the difference $\mathrm{EU}(\mathbf{x}) = \mathrm{TU}(\mathbf{x}) - \mathrm{AU}(\mathbf{x})$, which can be shown to yield the mutual information of the predictor h and outcome y (both considered as random variables). Intuitively, epistemic uncertainty thus represents the (expected) reduction of uncertainty that is achieved by revealing the (uncertain) predictor h [13].

4 Uncertainty in Medical Image Data

Medical image data introduce added sources of uncertainty compared to standard datasets (e.g. MNIST, CIFAR [6, 16]) usually used in UQ. Firstly, aleatoric uncertainty in medical images is connected to their immense complexity. These images are typically voluminous, containing various modalities with structures that vary in shape, size, and appearance. For instance, PET/CT image data consists of multi-modal 3D volumes, including whole-body scans with dimensions such as $400 \times 400 \times 300$ pixels, exhibiting diverse anatomical structures within and across patients. In contrast, MNIST features images of 28×28 pixels, focusing on a single object centered in the image. In addition, medical images are susceptible to noise, inconsistencies, and artifacts stemming from differences in equipment, patient motion, or physiological processes [3].

Secondly, epistemic uncertainty emerges from the choice of model type and a limited number of samples, which is often the case in medical image datasets, resulting in a lack of knowledge within the model. Similarly, missing or erroneous values in existing data reduces the amount of extractable knowledge, thus elevating aleatoric uncertainty.

Thirdly, interpreting medical images requires the expertise of physicians, introducing subjective assessment. Ambiguous labels may arise when multiple physicians interpret an image, thereby increasing aleatoric uncertainty stemming from the labeling process.

Quantifying uncertainty resulting from these diverse sources in medical image settings has multiple implications for trust and interpretability of predictions. A promoted benefit is selective prediction, wherein images with high uncertainty are excluded from automated predictions or redirected to physicians for further manual inspection. In contrast, predictions with low uncertainty can be trusted, thereby reducing costs and enhancing examination quality for patients.

Further, the distinction between aleatoric and epistemic uncertainty provides physicians with greater insights for informed decision-making. High aleatoric uncertainty identifies challenging images necessitating further patient examinations by the physician. Consequently, data scientists might need to adapt the model to additional data input. If further examinations are not feasible, physicians are informed that a reliable decision cannot be made based on the available data. High epistemic uncertainty indicates a lack of knowledge within the learning model, stemming either from inadequate model selection or encountering data beyond the learned distribution. Identifying such out-of-distribution (OOD) data is vital for medical applications to prevent the model from making predictions on unfamiliar image types. Addressing high epistemic uncertainty involves either enlarging the training data with additional images by a physician or refining the model architecture for the specific learning task by a data scientist.

In other words, understanding the source of uncertainty guides the identification of challenges to be addressed. Additionally, it clarifies the responsibilities and strengthens the collaboration between physicians and data scientists.

5 Method and Implementation

The Bayesian approach outlined in Sect. 3 is computationally intractable for complex model classes, due to the need for integrating over $\mathcal{H}$. In practice, ensemble methods are commonly used as an approximation [17]. Such methods train a set of M different predictors $h_1, \dots, h_M \in \mathcal{H}$, which are considered as a representative sample from the true distribution Q. Concretely, we instantiate this approach with deep neural networks as our model class, i.e., each ensemble members h_j is a deep network. Model architecture, hyperparameters, and training settings are chosen based on related work [28], with the model mainly derived from ResNet [12]. As the training cost of deep ensembles is prohibitively high, we limit the size of our ensemble to $M = 10$, which is a typical choice also used in other studies [17]. For further details and the full experimental code refer to Appendix B and our public repository[1].

[1] https://github.com/tloehrlmu/uq-in-mi.

Given a trained ensemble, the posterior predictive distribution for a query instance $\vec{x}$ is obtained by averaging (instead of integration):

$$\bar{p}(\cdot \mid \mathbf{x}) = \frac{1}{M}(p(\cdot \mid h_1, \mathbf{x}) + \ldots + p(\cdot \mid h_M, \mathbf{x})) .$$

Likewise, discrete versions of the uncertainty measures are given as follows:

$$\begin{aligned}
\mathrm{TU}(\mathbf{x}) = H\Big[\bar{p}(\cdot \mid \mathbf{x})\Big] &= H\Big[\tfrac{1}{M}\sum_{j=1}^{M} p(\cdot \mid h_j, \mathbf{x})\Big] \\
\mathrm{AU}(\mathbf{x}) &= \tfrac{1}{M}\sum_{j=1}^{M} H\big[p(\cdot \mid h_j, \mathbf{x})\big] \\
\mathrm{EU}(\mathbf{x}) &= \mathrm{TU}(\mathbf{x}) - \mathrm{AU}(\mathbf{x})
\end{aligned}$$

6 Empirical Evaluation

To evaluate uncertainty quantification in a real-world medical imaging application, we tackle the problem of tumor classification on Integrated Positron Emission Tomography/Computed Tomography (PET/CT) images. The clinical demand for quantitative analysis of PET/CT grows increasingly as it offers detailed anatomical insights through CT scans alongside functional and molecular information from PET scans [25]. We leverage a publicly accessible dataset consisting of 501 consecutive whole-body FDG-PET/CT scans from patients with malignant lymphoma, melanoma, and lung cancer, alongside 513 scans from patients without malignant lesions (negative controls) [9]. Examples of PET/CT scans are depicted in Fig. 1, with further details on the image retrieval provided in Appendix A. We assess the distinction between aleatoric and epistemic uncertainty qualitatively and quantitatively in selective prediction and the detection of out-of-distribution samples, which are highly important tasks in medical settings.

Qualitative Examples. Figure 1 depicts three PET/CT images alongside their corresponding predictions and quantified uncertainty (cf. Sect. 5). In the first row, we observe an instance with low uncertainty, accurately predicting the class "Lung Cancer" which aligns with the visual impression. A lung tumor is distinctly visible in both the CT and PET images. Such cases can be automatically predicted by the model without manual inspection. Moving to the second row, we encounter an instance with high aleatoric but low epistemic uncertainty, leading to a misclassification. This scenario presents a challenging image, as no evident signs of malignant lesions are apparent in either the CT or PET image, leading to high aleatoric uncertainty. The physician is alerted that reliable predictions cannot be made based solely on the provided images, necessitating further examinations. In the third row, an example with high epistemic uncertainty is presented, also resulting in misclassification. From a visual standpoint, this example likely falls outside the learned distribution due to the unusual, rather shifted positioning of the patient. The elevated epistemic uncertainty suggests the need for additional knowledge acquisition to furnish a reliable prediction. This could involve, for instance, augmenting the training data with more samples where the patient exhibits a similar shifted position.

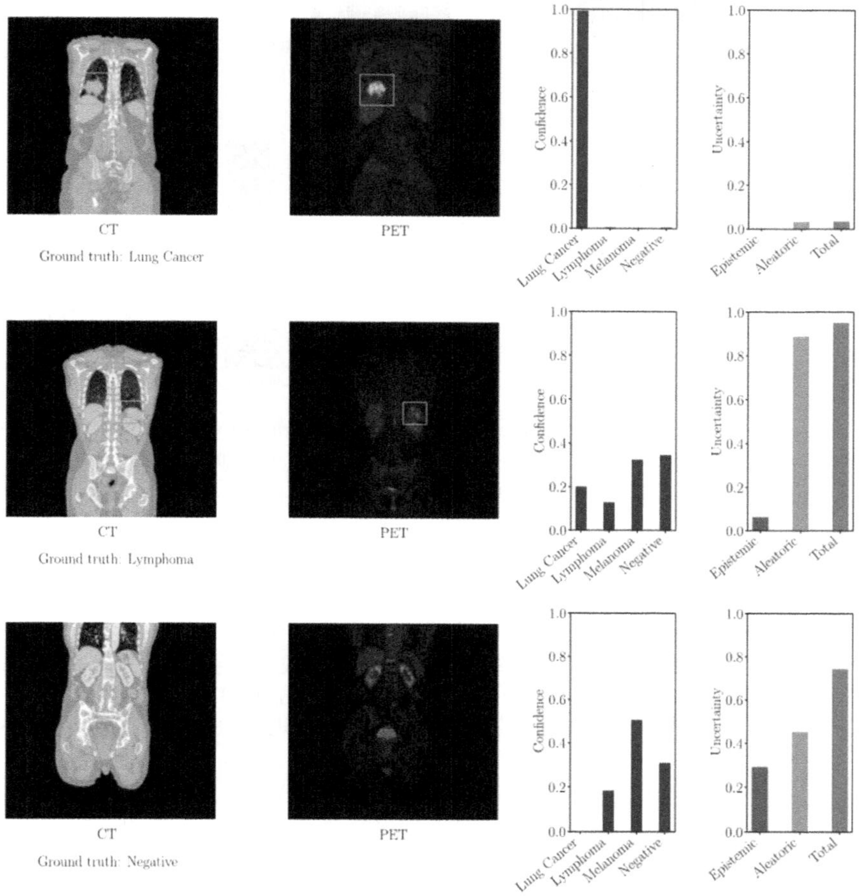

Fig. 1. Three PET/CT images with their corresponding probabilistic predictions and quantified uncertainty. Orange boxes surround the tumors identified by physicians.

Selective Prediction. Selective prediction refers to the rejection of uncertain predictions, thereby improving the overall prediction quality. Accuracy-Rejection Curves (ARCs) [20] assess the performance by sorting samples based on their uncertainty and discarding step-by-step those with highest uncertainty. One would anticipate an increase in accuracy for the remaining predictions, thus a monotonic increase of the ARC. Figure 2a illustrates a consistent increase of all curves based on the values for aleatoric, epistemic, and total uncertainty. This demonstrates that UQ enhances the model's predictive performance, enabling physicians to depend on the quantified uncertainty when assessing individual cases.

Out-of-Distribution (OOD) Detection. Accurate quantification of epistemic uncertainty is crucial in detecting OOD samples. We anticipate higher epistemic uncertainty for samples beyond the known distribution given the model's lack of knowledge for such images. In this experiment, we train our ensemble on three classes within our

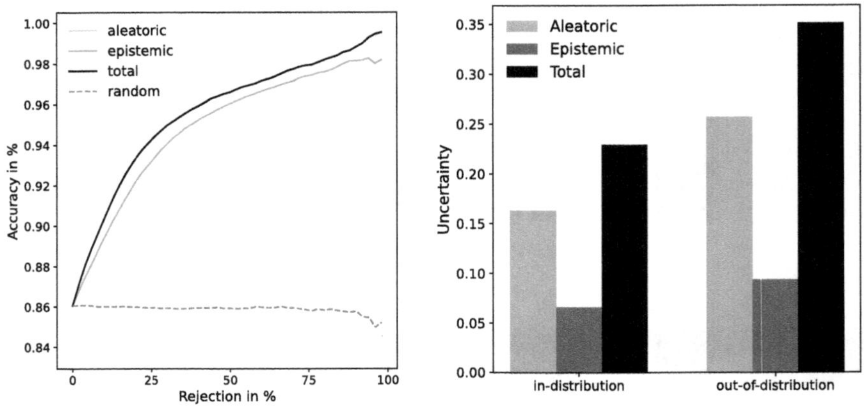

Fig. 2. (a) Accuracy-rejection curve based on the model predictions for the PET/CT test set. (b) Average uncertainty values split between in-distribution and out-of-distribution samples separated by aleatoric, epistemic and total uncertainty.

dataset, reserving samples from the Negative class as OOD data. Figure 2b depicts the average uncertainty for in- and out-of-distribution data, revealing higher aleatoric and epistemic uncertainty for OOD samples. Although the causes of a rise in aleatoric uncertainty may differ, the increased epistemic uncertainty facilitates the detection of OOD samples. Besides alerting physicians in medical practice, it enables data scientists to analyze OOD data, thereby refining the model and training process for enhanced data generalization.

Discussion. Our work highlights the higher complexity of medical images (e.g. PET/CT images) in comparison to standard images like CIFAR or MNIST. We demonstrate the importance of UQ in tumor classification within PET/CT images, particularly in enhancing trust and interpretability. Through a detailed analysis of model predictions, we facilitate the identification of challenging samples for physicians based on aleatoric uncertainty or out-of-distribution instances based on epistemic uncertainty. Hence, physicians can make informed and dependable decisions in clinical practice, while data scientists can analyze the model's knowledge and comprehension, thereby refining its adaptation to the learning task. This underlines the high importance of a collaboration in these two disciplines.

7 Conclusion and Outlook

The distinction between aleatoric and epistemic uncertainty holds critical importance in medical image analysis. We illustrated this distinction for tumor classification in PET/CT images, showcasing its potential to enhance clinical decision-making by identifying difficult and out-of-distribution cases.

Future research will expand the evaluation of our findings to include additional medical image types beyond PET/CT images. Moreover, we would like to explore other UQ methods as alternatives to the entropy-based approach used in this work. Finally, a clinical validation is still pending.

Appendix A: FDG-PET/CT Dataset

All examinations were conducted between 2014 and 2018 at the University Hospital Tübingen, utilizing a state-of-the-art PET/CT scanner (Siemens Biograph mCT). The imaging protocol involved a diagnostic CT scan, typically covering from the skull base to mid-thigh level, with intravenous contrast enhancement, except for patients with contraindications. The CT parameters utilized included a reference dose of 200 mAs, a tube voltage of 120 kV, iterative reconstruction with a slice thickness of 2–3 mm. Additionally, a whole-body FDG-PET scan was performed 60 min following the intravenous injection of 300–350 MBq 18F-FDG. PET data reconstruction was accomplished using an ordered-subset expectation maximization (OSEM) algorithm featuring 21 subsets and 2 iterations, a 2 mm Gaussian kernel, and a matrix size of 400 × 400 [9].

All the acquired images underwent comprehensive clinical evaluation, with radiologists and nuclear medicine physicians collaborating to identify primary tumors and metastases. Malignant FDG-avid lesions, as determined by patient history and previous examinations, were meticulously segmented on PET images, slice by slice, using specialized software (NORA imaging platform, University of Freiburg, Germany) by a single reader [9].

Appendix B: Deep Neural Network Training and Evaluation

In order to quantify the uncertainty for a deep learning approach for PET/CT images, we first train a deep neural network ensemble for classifying tumors in the previously described data set (see Sect. 4). We choose the model architecture, the hyperparameters as well as the general training setting based on related work from this field [28].

In particular, an input sample comprises a three-channel image containing the PET, CT and fusion image. Each image comes with a class label which is used as ground truth when computing the loss function or evaluating a model.

The model architecture is based on the commonly known ResNet and slightly adapted in order to match the dimensionality of our data [12]. We train in total 10 deep neural networks as an ensemble and average the probabilistic outputs in order to get the final prediction. As a prerequisite for the training, several data preprocessing steps have to be performed.

Preprocessing

Each PET/CT volume has to be prepared in order to get a homogeneous image dataset in terms of image size, intensity value range and noise. For this paper, we first scale the intensity range for each PET and CT image to [0, 1]. Second, we remove obvious artefacts from each volume (e.g. the examination table). For our training setting, we decided to use 2D images from a coronal view as this depicts the whole body at once and has a similar structure throughout all slices in the volume (see Fig. 1). Therefore, we are extracting in a third step for each patient 2D coronal slices from the 3D volume and resize to a matrix size of 400 × 400 pixels. Finally, the PET, CT and fusion image of those two will be stacked and saved as a three-channel image. As mentioned before, for each PET volume a segmentation of the existent tumors is available. Therefore, we extract a class label for each slice based on the segmented lesions. Each tumor and therefore also the slice which depicts the tumor is associated to one of three types, Melanoma, Lymphoma or Lung Cancer. Slices without tumor are associated with a fourth class called Negative.

As a result, we receive a dataset of 97014 2D images (100%) distributed as follows over the four classes: 9529 Lung Cancer (9.82%), 6433 Lymphoma (6.63%), 7096 Melanoma (7.31%), 73956 Negative (76.24%). It can be seen that the data set is highly imbalanced having a large number of slices without malignant lesions.

Model Training

For the model training, we first split the dataset into a train and test set in order to evaluate the model after training on an unseen data subset. In detail, we split into a 80% train set and 20% test set while maintaining the class distribution of the whole data set for the subsets by a weighted sampling. For the training itself, we use the cross-entropy loss, an Adam optimizer with a learning rate of 0.001 and a weight decay. The batch size is set to 50 and we train each network for 50 epochs as the training loss already converges before. In order to tackle the data imbalance in our data set, we use a weighted batch sampling which ensures a uniform distribution over all classes per batch.

Model Evaluation

After training, we evaluate the deep neural network ensemble based on commonly used metrics, for instance the accuracy metric as well as the false positive and false negative rate. We use this to be certain that the model learned to classify 2D PET/CT images correctly into one of the four tumor classes to a certain degree. We look at results at slice level, meaning we calculate all evaluation metrics for each slice and average over all samples.

The trained ensemble achieves an overall accuracy of 0.861. Our focus is not on achieving state-of-the-art performance but rather on studying uncertainty quantification in practical model applications.

References

1. Abdar, M., et al.: A review of uncertainty quantification in deep learning: techniques, applications and challenges. Inf. Fusion **76**, 243–297 (2021)
2. Ayhan, M.S., Berens, P.: Test-time data augmentation for estimation of heteroscedastic aleatoric uncertainty in deep neural networks. In: Medical Imaging with Deep Learning (2022)
3. Berti, V., Pupi, A., Mosconi, L.: PET/CT in diagnosis of movement disorders. Ann. N. Y. Acad. Sci. **1228**(1), 93–108 (2011)
4. Buddenkotte, T., et al.: Calibrating ensembles for scalable uncertainty quantification in deep learning-based medical image segmentation. Comput. Biol. Med. **163**, 107096 (2023)
5. Combalia, M., Hueto, F., Puig, S., Malvehy, J., Vilaplana, V.: Uncertainty estimation in deep neural networks for dermoscopic image classification. In: Proceedings of the IEEE/CVF Conference on Computer Vision and Pattern Recognition Workshops, pp. 744–745 (2020)
6. Deng, L.: The mnist database of handwritten digit images for machine learning research. IEEE Signal Process. Mag. **29**(6), 141–142 (2012)

7. Depeweg, S., Hernandez-Lobato, J.M., Doshi-Velez, F., Udluft, S.: Decomposition of uncertainty in Bayesian deep learning for efficient and risk-sensitive learning. In: International Conference on Machine Learning, pp. 1184–1193. PMLR (2018)
8. Filos, A., et al.: A systematic comparison of Bayesian deep learning robustness in diabetic retinopathy tasks. arXiv preprint arXiv:1912.10481 (2019)
9. Gatidis, S., et al.: A whole-body FDG-PET/CT dataset with manually annotated tumor lesions. Sci. Data **9**(1), 601 (2022)
10. Glocker, B., Jones, C., Roschewitz, M., Winzeck, S.: Risk of bias in chest radiography deep learning foundation models. Radiol. Artif. Intell. **5**(6), e230060 (2023)
11. Gruber, C., Schenk, P.O., Schierholz, M., Kreuter, F., Kauermann, G.: Sources of uncertainty in machine learning–a statisticians' view. arXiv preprint arXiv:2305.16703 (2023)
12. He, K., Zhang, X., Ren, S., Sun, J.: Deep residual learning for image recognition. In: Proceedings of the IEEE Conference on Computer Vision and Pattern Recognition, pp. 770–778 (2016)
13. Hüllermeier, E., Waegeman, W.: Aleatoric and epistemic uncertainty in machine learning: an introduction to concepts and methods. Mach. Learn. **110**, 457–506 (2021)
14. Jungo, A., Balsiger, F., Reyes, M.: Analyzing the quality and challenges of uncertainty estimations for brain tumor segmentation. Front. Neurosci. **14**, 501743 (2020)
15. Kendall, A., Gal, Y.: What uncertainties do we need in Bayesian deep learning for computer vision? Adv. Neural Inf. Process. Syst. **30** (2017)
16. Krizhevsky, A., Hinton, G., et al.: Learning multiple layers of features from tiny images (2009)
17. Lakshminarayanan, B., Pritzel, A., Blundell, C.: Simple and scalable predictive uncertainty estimation using deep ensembles. Adv. Neural Inf. Process. Syst. **30** (2017)
18. Lambert, B., Forbes, F., Doyle, S., Dehaene, H., Dojat, M.: Trustworthy clinical AI solutions: a unified review of uncertainty quantification in deep learning models for medical image analysis. Artif. Intell. Med. **150**, 102830 (2024)
19. Mehrtens, H.A., Kurz, A., Bucher, T.C., Brinker, T.J.: Benchmarking common uncertainty estimation methods with histopathological images under domain shift and label noise. Med. Image Anal. **89**, 102914 (2023)
20. Nadeem, M.S.A., Zucker, J.D., Hanczar, B.: Accuracy-rejection curves (arcs) for comparing classification methods with a reject option. In: Machine Learning in Systems Biology, pp. 65–81. PMLR (2009)
21. Senge, R., et al.: Reliable classification: learning classifiers that distinguish aleatoric and epistemic uncertainty. Inf. Sci. **255**, 16–29 (2014)
22. Seoni, S., Jahmunah, V., Salvi, M., Barua, P.D., Molinari, F., Acharya, U.R.: Application of uncertainty quantification to artificial intelligence in healthcare: a review of last decade (2013–2023). Comput. Biol. Med. **165**, 107441 (2023)
23. Shehab, M., et al.: Machine learning in medical applications: a review of state-of-the-art methods. Comput. Biol. Med. **145**, 105458 (2022)
24. Shelmanov, A., et al.: Active learning for sequence tagging with deep pre-trained models and Bayesian uncertainty estimates. arXiv preprint arXiv:2101.08133 (2021)
25. Tan, H., et al.: Total-body PET/CT: current applications and future perspectives. Am. J. Roentgenol. **215**(2), 325–337 (2020)
26. Wang, G., Li, W., Aertsen, M., Deprest, J., Ourselin, S., Vercauteren, T.: Aleatoric uncertainty estimation with test-time augmentation for medical image segmentation with convolutional neural networks. Neurocomputing **338**, 34–45 (2019)
27. Woo, J.O.: Analytic mutual information in Bayesian neural networks. In: 2022 IEEE International Symposium on Information Theory (ISIT), pp. 300–305. IEEE (2022)
28. Yang, J., et al.: MedMNIST v2-a large-scale lightweight benchmark for 2D and 3D biomedical image classification. Sci. Data **10**(1), 41 (2023)

Ultrasound Image Segmentation via a Multi-scale Salient Network

Abdalrahman Alblwi(✉) and Kenneth E. Barner

University of Delaware, Newark, DE 19716, USA
{alblwi,Barner}@udel.edu

Abstract. Deep learning methods have significantly advanced the field of ultrasound imaging analysis. However, the low quality of ultrasound images remains a bottleneck, especially in the case of tumor segmentation. Motivated by these facts, this paper introduces a novel method called the Deep Saliency Attention Network (DSA-net) to segment breast tumors in ultrasound images. The DSA-net is developed to learn multi-scale characteristics of tumors in a noisy environment and with limited data. The proposed method achieves accurate segmentation performance by combining saliency and attention mechanisms. The effectiveness of DSA-net is evaluated on three benchmark databases of breast ultrasound images, benchmarked against various widely used and state of-the-art (SOTA) methods (*e.g.* U-net, AAU-net, U^2-net, and Attention U-net). The proposed DSA-net method demonstrates SOTA visual and quantitative improvements evaluated on the breast cancer databases.

Keyword: Tumor segmentation · Saliency · Attention mechanism · Residual U-shape blocks

1 Introduction

According to the World Health Organization (WHO), breast cancer is a pressing global health issue, particularly among women. Early detection can significantly improve treatment success [5]. Breast cancer diagnosis typically involves a variety of tools, such as MRI, ultrasound imaging, and mammography, to detect breast tumors [9]. The increasing preference for ultrasound imaging in diagnosing breast tumors can be attributed to its cost-effectiveness and the absence of ionizing radiation exposure. However, ultrasound images often exhibit noise, distortions, and artifacts, resulting in images with a comparatively lower resolution. As a consequence, these factors affect image quality and potentially undermine diagnostic interpretations. Segmenting breast ultrasound images, especially for malignant lesions, poses challenges due to their intricate structures resulting from invading neighboring tissues [9]. Due to these challenges, deep learning algorithms have become indispensable in processing ultrasound images, particularly in noisy data tasks. This work aims to leverage saliency and attention mechanisms to enhance the learning of complex ultrasound imaging samples.

J. Finkelstein et al. (Eds.): AIME 2024, LNAI 14845, pp. 156–161, 2024.
https://doi.org/10.1007/978-3-031-66535-6_18

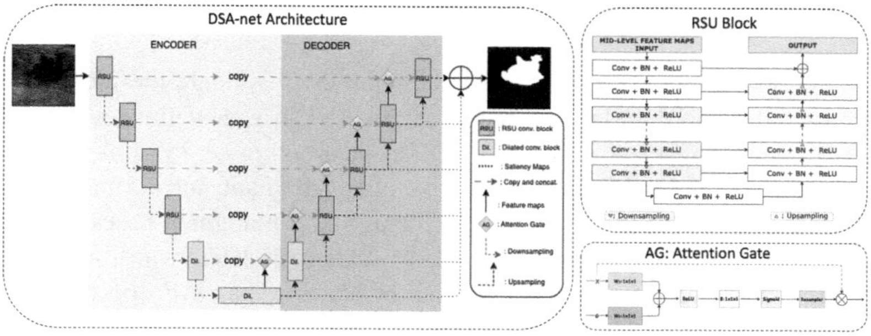

Fig. 1. The DSA-net architecture: The left side shows the pipeline, while the right side details its components, including RSU blocks and attention gates.

2 Methodology

Attention gates [4], a well-regarded variant of U-net as depicted in Fig. 1, comprises two inputs: the first input corresponds to the feature maps from an intermediate encoding layer through a skip-connection, denoted by **X**, and the second input symbolizes the coarse-scale feature maps from the lower decoder block, represented by **G**. A stride convolution layer is applied to **X** before performing an element-wise addition with **G** to align the dimensions. Linear transformations utilizing weights $\mathbf{W_G}$ and $\mathbf{W_X}$ are then followed by ReLU activation and a $1 \times 1 \times 1$ convolution, culminating in a single feature representation. Subsequently, a sigmoid activation function is employed to produce attention coefficients. These coefficients are resampled and multiplied element-wise with **X** to emphasize relevant features. Qin *et al.* [7] presented a novel nested network that utilizes RSU blocks instead of traditional convolutional settings commonly seen in segmentation models like U-Net. The RSU block comprises an initial convolutional layer for extracting local features, followed by a U-Net-inspired symmetric encoder-decoder structure capable of capturing multi-scale contextual information. As depicted in Fig. 1, the RSU block is designed with an inner encoder-decoder shape at each convolutional stage, facilitating local and global salient features extraction through mid-level representation and downscale pooling. By preserving feature resolution and network depth, RSU facilitates extracting multi-level features within a heterogeneous environment. During training process, RSU blocks help assimilate features from various receptive fields, effectively capturing contextual information across scales. As shown in Fig. 1, the red and blue blocks represent the RSU and the dilated convolution layers, respectively. Using dilated convolutions within low-level feature blocks, rather than inner downsampling and upsampling operations in RSU, is essential to maintain high resolution features, ensuring their retention despite substantial downsampling [7]. Furthermore, to improve the training as discussed in [2, 7], saliency maps are generated and upsampled to the original input size using nearest-neighbor interpolation. The final model mask is derived by concatenating these upsampled saliency maps and convolving them with a (1×1) kernel.

2.1 Datasets and Experiments Settings

This study employs three breast ultrasound databases to validate the proposed DSA-net approach. The first dataset, referred to as Breast Ultrasound Images (BUSI), was collected by Al-Dhabyani *et al.* [1]. The BUSI data comprise 780 images from 600 female patients, encompassing three classes: benign, malignant, and normal. The second dataset, Dataset B, includes 163 images of benign and malignant cases collected by Yap *et al.* [10]. The third breast cancer dataset, called BrEaST, with three classes of benign, malignant, and normal, was collected by Pawłowska *et al.* [6] and these data were acquired from 256 women. Comprehensive experiments have been conducted to validate the proposed DSAnet and compare its performance with existing approaches. Three variations of training scenarios are considered: (1) BUSI dataset, (2) Dataset B, and (3) BrEaST dataset. The models have been validated via k-fold cross-validation, with $k = 4$ for a fair comparison. The training is set for 50 epochs, with a batch size 12, utilizing the Adam optimizer and the binary cross-entropy loss function for network optimization. The testing results are averaged across all k-fold validation procedures.

2.2 Results and Discussion

Table 1 presents segmentation results across diverse datasets with varying classes, including normal, benign, and malignant. For the BUSI dataset, the results show that the proposed DSA-net outperforms all SOTA models, except for the recall metric. Surpassing the closest comparable model, (*e.g.* U^2-net [7]), the DSA-net demonstrates a precision accuracy enhancement of about 5.2%, with a value of 0.7867 compared to 0.7477. For the case of dataset B, the DSA-net achieves the best performance for most of the evaluation metrics, as seen in Table 1. Interestingly, compared to other models, the proposed DSA-net shows promising efficacy in learning from smaller datasets, evidenced by its Jaccard score of 0.4378 compared to the closest comparable model, the AAU-net, which achieved a score of 0.3549. This represents a percentage improvement of approximately 23.3% in Jaccard score. Despite the constraints of limited data, which constitutes the smallest dataset size in our study, this superior performance underscores the potential of the DSA-net for efficient learning from small datasets. In the BrEaST dataset, our model shows comparable or superior performance compared to U^2net. As highlighted in Table 1, the U^2-net and our DSA-net achieve Dice scores of 0.7271 and 0.7296, respectively, showing an improvement in the range of approximately 8.7% to 9.1% compared to the AAU-net. Furthermore, the Jaccard scores for U^2-net and our DSA-net, 0.5728 and 0.5752 respectively, reflect an improvement in the range of approximately 13% to 14% compared to the AAUnet's score of 0.5031. In composite, these results indicate the generality and efficiency of DSA-net. As shown in Fig. 2, a side-by-side visualization highlights the considerable success of DSA-net in identifying tumors amidst the noisy backdrop of ultrasound images. Further experiments are highlighted in Fig. 3, where the ROC curves demonstrate, by a significant margin in all cases, the superior performance of our DSA-net compared to other SOTA approaches.

Table 1. The segmentation results (mean ± std) on several breast ultrasound datasets.

Model	BUSI				
	Precision	Specificity	Dice	Recall	Jaccard
U-net [8]	0.7334 ± 0.0558	0.9815 ± 0.0059	0.6530 ± 0.0187	0.5917 ± 0.0258	0.4851 ± 0.0209
Att. U-net [4]	0.7781 ± 0.0240	0.9844 ± 0.0025	0.7081 ± 0.0119	0.6517 ± 0.0357	0.5482 ± 0.0141
U^2-net [7]	0.7477 ± 0.0255	0.9804 ± 0.0030	0.7173 ± 0.0163	0.6896 ± 0.0141	0.5595 ± 0.0199
AAU-net [3]	0.6560 ± 0.0613	0.9728 ± 0.0092	0.6288 ± 0.1223	0.6296 ± 0.1791	0.4692 ± 0.1200
Ours	0.7867 ± 0.0382	0.9844 ± 0.0038	0.7267 ± 0.0163	0.6779 ± 0.0373	0.5709 ± 0.0201
Model	Dataset B				
	Precision	Specificity	Dice	Recall	Jaccard
U-net [8]	0.5309 ± 0.3737	0.8907 ± 0.1203	0.4133 ± 0.1566	0.6000 ± 0.1923	0.2730 ± 0.1263
Att. U-net [4]	0.7379 ± 0.3154	0.9553 ± 0.0748	0.3938 ± 0.1791	0.4046 ± 0.2287	0.2605 ± 0.1378
U^2-net [7]	0.6926 ± 0.2496	0.9774 ± 0.0321	0.3904 ± 0.2157	0.3829 ± 0.2536	0.2666 ± 0.1809
AAU-net [3]	0.6069 ± 0.2046	0.9733 ± 0.0209	0.5233 ± 0.0302	0.5423 ± 0.1460	0.3549 ± 0.0274
Ours	0.7425 ± 0.2046	0.9863 ± 0.0121	0.6056 ± 0.0686	0.5604 ± 0.1379	0.4378 ± 0.0701
Model			BrEaST		
	Precision	Specificity	Dice	Recall	Jaccard
U-net [8]	0.6922 ± 0.0162	0.9865 ± 0.0012	0.5277 ± 0.0468	0.4296 ± 0.0592	0.3598 ± 0.0427
Att. U-net [4]	0.6785 ± 0.1737	0.9756 ± 0.0303	0.4331 ± 0.0852	0.4116 ± 0.2355	0.2802 ± 0.0695
U^2-net [7]	0.8125 ± 0.0132	0.9891 ± 0.0011	0.7271 ± 0.0407	0.6622 ± 0.0693	0.5728 ± 0.0488
AAU-net [3]	0.7251 ± 0.0525	0.9825 ± 0.0062	0.6684 ± 0.0382	0.6209 ± 0.0365	0.5031 ± 0.0419
Ours	0.8089 ± 0.0489	0.9883 ± 0.0046	0.7296 ± 0.0314	0.6728 ± 0.0763	0.5752 ± 0.0380

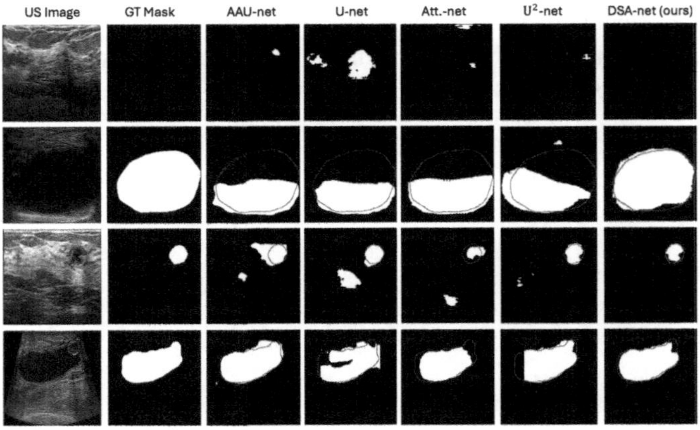

Fig. 2. Results of the predicted mask of our DSA-net and other existing methods.

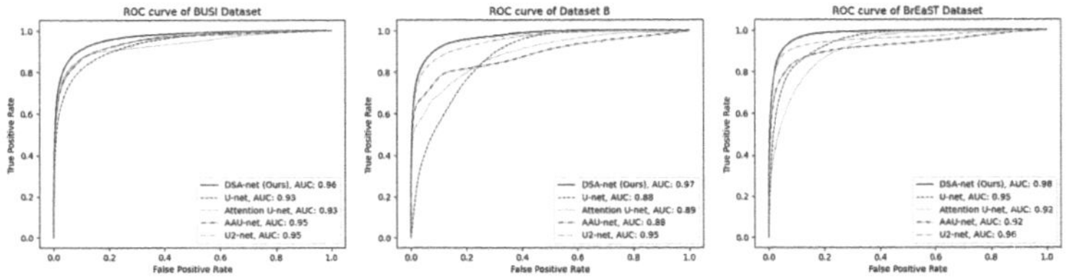

Fig. 3. ROC Curves of BUSI, Dataset B, and BrEaST databases.

3 Conclusion

This paper presents a model, DSA-net, that synergistically integrates attention and saliency mechanisms that address the limitations of current ultrasound based tumor segmentation approaches. The architecture facilitates advanced multi-scale feature learning by exploiting the synergies between saliency and attention modules. The DSA-net framework effectively delineates complex tumor shapes and edges within inherently noisy ultrasound imaging environments. Validation in three different breast ultrasound datasets underscores the efficacy of the proposed approach. Our results suggest that DSA-net represents an advancement in the critically important application of breast tumor segmentation in ultrasound imaging.

References

1. Al-Dhabyani, W., Gomaa, M., Khaled, H., Fahmy, A.: Dataset of breast ultrasound images. Data Brief **28**, 104863 (2020)
2. Alblwi, A., Barner, K.E.: Optimizing feature representation via a nested network for object segmentation. In: 2022 8th International Conference on Optimization and Applications (ICOA), pp. 1–6. IEEE (2022)
3. Chen, G., Li, L., Dai, Y., Zhang, J., Yap, M.H.: Aau-net: An adaptive attention u-net for breast lesions segmentation in ultrasound images. IEEE Trans. Med. Imaging **42**(5), 1289–1300 (2022)

4. Oktay, O., et al.: Attention u-net: learning where to look for the pancreas. arXiv preprint arXiv:1804.03999 (2018)
5. World Health Organization et al.: WHO Position Paper on Mammography Screening. World Health Organization (2014)
6. Pawłowska, A., et al.: Curated benchmark dataset for ultrasound based breast lesion analysis. Sci. Data **11**(1), 148 (2024)
7. Qin, X., Zhang, Z., Huang, C., Dehghan, M., Zaiane, O.R., Jagersand, M.: U2net: going deeper with nested u-structure for salient object detection. Pattern Recogn. **106**, 107404 (2020)
8. Ronneberger, O., Fischer, P., Brox, T.: U-net: convolutional networks for biomedical image segmentation. In: Medical Image Computing and Computer-Assisted Intervention–MICCAI 2015: 18th International Conference, Munich, 5–9 October 2015, Proceedings, Part III 18, pp. 234–241. Springer (2015)
9. Wasif, N., Garreau, J., Terando, A., Kirsch, D., Mund, D.F., Giuliano, A.E.: MRI versus ultrasonography and mammography for preoperative assessment of breast cancer. Am. Surg. **75**(10), 970–975 (2009)
10. Yap, M.H., et al.: Automated breast ultrasound lesions detection using convolutional neural networks. IEEE J. Biomed. Health Inform. **22**(4), 1218–1226 (2017)

Data Integration and Multimodal Analysis

A 360° View for Large Language Models: Early Detection of Amblyopia in Children Using Multi-view Eye Movement Recordings

Dipak P. Upadhyaya[1], Aasef G. Shaikh[2], Gokce Busra Cakir[3], Katrina Prantzalos[1], Pedram Golnari[1], Fatema F. Ghasia[3], and Satya S. Sahoo[1](✉)

[1] Department of Population and Quantitative Health Sciences, School of Medicine, Case Western Reserve University, Cleveland, OH, USA
{dipak.upadhyaya,katrina.prantzalos,pedram.golnari, satya.sahoo}@case.edu

[2] National VA Parkinson's Consortium Center, Louis Stokes Cleveland VA Medical Center, Cleveland, Cleveland, OH, USA
aasef.shaikh@case.edu

[3] Visual Neuroscience Laboratory, Cole Eye Institute, Cleveland Clinic, Cleveland, OH, USA
{cakrig,ghasiaf}@ccf.org

Abstract. Amblyopia is a neurodevelopmental visual disorder that affects approximately 3–5% of children globally and it can lead to vision loss if it is not diagnosed and treated early. Traditional diagnostic methods, which rely on subjective assessments and expert interpretation of eye movement recordings presents challenges in resource-limited eye care centers. This study introduces a new approach that integrates the Gemini large language model (LLM) with eye-tracking data to develop a classification tool for diagnosis of patients with amblyopia. The study demonstrates: (1) LLMs can be successfully applied to the analysis of fixation eye movement data to diagnose patients with amblyopia; and (2) Input of medical subject matter expertise, introduced in this study in the form of medical expert augmented generation (MEAG), is an effective adaption of the generic retrieval augmented generation (RAG) approach for medical applications using LLMs. This study introduces a new multi-view prompting framework for ophthalmology applications that incorporates fine granularity feedback from pediatric ophthalmologist together with in-context learning to report an accuracy of 80% in diagnosing patients with amblyopia. In addition to the binary classification task, the classification tool is generalizable to specific subpopulations of amblyopic patients based on severity of amblyopia, type of amblyopia, and with or without nystagmus. The model reports an accuracy of: (1) 83% in classifying patients with moderate or severe amblyopia, (2) 81% in classifying patients with mild or treated amblyopia; and (3) 85% accuracy in classifying patients with nystagmus. To the best of our knowledge, this is the first study that defines a multi-view prompting framework with MEAG to analyze eye tracking data for the diagnosis of amblyopic patients.

Keywords: Transfer learning · large language models · in-context learning · fixation eye movement · amblyopia diagnosis

J. Finkelstein et al. (Eds.): AIME 2024, LNAI 14845, pp. 165–175, 2024.
https://doi.org/10.1007/978-3-031-66535-6_19

1 Introduction

Amblyopia is a neurodevelopmental disorder of the eye affecting 3–5% of the children's population, that is, an estimated 99.2 million children are affected by amblyopia worldwide [1]. This condition leads to decreased vision in one eye due to improper visual stimulation during early childhood [1]. If left untreated, amblyopia can result in irreversible visual impairment, and it contributes to lifelong consequences for the affected individuals. In the United States, amblyopia is the most common cause of monocular vision loss among children, adolescents, and middle-aged adults, underscoring the substantial public health challenge it poses across age groups [1]. The diagnosis of amblyopia has traditionally relied on subjective visual acuity assessments and the interpretation of eye movement recordings, which requires specialized equipment and expert clinical analysis. These diagnostic challenges are exacerbated in resource-limited settings, where access to specialized pediatric ophthalmology services is often unavailable. The increasing use of high-fidelity eye tracking instruments, such as video oculography tracker (EyeLink1000 plus), has enabled the use of eye tracking data for the detection of amblyopia. In this study, we leverage the powerful transfer learning capabilities of the Gemini large language model (LLM) (Google, Inc.) to analyze eye tracking data in a cohort of 135 subjects to develop a high accuracy classification tool that can be deployed in remote, resource scarce eye care clinics.

LLMs have demonstrated state of the art performance in a variety of applications [2–5]; however, their performance in medical applications have highlighted multiple challenges due to the complexity of the medical domain [6–9]. Specifically, customized LLMs such as Med-PaLM, which extends the Pathway Language Model (PaLM) [3] and MedGemini [10], require the development of fine-tuned prompting methods to answer medical domain questions (e.g., questions in the MultiMedQA multiple-choice benchmark dataset). Building on the PaLM model, Gemini is a state-of-the-art language model that has achieved high accuracy in a range of natural language processing tasks [2, 11]. However, initial results from our approach to use the Gemini model to classify patients with amblyopia using eye movement data visualized as a Cartesian plot resulted in low accuracy with multiple errors. Specifically, as a baseline prompt, we used eye tracking plots (Fig. 1) corresponding to a single eye together with a textual description of the plot for the Gemini model. The model reported an accuracy of 50% or less for classifying patients with amblyopia with errors in classifying patients based on the severity of amblyopia and amblyopic patients with or without nystagmus.

To address these limitations, we built on the few shot learning capabilities of LLMs together with the development of a new multi-view prompting framework for ophthalmology applications. Specifically, we combine eye tracking data recorded under multiple viewing conditions to create a collage of eye movement recordings for each patient. In the next step, we systematically label the images in the collage with input from a pediatric ophthalmologist (co-author FG) and construct a structured prompt expression for each patient. Together with few-shot learning approach, this new multiple view prompt with medical expert feedback demonstrates:

1. The feasibility of using LLMs to classify patients with amblyopia using recordings from eye tracking instruments.

2. The increase in contextual information (represented as a multi-view collage of eye movement plots in this study) is effective in improving the accuracy of results generated by language models in a specialized medical domain.

In contrast to fine tuning LLMs, which requires retraining the LLM to modify gradient weights, prompt tuning is a low-cost approach to improve the accuracy of LLM generated results [6]. However, prompt tuning for medical applications requires customizations that reflect the specific attributes of the medical data. Therefore, medical expertise augmented retrieval (MEAG), which is a form of retrieval augmented generation (RAG) [12] is essential for effective use of language models to analyze patient data. We discuss the characteristics of eye tracking data and its use in diagnosis of patients with amblyopia.

1.1 Background

Eye movement abnormalities are being increasingly used as diagnostic markers in amblyopia research as they can be recorded in a non-invasive manner [13]. Individuals with amblyopia demonstrate notable fixation instability, particularly in the amblyopic eye as compared to its fellow eye (Fig. 1), which highlights the close link between fixation eye movement (FEM) abnormalities and the sensory deficits in amblyopia [14, 15]. The use of modern eye trackers provides an objective dataset crucial for assessing visual function and understanding the nuance of impairments in eye movement control associated with the disorder [16]. In parallel, the development of LLMs with capabilities to analyze multimodal data including text and images, presents new avenues for ophthalmology applications. Therefore, this study uses the Gemini LLM to analyze eye tracking data for classification of patients with amblyopia.

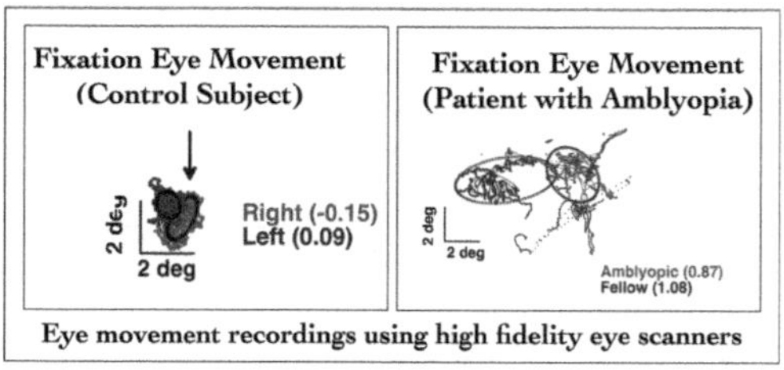

Fig. 1. The visualization of eye movement recordings in children with amblyopia and control subjects with dense, focused saccades seen in control subjects (left); and dispersed, wide-ranging saccades seen in amblyopic patient (right)

1.2 Related Work

The application of machine learning algorithms in ophthalmology, particularly for amblyopia detection, has evolved over time [17]. Convolutional neural networks (CNNs) have been employed to analyze retinal images for amblyogenic risk factors and they have shown greater accuracy as compared to traditional methods [18]. Other machine learning methods have been explored for classification of visual field defects and assessing binocular vision, which is important for amblyopia diagnosis [19]. However, traditional approaches require extensive training data and their effectiveness have been affected due to the difficulty in identifying subtle features associated with amblyopia [20]. We have addressed some of these challenges in this study by leveraging the powerful capabilities of LLMs, including transfer learning that rely on few-shot based in-context learning for generating accurate results. In-context learning refers to the ability of language models to adapt to new tasks or domains without explicit retraining of the LLM through the use of one or more examples [2, 21].

2 Method

Figure 2 gives an overview of the method used in this study that consists of a multi-stage workflow, including: (i) acquisition of eye movement data using high fidelity eye tracking instrument; (ii) signal pre-processing methods to enhance data quality by reducing noise in the recordings; (iii) visualization of eye tracking recordings with appropriate scaling measures as Cartesian plots; and (iv) the use of multi-view prompting with input from pediatric ophthalmologist.

2.1 Study Cohort and Data Collection

This study uses eye movement recordings from 135 participants collected at the Cleveland Clinic Eye Institute (the study was reviewed and approved by the Cleveland Clinic Institutional Review Board). The study cohort consisted of 95 patients with amblyopia and 40 control subjects. The age of the participants ranged between 3 years and 72 years. Control subjects were selected based on the absence of ocular or systemic abnormalities affecting visual acuity, except for refractive errors. The 95 patients with amblyopia included 51 females and 44 males. The 40 control subjects included 23 females and 17 males. The control group consisted of 24 White Caucasian, 6 African American, 1 Hispanic, 5 Asians, and 4 multi-cultural patients. The amblyopic subjects included 61 Whites Caucasians, 12 African American, 10 Hispanics, 3 Asians, and 9 multi-cultural patients.

The EyeLink1000 plus video-based eye tracker was used with a sampling rate of 500 Hz with participants wearing corrective lenses. In a dimly lit room, participants faced an LCD screen displaying various stimuli, with calibration involving a cartoon animal accompanied by sound cues to focus attention. The calibration process adhered to a 5-point manufacturer guideline [22]. The fixation measurements were taken as subjects focused on a white dot against a black screen, employing animal sounds to redirect attention if needed. Each session was recorded with both eyes open (labeled as binocular vision, BV) or with left/right eye viewing while the other was covered (labeled as monocular viewing, MV), in a sequence determined randomly. The MV sessions were recorded from the active viewing eye, distinguishing between fellow eye (FE) and amblyopic eye viewing (AE) conditions. The covered eye was monitored using an infrared filter to avoid exposure to visible light.

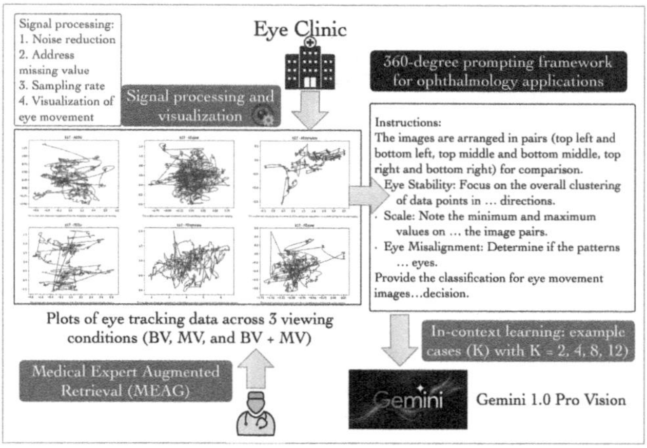

Fig. 2. An overview of this study that introduced a new Multiview imaging LLM prompting techniques for analyzing pediatric ophthalmology data.

2.2 A Multi-view Prompting Framework with Few-Shot Learning

Following the data pre-processing steps, the eye tracking data were visualized as Cartesian plots using the Pandas library for data processing and the Python Imaging Library (PIL). As described in the previous section, the plots of eye tracking data were combined into a collage of six images with recording of each eye under three viewing conditions, that BV, MV, and BV + MV. In the next step, the collage was annotated with text prepared by pediatric ophthalmologist to describe the characteristics of the eye tracking data (Fig. 3). In the final stage, the prompt expression was constructed using this multi-view expert annotated images together with structured input that described the context of the task and added example query examples as part of the few-shot learning approach.

2.3 In-context Learning for Eye Movement Data

LLMs have demonstrated a notable capability to learn specific contexts using few examples (few-shot or in-context learning) that was highlighted in the Generative Pre-trained Transformer (GPT 3.5) model and many other subsequent studies [2, 3, 6]. In this study, we evaluated the performance of the Gemini LLM by combining the multi-view prompt expressions with medical expertise together with in-context learning using a different number of examples for in-context learning. Figure 3 shows an overview of the multi-view prompting approach developed and implemented in this study for the Gemini LLM. This prompting framework includes several constituent tasks, including the system prompt together with multi-view expert annotated text and examples for few-shot in-context learning. Using this new prompting framework, the classification tool was evaluated using held out test dataset with 123 subjects (26 control subjects and 87 patients with amblyopia). We evaluated the effectiveness of the new multi-view prompting approach with and without feedback from the medical expert together with a systematic evaluation of the impact of different number of examples on the accuracy of the model.

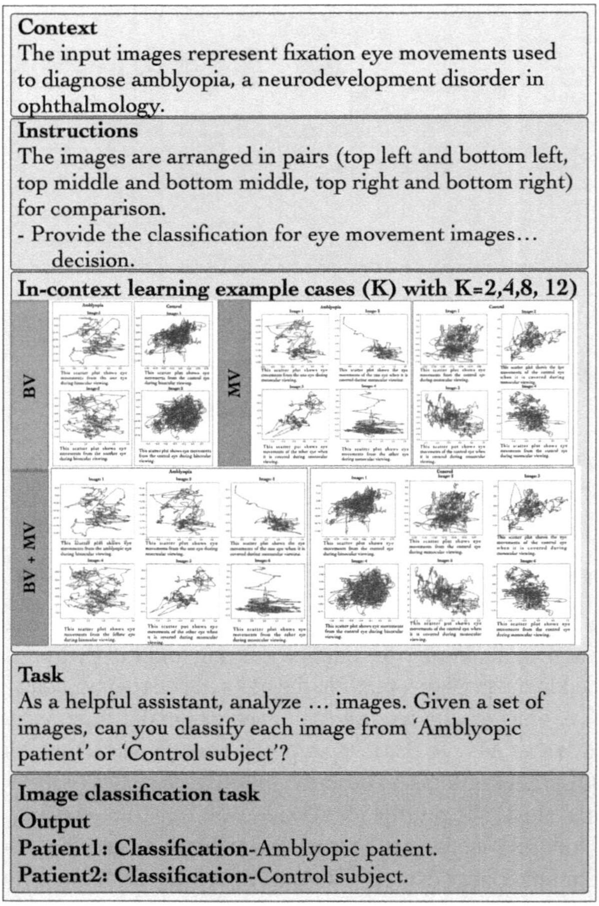

Fig. 3. Overview of multi-view prompting framework developed for diagnosis of amblyopia using structured feedback from pediatric ophthalmologist.

3 Results

3.1 Experimental Setup

This study used Gemini 1.0 Pro Vision model that accepts both text and images as input and generates text as output [4]. The evaluation was structured into four few-shot prompting scenarios with multi-view collages corresponding to recordings from 2, 4, 8 or 12 subjects under the three viewing conditions. The subjects used in the few-shot prompting were selected to be representative of amblyopic patients and control subjects. This evaluation process was performed for each of the three viewing conditions (BV, MV, and BV + MV).

To demonstrate the effectiveness of prompt expression created using fine granularity input from pediatric ophthalmologist, we performed a comparative evaluation of the LLM with and without the inclusion of medical expert generated text. The results of this evaluation demonstrated the importance of feedback from medical experts in constructing effective prompts. Based on these results, all subsequent prompts incorporate the input of ophthalmologist in labeling the eye movement plots. In the second phase of the evaluation, the prompt-tuned Gemini model was used to classify patients with amblyopia based on the severity of amblyopia, that is, across the three categories of: (1) patients with severe or moderate amblyopia and control subjects; (2) patients with mild or treated amblyopia and control subjects; and (3) amblyopic patient with nystagmus and control subjects. A key feature of the multi-view prompt development was an iterative feedback mechanism that tuned the prompt to reflect the reasoning approach used by pediatric ophthalmologists in diagnosing patients with amblyopia.

The evaluation experiments were conducted on a server with 32 GB RAM and Intel® Core™ i7-13700K CPU, using Ubuntu 22.04.1 LTS (64-bit). The 123 subjects constituting the test dataset did not have any overlap with the 12 subjects used for in-context learning. Temperature is a tunable parameter of LLMs that corresponds to increased randomness in results with higher temperature value and more deterministic results with lower temperature values [7]. In this study, a temperature parameter of 0.7 was found to be optimal for achieving the highest accuracy.

3.2 Evaluation Results

Table 1 lists the performance of the LLM in phase 1 using a single image corresponding to eye movement recordings under a specific viewing condition. The results generated by the Gemini model have low accuracy due to the limited context provided in the prompt, including a limited "view" of the eye movement recording and lack of fine granularity medical expert feedback in the prompt. Specifically, the model reports an accuracy of 0.54 when distinguishing between patients with amblyopia (irrespective of their severity) and control subjects; an accuracy of 0.58 for diagnosing patients with mild or treated amblyopia; and an accuracy of 0.62 for diagnosing patients with moderate and severe amblyopia.

Table 1. Performance of the classification tool using a baseline prompt to: (1) classify patients with amblyopia and control subjects; (2) diagnosing patients based on their severity.

Model	Amblyopia versus control (BV)	Mild and treated amblyopia versus control (BV)	Moderate and severe amblyopia versus control (BV)
Few shot (K = 12)	0.54	0.58	0.62

To address the limitations of the baseline prompting approach, we used the multi-view prompting approach as described in the previous section (Sect. 2.2). Table 2 shows the evaluation results that compares the performance of the Gemini model with and without feedback from the medical expert. Table 2 lists the results across the three different viewing conditions with few-shot learning approach. The results validate the effectiveness of few-shot learning in LLMs with improvement in performance across the three viewing conditions with increasing number of examples. Further, the results show that there is a significant improvement in the performance of the model with input of the medical expert.

Table 3 lists the performance of the models across different classification tasks based on: (1) the severity of amblyopia, including mild and treated amblyopia, moderate and severe amblyopia, and (2) amblyopia with nystagmus as compared to the control group using the few-shot in-context learning approach. Table 3 also includes the results for combined binocular and monocular viewing (BV + MV), MV, and BV models, which demonstrates the improvement in accuracy of the model with an increased number of examples for in-context learning across all viewing conditions. We also note that the classification accuracy is the highest for eye tracking data recorded under the binocular viewing condition.

Table 2. Comparative evaluation of the classification tool with and without MEAG prompting approach for classifying amblyopic patients and control subjects across the three viewing conditions.

Model	BV + MV viewing condition (baseline prompt)	BV + MV viewing condition (MEAG prompt)	MV viewing condition (baseline prompt)	MV viewing condition (MEAG prompt)	BV viewing condition (baseline prompt)	BV viewing condition (MEAG prompt)
	Accuracy	Accuracy	Accuracy	Accuracy	Accuracy	Accuracy
Few shot (K = 2)	0.55	0.62	0.55	0.63	0.6	0.65
Few shot (K = 4)	0.55	0.64	0.55	0.67	0.6	0.6
Few shot (K = 8)	0.57	0.74	0.57	0.79	0.62	0.79
Few shot (K = 12)	0.57	0.78	0.57	0.78	0.62	0.8

BV: Binocular viewing (Two images per collage), MV: Monocular viewing (Four images per collage), BV + MV (6 images per collage)

Table 3. An evaluation of the generalizability capabilities of the classification tool using multi-view prompting framework with MEAG across different subpopulations of amblyopic patients.

Model	BV + MV viewing condition	MV viewing condition	BV viewing condition
	Accuracy	Accuracy	Accuracy
Mild and treated amblyopia versus control			
Few shot (K = 2)	0.6	0.65	0.7
Few shot (K = 4)	0.63	0.69	0.68
Few shot (K = 8)	0.76	0.78	0.8
Few shot (K = 12)	0.77	0.81	0.81
Moderate and severe amblyopia versus control			
Few shot (K = 2)	0.68	0.69	0.7
Few shot (K = 4)	0.71	0.74	0.77
Few shot (K = 8)	0.78	0.8	0.81
Few shot (K = 12)	0.8	0.82	0.83
Amblyopia with nystagmus versus control			
Few shot (K = 2)	0.7	0.72	0.73
Few shot (K = 4)	0.74	0.74	0.76
Few shot (K = 8)	0.8	0.81	0.83
Few shot (K = 12)	0.8	0.83	0.85

4 Discussion and Conclusion

The study defines a new prompting framework for use with foundational language models in ophthalmology applications. The multi-view prompting framework that provides an enhanced visual context to language model is effective in leveraging the transfer learning capabilities of language model to accurately analyze eye tracking data. Further, the results demonstrate that notwithstanding the few shot learning capability of language model, integration of high quality, fine granular feedback from medical experts results in significant improvement in the performance of the language model.

The integration of expert feedback together with reinforcement learning method or reinforcement learning with human feedback (RHLF) has been successfully used to improve instruction following in LLMs [23]. Our approach to incorporate expert feedback in prompt tuning results in significant improvement in the quality of results however, the study does not evaluate the specific type of feedback from medical experts that is most effective in improving the accuracy of results. Specifically, medical expert input in the form of a structured, stepwise description may further improve the performance of the LLM as demonstrated in other domains using the Chain of Thought (CoT) approach. Further, the size of the study cohort in this paper is small; therefore, we are unable to draw broader conclusions regarding the effectiveness of the combined multi-view and

MEAG prompting framework. As part of our ongoing work, we aim to increase the size of the study cohort and analyze data from a broader spectrum of ophthalmology disorders using the new prompting framework introduced in this study.

Acknowledgement. This research was funded in part by grants from the US National Institutes of Health (NIH): U24EB029005, R01DA053028, the US Department of Defense (DoD) grant W81XWH2110859, and the Clinical and Translational Science Collaborative of Cleveland, which is funded by the NIH, National Center for Advancing Translational Sciences, Clinical and Translational Science Award grant, UL1TR002548. The content is solely the responsibility of the authors and does not necessarily represent the official views of the NIH. Additional funding came from NEI T32: 5 T32 EY 24236-4, Case Western Reserve University Biomedical Research Fellowship-Hartwell Foundation (F.G.), Blind Children's Foundation (F.G.), Research to Prevent Blindness Disney Amblyopia Award (F.G.), and Cleveland Clinic RPC Grant (F.G.). Support also included a Cleveland VA Medical Center grant 5I21RX003878-02 (AS and FG), Research to Prevent Blindness, CCLCM Unrestricted Block Grant, NIH-NEI P30 Core Grant, and Cleveland Eye Bank.

References

1. McKean-Cowdin, R., et al.: Prevalence of amblyopia or strabismus in Asian and non-Hispanic white preschool children: multi-ethnic pediatric eye disease study. Ophthalmology **120**(10), 2117–2124 (2013)
2. Brown, T., et al.: Language models are few-shot learners. Adv. Neural. Inf. Process. Syst. **33**, 1877–1901 (2020)
3. Chowdhery, A., et al.: Palm: scaling language modeling with pathways. J. Mach. Learn. Res. **24**(240), 1–113 (2023)
4. Reid, M., et al.: Gemini 1.5: Unlocking multimodal understanding across millions of tokens of context. arXiv preprint arXiv:2403.05530 (2024)
5. Touvron, H., et al.: Llama 2: open foundation and fine-tuned chat models. arXiv preprint arXiv:2307.09288 (2023)
6. Singhal, K., et al.: Large language models encode clinical knowledge. Nature **620**(7972), 172–180 (2023)
7. Sahoo, S.S., et al.: Large language models for biomedicine: foundations, opportunities, challenges, and best practices. J. Am. Med. Inform. Assoc. (2024)
8. Shah, N.H., Entwistle, D., Pfeffer, M.A.: Creation and adoption of large language models in medicine. JAMA **330**(9), 866–869 (2023)
9. Thirunavukarasu, A.J., Ting, D.S.J., Elangovan, K., Gutierrez, L., Tan, T.F., Ting, D.S.W.: Large language models in medicine. Nat. Med. **29**(8), 1930–1940 (2023)
10. Saab, K., et al.: Capabilities of Gemini models in medicine. arXiv preprint arXiv:2404.18416 (2024)
11. Team, G., et al.: Gemini: a family of highly capable multimodal models. arXiv preprint arXiv: 2312.11805 (2023)
12. Guu, K., et al.: Retrieval augmented language model pre-training. In: International Conference on Machine Learning. PMLR (2020)
13. Ghasia, F., Wang, J.: Amblyopia and fixation eye movements. J. Neurol. Sci. **441**, 120373 (2022)
14. Subramanian, V., Jost, R.M., Birch, E.E.: A quantitative study of fixation stability in amblyopia. Invest. Ophthalmol. Vis. Sci. **54**(3), 1998–2003 (2013)

15. Shi, X.F., et al.: Fixational saccadic eye movements are altered in anisometropic amblyopia. Restor. Neurol. Neurosci. **30**(6), 445–462 (2012)
16. Niechwiej-Szwedo, E., Colpa, L., Wong, A.M.: Visuomotor behaviour in amblyopia: deficits and compensatory adaptations. Neural Plasticity **2019**, 1 (2019)
17. Vadhera, R., Sharma, M.: Review of amblyopia and artificial intelligence techniques used for its detection. In: Proceedings of the Second International Conference on Information Management and Machine Intelligence: ICIMMI 2020. Springer (2021)
18. Ting, D.S.W., et al.: Development and validation of a deep learning system for diabetic retinopathy and related eye diseases using retinal images from multiethnic populations with diabetes. JAMA **318**(22), 2211–2223 (2017)
19. Holmes, J.M., Clarke, M.P.: Amblyopia. Lancet **367**(9519), 1343–1351 (2006)
20. Webber, A.L., Wood, J.: Amblyopia: prevalence, natural history, functional effects and treatment. Clin. Exp. Optom. **88**(6), 365–375 (2005)
21. Wang, Y., et al.: Generalizing from a few examples: a survey on few-shot learning. ACM Comput. Surv. **53**(3), 1–34 (2020)
22. Research, S.: EyeLink 1000 Plus: A Highly Accurate, Precise, and Versatile Eye Tracker (2024). https://www.sr-research.com/eyelink-1000-plus/
23. Ouyang, L., et al.: Training language models to follow instructions with human feedback. Adv. Neural. Inf. Process. Syst. **35**, 27730–27744 (2022)

Enhancing Anti-VEGF Response Prediction in Diabetic Macular Edema Through OCT Features and Clinical Data Integration Based on Deep Learning

Sung Ho Jung[1], Soon Jung Oh[1], Hakje Yoo[1,2](✉), and Je Moon Yoon[3]

[1] Medical AI Research Center, Research Institute for Future Medicine, Samsung Medical Center, Gangnam-gu, Seoul, Republic of Korea
waihjei11@gmail.com

[2] Department of Bio-Mechatronic Engineering, Sungkyunkwan University College of Biotechnology and Bioengineering, Jangan-gu, Suwon, Gyeonggi, Republic of Korea

[3] Department of Ophthalmology, Samsung Medical Center, Sungkyunkwan University School of Medicine, Seoul, Republic of Korea

Abstract. This study presents a deep learning model that combines OCT image features and clinical data to predict responses to anti-VEGF treatment in DME patients. Using data from 107 patients, the model integrates EfficientNetB2 for OCT image analysis and a DNN for clinical data, achieving a notable performance with an AUROC of 0.830. This suggests the model's effectiveness in facilitating personalized DME treatment strategies by leveraging multimodal data analysis.

Keywords: DME · Anti-VEGF Treatment Response · OCT · Multimodal Deep Learning · Lesion Feature Extraction

1 Introduction

Diabetes and its complication, diabetic macular edema (DME), are key concerns in modern medicine. Over 422 million people worldwide suffer from diabetes, which can lead to various health issues, including DME, a major cause of vision loss in diabetic patients [1]. Optical coherence tomography (OCT) is the gold standard for diagnosing and monitoring DME, providing detailed information on retinal health. Treatment for DME has evolved, with intravitreal injections targeting Vascular Endothelial Growth Factor (VEGF) proving more effective than traditional laser therapy [2]. The primary treatments include anti-vascular endothelial growth factor (anti-VEGF) drugs (bevacizumab, ranibizumab, aflibercept) and steroids, although steroids have side effects like cataract formation and increased eye pressure [3, 4]. However, there's a lack of clear guidelines for DME treatment, leading to variability in treatment choices by ophthalmologists [5]. This highlights the need for personalized treatment approaches. Deep learning technology offers promising advancements in this area. Studies by Xu et al. (2022) show the potential of deep learning in predicting treatment outcomes and detecting DME from

J. Finkelstein et al. (Eds.): AIME 2024, LNAI 14845, pp. 176–180, 2024.
https://doi.org/10.1007/978-3-031-66535-6_20

OCT images [6]. This could greatly assist ophthalmologists in decision-making and provide optimal patient care, contributing to both academic and clinical advancements in DME treatment. In this study, we have developed a deep learning-based model to predict the treatment response for DME, utilizing clinical data, OCT images, and lesion characteristic data extracted therefrom, collected at the patient's initial visit.

2 Methods

Figure 1 is an overview for multimodal model development process of this study. At step (a), raw clinical data and OCT images are processed. At step (b), Training multimodal models with feature fusion techniques. At step (c), the trained model evaluates patient-wisely with 5 oct slices.

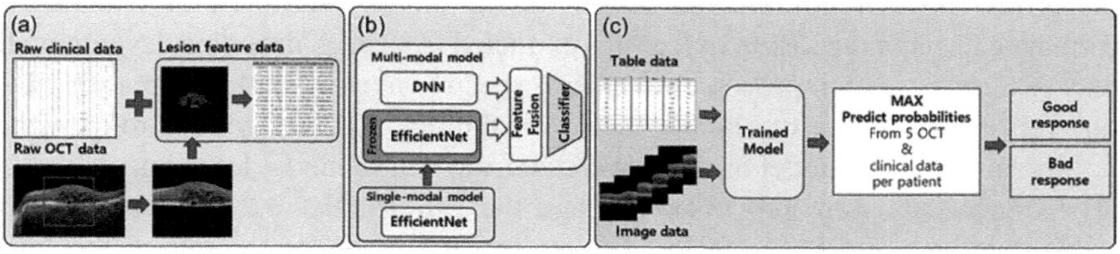

Fig. 1. Overview for development of multimodal prediction model. (a) Data preprocessing of table and image data, (b) Model training with feature fusion, (c) Patient-wise Evaluation.

This study collected clinical data and OCT images from 107 patients. The data collection spanned from 2010 to 2022 at Samsung Medical Center, Seoul, focusing on patients diagnosed with DME and treated with three rounds of anti-VEGF therapy. Treatment response was classified as 'good' if central macular thickness reduced by more than 50 μm, and 'bad' otherwise. OCT images were captured using Spectral domain-optical coherence tomography (Spectralis HRA + OCT, Heidelberg Engineering, Germany). Clinical data included basic patient information and medical history.

The preprocessing of collected OCT images was conducted to prepare for analysis. Figure 2(a)–(c) shows the extraction of OCT images from B-scan images. From B-scan images consisting of 31 frames, five slices were extracted, focusing on the center of the macular edema. Two slices on each side of this central slice were included, making a total of five slices per patient. The process of isolating the regions of interest (ROI) led to the removal of the surrounding macular area. It involved the exclusion of the left and right areas diverging from the macula and the areas where optic nerve fibers converge. Moreover, the areas above and below the retina, including the vitreous and choroid, were also eliminated.

Post-preprocessing, clinicians manually masked key lesions in the OCT images, which included intraretinal fluid (IRF), and subretinal fluid (SRF), hyperreflective dots (HRD) as Fig. 2(d). Lesion areas were excised from OCT images for further analysis. Image processing techniques extracted shape features like area and perimeter, and image characteristics such as pixel intensity and texture. These features were combined with clinical data for training as table data.

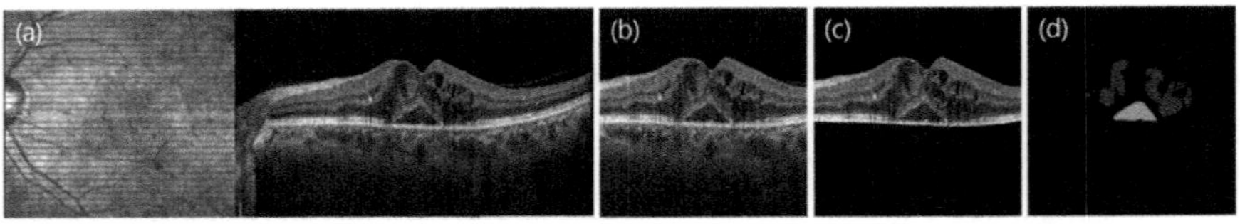

Fig. 2. Preprocess of OCT images. (a) Initial OCT image slice, (b) Exclude non-edematous areas, (c) Eliminate vitreous and choroidal areas, (d) Highlight SRF, IRF, and HRD lesion areas of image.

For deep learning model training, deep neural network (DNN) model used for the table data. The model structure consisted of two fully connected layers (FCL) with 32 units each using the ReLU activation function, followed by a binary classification FCL using the Sigmoid activation function. EfficientNet-B2 model used for OCT image data [7], pre-trained with ImageNet dataset weights, and fine-tunned for binary classification output probability. The multimodal model was developed with the aim of enhancing performance through the fusion of features extracted from each data type, leveraging the distinct feature extraction processes of their respective models, and incorporating FCL for the refined prediction of classification probabilities. The image feature extraction part of the multimodal model used the weights from single-modal learning.

The image data were processed through the EfficientNet-B2 architecture, while the table data were fed into a DNN. The multimodal model training used Stochastic Gradient Descent (SGD) as the optimizer, with a learning rate of 0.001, batch size of 16, and 200 epochs. Data augmentation techniques like rotation, flipping, and shifting were applied for performance enhancement and generalization. The table data were standardized before training.

For each patient, the treatment response prediction was determined by the highest probability class among the five extracted OCT images and table data. Performance metrics included area under the receiver operating characteristic (AUROC), Accuracy, Sensitivity, Specificity, Precision and F1-score. 5-Fold Cross Validation method was used for evaluation. Performance metrics of each fold were Averaged. Gradient-weighted Class Activation Mapping (Grad-CAM) visualized to understand the regions of the images the model was focusing on during classification.

3 Results

All OCT images were processed by extracting and cropping only the ROI. Following this, lesion masking was carried out to extract features, which were then combined with clinical data to form a comprehensive table dataset. The dataset was divided using a 5-fold cross-validation method with a split ratio of 8:2, allocating 425 instances for training and 110 for evaluation in each fold. The dataset was arranged so that data from the same patient were not included in both the training and evaluation sets. The Receiver Operating Characteristic (ROC) curves, as showed in Fig. 3, provide a comparative analysis of the performance across three models. The DNN model, trained with table data, manifested an AUROC of 0.790. In comparison, the AUROC for the EfficientNet-B2 model, trained with image data, was observed at 0.751. Notably, the multimodal model, integrating both table and image data, demonstrated higher performance with an AUROC of 0.830.

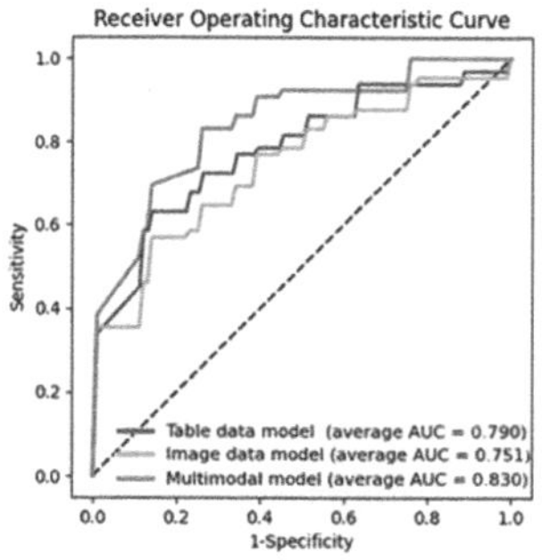

Fig. 3. Comparative ROC curves for single-modal and multimodal learning approaches using OCT images and table data.

Table 1 compiles the additional performance metrics across the validation folds. The model trained on table data exhibited the highest accuracy of 0.757. Both the table and image data models achieved a sensitivity of 0.785, surpassing the multimodal model, which registered a sensitivity of 0.738. The multimodal model, however, attained the highest specificity of 0.778 and precision of 0.863, culminating in the most favorable F1-score of 0.792.

Table 1. Average performance of 5-fold cross validation for trained models

Data type	Accuracy	Sensitivity	Specificity	Precision	F1
Clinical & Lesion features	**0.757**	0.785	0.719	0.775	0.755
OCT images	0.737	0.785	0.656	0.796	0.781
Multi-modal	0.755	0.738	**0.778**	**0.863**	**0.792**

Figure 4 provides Grad-CAM heatmaps of sample OCT images, as classified by the multimodal model. These heatmaps offer a visual representation of the model's predictive focus, with image (a) correctly identified as 'Bad response' and image (b) as 'Good response'. Both heatmaps highlight the regions of macular edema within the retina, corroborating the model's ability to localize clinically significant areas that influence its predictive accuracy.

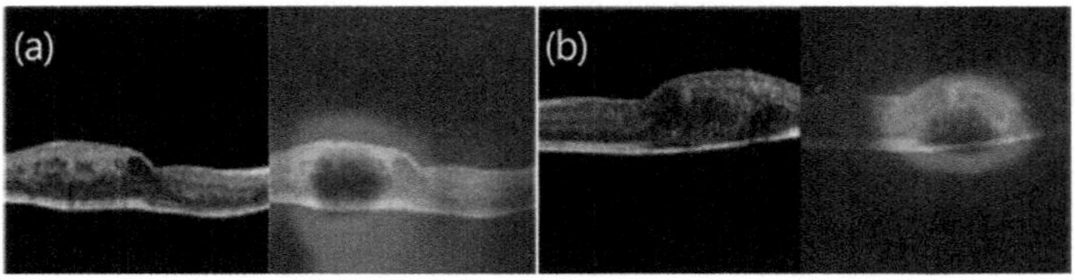

Fig. 4. Grad-CAM heatmap visualization of multimodal model predictions (a) bad response (b) good response.

4 Discussion and Conclusion

In this study, a deep learning model was developed to predict early responses to anti-VEGF treatment in DME. The model was trained by integrating OCT images, clinical data, and features extracted from key lesion areas collected during the initial patient visit. Compared to training with only OCT images or table data, the multimodal approach, which combined both data types, showed higher performance in terms of AUROC. Although its sensitivity, an indicator of accurately identifying good responses, was slightly lower, the model exhibited higher specificity, which is indicator of accurately identifying bad responses. This is expected to aid in determining the appropriate treatment direction for patients. However, this study has several limitations. Firstly, the dataset originated from a single institution, limiting the scope and diversity of the patient data. Secondly, the process required clinicians to manually mask lesions to extract features, potentially complicating the immediate assessment of predictive results at the point of patient visitation in a clinic. Future research will seek to expand the dataset, thereby enhancing the model's generalizability. Efforts will be directed towards the development of an automated segmentation model for OCT images, aiming to expedite the prediction process during clinical consultations. Such advancements are expected to refine DME treatment protocols and elevate the standard of patient care.

Acknowledgments. This work was also supported by the Technology Innovation Program (20014111) funded by the Ministry of Trade, Industry & Energy (MOTIE,Korea) and Samsung Medical Center Grant #SMO1230381.

References

1. Cheung, N., Mitchell, P., Wong, T.Y.: Diabetic retinopathy. Lancet **376**, 124–136 (2010)
2. Nguyen, Q.D., Brown, D.M., Marcus, D.M., et al.: Ranibizumab for diabetic macular edema: results from 2 phase III randomized trials: RISE and RIDE. Ophthalmology **119**, 789–801 (2012)
3. Wells, J.A., Glassman, A.R., Ayala, A.R., et al.: Aflibercept, bevacizumab, or ranibizumab for diabetic macular edema: two-year results from a comparative effectiveness randomized clinical trial. Ophthalmology **123**, 1351–1359 (2016)
4. Jonas, J.B., Jonas, R.A., Neumaier, M., Findeisen, P.: Cytokine concentration in aqueous humor of eyes with diabetic macular edema. Retina **32**, 2150–2157 (2012)
5. Sim, D.A., Keane, P.A., Tufail, A., Egan, C.A., Aiello, L.P., Silva, P.S.: Automated retinal image analysis for diabetic retinopathy in telemedicine. Curr. Diab. Rep. **15**, 14 (2015)
6. Xu, F., Liu, S., Xiang, Y., et al.: Prediction of the short-term therapeutic effect of anti-VEGF therapy for diabetic macular edema using a generative adversarial network with OCT images. J. Clin. Med. **11**, 2878 (2022)
7. Tan, M., Le, Q.V.: EfficientNet: rethinking model scaling for convolutional neural networks. arXiv preprint arXiv:1905.11946 (2019)

Expert Insight-Enhanced Follow-Up Chest X-ray Summary Generation

Zhichuan Wang[1,2], Kinhei Lee[1], Qiao Deng[1,2], Tiffany Y. So[1], Wan Hang Chiu[3], Bingjing Zhou[1], and Edward S. Hui[1,2,4](✉)

[1] Department of Imaging and Interventional Radiology, The Chinese University of Hong Kong, Hong Kong SAR, China
edward.s.hui@gmail.com
[2] CU Lab for AI in Radiology (CLAIR), The Chinese University of Hong Kong, Hong Kong SAR, China
[3] Hospital Authority, Hong Kong SAR, China
[4] Department of Psychiatry, The Chinese University of Hong Kong, Hong Kong SAR, China

Abstract. A chest X-ray radiology report describes abnormal findings not only from X-ray obtained at current examination, but also findings on disease progression or change in device placement with reference to the X-ray from previous examination. Majority of the efforts on automatic generation of radiology report pertain to reporting the former, but not the latter, type of findings. To the best of the authors' knowledge, there is only one work dedicated to generating summary of the latter findings, i.e., follow-up summary. In this study, we therefore propose a transformer-based framework to tackle this task. Motivated by our observations on the significance of medical lexicon on the fidelity of summary generation, we introduce two mechanisms to bestow expert insight to our model, namely expert soft guidance and masked entity modeling loss. The former mechanism employs a pretrained expert disease classifier to guide the presence level of specific abnormalities, while the latter directs the model's attention toward medical lexicon. Extensive experiments were conducted to demonstrate that the performance of our model is competitive with or exceeds the state-of-the-art.

Keywords: Follow-up chest X-ray · follow-up summary generation · multimodal representation learning

1 Introduction

Chest X-ray radiology report generation can automatically report radiological findings on radiogram in an end-to-end manner [1, 2], potentially providing a means to alleviate the workloads of radiologists. A major shortcoming of the majority of these models is that they are limited to reporting abnormalities in a given chest X-ray examination. Typically, radiologist writes reports not only based on a single (follow-up) X-ray examination but also by incorporating information from the X-ray from previous check-up to discern disease progression and/or change in device placement. Automatic generation of this

J. Finkelstein et al. (Eds.): AIME 2024, LNAI 14845, pp. 181–193, 2024.
https://doi.org/10.1007/978-3-031-66535-6_21

type of report is known as follow-up report generation, a clinically critical yet largely understudied area. To the best of the knowledge, there is only one work [3] focusing on this field. We therefore aim to develop an end-to-end deep learning model to generate textural summary of disease progression from chest X-ray obtained at follow-up and baseline examinations (see Fig. 1).

However, majority of the current models, namely follow-up summary generation [3], single radiogram report generation [1, 2], or natural image difference caption [4, 5], overlook the importance of meaningful entities, *i.e.*, the medical lexicon within the domain of follow-up summary generation. Specifically, these models typically generate a sentence word by word under the supervision of the masked language modeling loss. In this paradigm, these models pay equal attention to each word, which may not be optimal, as words related to abnormalities should deserve more attention than other words. This asymmetry results in a relatively lower accuracy in predicting abnormality-related words, thereby necessitating additional mechanism to improve the fidelity of follow-up summary generation.

Toward this end, we propose two mechanisms to bestow expert insight on follow-up chest X-ray summary generation, namely expert soft guidance and masked entity modeling loss. The former employs a pretrained expert disease classifier to provide complementary guidance on the presence level of certain diseases, while the latter directs the model's attention toward abnormality-related words. Our contributions are:

- A framework is proposed to address the follow-up chest X-ray summary generation task, a clinically critical but largely understudied area.
- wo mechanisms, namely expert soft guidance and masked entity modeling loss, are proposed to bestow expert insight on disease-related words.
- Comprehensive experiments were conducted to demonstrate that the performance of our model is competitive with or exceeds the state-of-the-art.

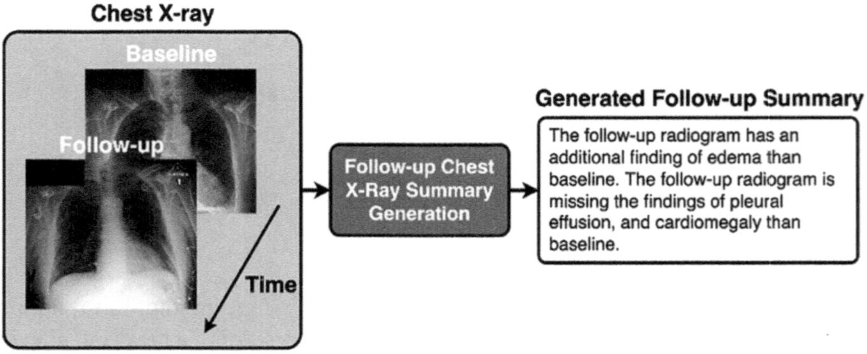

Fig. 1. A schematic of our proposed model that generates a textural summary of disease progression from a pair of chest X-rays taken at follow-up and baseline examinations.

2 Related Work

The work of Hu et al. [3] is arguably the inaugural work to tackle the task of follow-up summary generation. The authors have assembled a dataset, dubbed MIMIC-Diff-VQA, of radiogram pairs and their corresponding follow-up chest X-ray summaries using the MIMIC-CXR dataset [6]. They proposed EKAID, a graph-based network, to extract relationships between anatomical regions and diseases, with commendable performance.

The most related field in natural images is image difference caption, which aims to analyze the semantic differences in a given pair of images. The Spot-the-diff dataset was collected by [7] in 2018, marking the introduction of the image difference caption task. Later, Qiu et al. introduced a transformer-based model, MCCformers, to capture the intra and inter-image relations, further improving the performance [4]. With the pretraining paradigm demonstrating a powerful ability in vision-language tasks, Yao et al. devised three pretraining tasks, IDCPCL, to align visual and textural features, achieving superior results compared to previous works [5].

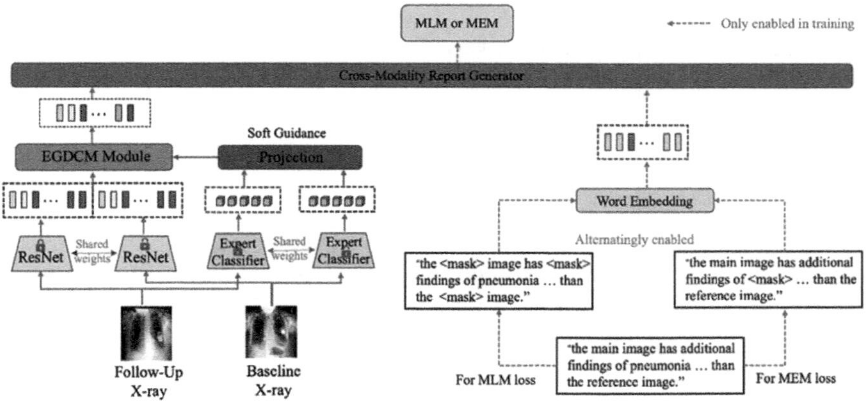

Fig. 2. Illustration of our proposed framework, expert insight-enhanced (EIE) follow-up chest X-ray summary generation. The dotted arrows indicate the branches that are only enabled in the training phase, and the symbol of lock indicates module that are frozen.

3 Method

3.1 Preliminaries

Given an X-ray pair (I^f, I^b) from the same patient, where I^f and I^b respectively represent X-ray images obtained at follow-up and baseline examinations. Our goal is to generate a summary of radiological findings s that describes disease progression and/or change in device placement at the time of follow-up X-ray examination. Such summary is dubbed as follow-up summary from hereon.

During the training phase, X-ray pair $\{\left(I^{f^{(i)}}, I^{b^{(i)}}\right)\}_{i\in D}$ and the corresponding ground-truth follow-up summary $\{s^{(i)}\}_{i\in D}$ from training dataset D were used to train the model. Building upon prior research [4, 5], every word in s underwent tokenization and was subsequently mapped to word embeddings via an embedding layer initiated with training from the ground up. Three special tokens were added to facilitate the training process, namely [CLS], [BOS] and [EOS]. [CLS] was added to merge the information of all word tokens, and [BOS] and [EOS] were used to identify the beginning or ending of a sentence. The text input can be represented as:

$$\bar{s} = \left([\text{CLS}], [\text{BOS}], w_1, \ldots, w_{N_s}, [\text{EOS}]\right), \tag{1}$$

where w_i represents the token of the ith word in s. Similar to the approach in [5], a pretrained ResNet101 [8] was adopted to obtain the image features from the X-ray pair, which were then flattened into a sequence of image tokens:

$$V^b = \left(\left[\text{Xray1}\right], V_0^b, \ldots, V_{N_v}^b\right), \quad V^f = \left(\left[\text{Xray2}\right], V_0^f, \ldots, V_{N_v}^f\right), \tag{2}$$

where $\left[\text{Xray1}\right]$ and $\left[\text{Xray2}\right]$ are two tokens for capturing the comprehensive representations of individual X-ray images, and V_i^b and V_i^f represent the ith image token of the baseline and follow-up X-ray, respectively. Furthermore, fixed positional embeddings and type embeddings were incorporated, with the former designed to denote token positions and the latter employed to distinguish the source of a particular token, such as V^f, V^b, or $\bar{s}$.

3.2 Model Architecture

The overall architecture of our proposed model, expert insight-enhanced (EIE) follow-up chest X-ray summary generation, is illustrated in Fig. 2. Illustration of our proposed framework, expert insight-enhanced (EIE) follow-up chest X-ray summary generation. The dotted arrows indicate the branches that are only enabled in the training phase, and the symbol of lock indicates module that are frozen. EGDCM: expert guided difference capture module; MLM: masked language modelling; MEM: masked entity modelling.. A pair of X-ray images are fed into a pretrained ResNet to extract image features, serving as image tokens after flattening. The same pair of X-ray are also input into a pretrained expert disease classifier to obtain the probability of the presence of certain diseases, which is then projected into the same dimension as image tokens. Image and soft-guidance tokens are fed into the transformer-based expert guided difference capture module (EGDCM) to obtain the difference representation. The output tokens of EGDCM and word tokens are input into the cross-modality summary generator to produce the follow-up chest X-ray summary. The generation is supervised by the masked language modeling (MLM) loss and masked entity modeling (MEM) loss.

Expert Soft Guidance. To bestow expert insight to the model for enhancing the prediction of abnormality-related words, known as entities, complementary guidance on the presence level of certain diseases was provided by a pretrained multi-class expert classifier.

Given an X-ray pair $\left(\boldsymbol{I}^f, \boldsymbol{I}^b\right)$, a pretrained multi-class disease classifier was employed to generate the probability of the presence of a number of abnormalities, denoted as

(p^f, p^b). Rather than using a hard threshold to binarize these probabilities, they were directly mapped into the same feature space of image tokens through a projection module. Subsequently, the guidance tokens were obtained and used to guide the EGDCM in capturing a more disease-related representation. Formally, the process of obtaining the soft guidance can be denoted as:

$$p^j = \mathrm{F}_e\left(I^j\right) \tag{3}$$

$$g^j = \mathrm{Proj}\left(p^j\right) \quad \text{for } j \in \{f, b\}, \tag{4}$$

where F_e denotes the expert classifier, $\mathrm{Proj}(\cdot)$ denotes the projection layer, and $\mathbf{g^j}$ represents the soft-guidance token. Similar to image tokens, the soft-guidance tokens $\mathbf{g^j}$ are appended with fixed positional embeddings and type embeddings.

Expert-Guided Difference Capture Module. We propose an EGDCM, $\mathrm{F_{diff}}$, to extract the representations of the difference in the image features of two different examinations. To better model long-distance dependencies and merge information from image and texts, EGDCM is composed of a two-layer transformer, which takes all image tokens and guidance tokens as input. Formally, the process can be written as:

$$V_E^b = \left([\mathrm{Xray1}], V_0^b, \ldots, V_{N_v}^b, g^b\right), V_E^f = \left([\mathrm{Xray2}], V_0^f, \ldots, V_{N_v}^f, g^f\right) \tag{5}$$

$$\{\overline{V^f}, \overline{V^b}\} = F_{\mathrm{diff}}\left(\{V_E^f, V_E^b\}\right). \tag{6}$$

Cross-Modality Follow-Up Summary Generator. A three-layer transformer was employed for the cross-modality report generator $\mathrm{F_{gen}}$, which took the image difference tokens and summary tokens as inputs:

$$\left(\widehat{V^f}, \widehat{V^b}, \hat{s}\right) = F_{\mathrm{gen}}\left(\overline{V^f}, \overline{V^b}, \bar{s}\right) \tag{7}$$

3.3 Loss Functions

To endow the model with the ability to generate follow-up chest X-ray summary, the widely used MLM loss [5, 9] was adopted. Furthermore, we propose a new MEM loss, an extension of MLM, to enhance the prediction of the abnormality-related words.

Masked Language Modeling Loss. In contrast to many vision-language pretraining works, the uni-directional MLM loss was employed to emulate the sentence generation process of inference. In other words, instead of considering all text tokens, the prediction of the ith (masked) word was based solely on its preceding words $w_{<i}$ and other image-related tokens $\left(\widehat{V^f}, \widehat{V^b}\right)$. The same masking strategy as BERT [9] was used, where 15% of the input text tokens were randomly selected, with 80% of which masked by the special token [MASK], 10% substituted with words randomly chosen in vocabulary, and the remaining 10% left unchanged. To obtain the final predictions, the masked tokens

were input into a linear classifier. Formally, the uni-directional MLM loss over D can be written as:

$$L_{\text{MLM}} = E_D\Big[-\log P_\theta\Big(w_i|w_{<i}, \widehat{V^f}, \widehat{V^b}\Big)\Big] \quad \text{for } \forall i \in \{1, \cdots, N_s\}, \tag{8}$$

where θ is the trainable parameters of the model.

Masked Entity Modeling Loss. Prediction of entities, can be enhanced by directing model's attention toward these words through our proposed MEM. Masked entity modeling loss is formulated similar to the MLM loss, but with the masking of the selected entity words:

$$L_{\text{MEM}} = E_D\Big[-\log P_\theta\Big(w_i|w_{<i}, \widehat{V^f}, \widehat{V^b}\Big)\Big] \quad \text{for } w_i \in \mathcal{E}, \tag{9}$$

where $\mathcal{E}$ is the selected entity set.

Given that expert soft guidance can only offer soft guidance for certain abnormalities, the MEM loss serves as a complement to expert soft guidance by enhancing the prediction of the remaining common and crucial abnormality-related words. For more details on entity selection, please refer to the appendix.

It is worth noting that, thanks to the parallel computation of transformer blocks, the soft guidance tokens g^f and g^b can be easily injected into or removed from the difference capturing process. When the model was trained solely with the MLM loss and without enabling expert soft guidance, we name the model as **EIE-base**. When **EIE-base** was combined with expert soft guidance or MEM loss, we refer to it as **EIE-esg** and **EIE-mem**, respectively. When **EIE-base** was combined with both proposed mechanisms, it is denoted as **EIE-all**.

3.4 Training Strategy

When both of the MLM and MEM losses were enabled, an alternating optimization strategy was adopted. Specifically, an enabling probability α was assigned to the MEM loss. In a given training iteration, a random sample r was drawn from a uniform distribution between 0 and 1. The MEM loss was employed if $r > \alpha$; the MLM loss was enabled otherwise. The optimization direction was controlled by adjusting the hyperparameter α.

4 Experiments

4.1 Dataset and Metrics

Our model was evaluated on the MIMIC-Diff-VQA [3] dataset, which contains 164,324 X-ray pairs and the corresponding follow-up reports. Following [3, 5], widely used natural language generation metrics including BLEU [10], METEOR [11, 12]ROUGE$_L$ and CIDEr [13] were adopted to evaluate the model on summary generation. Following [2], our method was also evaluated using accuracy Acc5 and Acc14 to demonstrate its ability on abnormality recognition, where Acc5 denotes the micro average accuracy over 5 most common abnormalities and Acc14 is that over 14 abnormalities of CheXpert [14]. Please refer to the appendix for more details on the abnormalities, the metrics and their computation.

4.2 Implementation Details

Consistent with previous works [5], a pretrained ResNet101 was employed to extract grid features with a shape of (7, 7, 1024), which were then flattened into a sequence of image tokens with a shape of (49, 1024). For all transformer-based modules, the hidden feature size was 512, the number of multi-heads was 8, and the layer numbers were 2 and 3 for the EGDCM and cross-modality follow-up summary generator, respectively. The word embedding layers were learnt from scratch, and the feature dimension was set to 512. The hyperparameter α in the alternating optimization strategy was set to 0.8. The Adam optimizer was used with a learning rate of 3e-5, and the total training iteration was set to 100k. Our expert classifier was based on PCAM [14]. For more details, please refer to the appendix.

Table 1. Comparison with State-of-the-Art (SOTA): **EIE-all** surpasses existing SOTA by a substantial margin across all metrics. Furthermore, each proposed mechanism notably enhances the performance of the base model. The combination of both mechanisms leads to further improvement. † Results were reported from EKAID. ‡ Results were reproduced using the official codes of EKAID.

Method	BLEU1	BLEU2	BLEU3	BLEU4	METEOR	$ROUGE_L$	CIDEr	Acc5	Acc14
MCCformers†	0.214	0.190	0.170	0.153	0.319	0.340	0.000	-	-
IDCPCL†	0.614	0.541	0.474	0.414	0.303	0.582	0.703	-	-
EKAID	0.628	0.553	0.491	0.434	0.339	0.577	1.027	68.21‡	81.00‡
EIE-base	0.574	0.510	0.456	0.407	0.416	0.590	0.968	59.94	77.57
EIE-mem	0.614	0.547	0.492	0.441	0.405	0.616	1.421	65.89	79.14
EIE-esg	0.638	0.573	0.519	0.469	0.399	0.636	1.615	68.33	81.63
EIE-all	0.646	0.583	0.528	0.477	0.401	0.653	1.698	70.98	81.47

4.3 Comparison with the State-of-the-Art

As depicted in Table 1, with the exception of the **EIE-base** model, other versions of the proposed methods outperformed existing SOTA by a considerable margin, with **EIE-all** achieving the best performance. Taking **EIE-all** as an example, there were improvements of 1.8%, 3.0%, 3.7%, and 4.3% on the BLEU metrics. For METEOR, a performance gain up to 6.2% was achieved. The increase was more pronounced on $ROUGE_L$ and CIDEr, with **EIE-all** boosting the performance by 5.8% and 0.671, respectively. The relative increase in CIDEr was particularly large, at 65.3% compared with EKAID. For Acc5 and Acc14, a 2.77% and 0.47% accuracy performance gain was obtained. These results demonstrate that our method performs better on both the report level and the clinical level.

4.4 Ablation Study

As illustrated in Table 1, each proposed mechanism notably enhanced the performance of our base model, both at the generation and clinical levels. **EIE-mem** achieved improvements of 4.0%, 3.7%, 3.6%, and 3.4% on the BLEU metric, with a 2.6% and 0.453 increase in $ROUGE_L$ and CIDEr, respectively. There were improvements in both Acc5 and Acc14, at 5.95% and 1.57% respectively. For **EIE-esg**, the performance gains were 6.4%, 6.3%, 6.3%, and 6.2% on BLEU, 8.39% on Acc5, and 4.06% on Acc14. $ROUGE_L$ increased from 59.0% to 63.6%, and CIDEr increased from 0.968 to 1.615. Notably, the relative increase in CIDEr was 66.84%. Combining the MEM loss and expert soft guidance further improved performance, with **EIE-all** achieving gains of 7.2%, 7.3%, 7.2%, 7.0%, 4.5%, 0.73, 11.04%, and 3.9% on BLEU, $ROUGE_L$, CIDEr, Acc5, and Acc14 respectively.

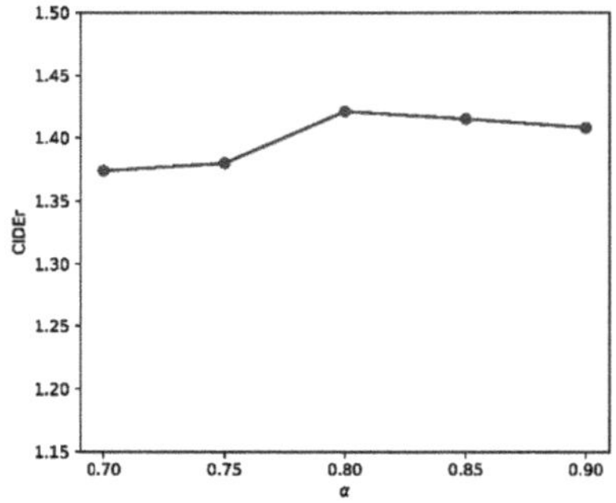

Fig. 3. Hyperparameter α sensitivity study of EIE-mem.

4.5 Hyper-parameter Sensitivity Study

Several experiments were conducted to analyze how the performance of **EIE-mem** varied with the hyper-parameter α, which controls the enable probability of the MEM loss. The results are presented in Fig. 3. The CIDEr score ranged from 1.37 to 1.42, and surpassed 1.027, which was that of the existing state-of-the-art [3]. The best result of 1.42 was obtained when $\alpha = 0.8$. Consequently, we set α to 0.8 in all experiments.

4.6 Comparison with Hard Guidance

Several experiments were conducted to compare the performance of hard guidance with that of soft guidance. To obtain hard guidance, a sigmoid function was applied to the output of the expert classifier, which was subsequently binarized using a threshold. The results are presented in Table 2. Soft guidance outperformed hard guidance, particularly in METEOR and $ROUGE_L$ metrics. Setting the threshold at 0.5 yields the best results among all hard guidance experiments. Soft guidance achieved 6.0%, 5.6%, 4.9%, and 4.3% improvement in BLEU metric, 8.2% in METEOR, 11.3% in $ROUGE_L$, and 0.036

in CIDEr. The relative performance increases in METEOR and $ROUGE_L$ were 25.87% and 21.6%, respectively. These experiments affirm that soft guidance is more effective and robust, as it eliminates the need for careful hyperparameter selection.

Table 2. Comparison with hard versus soft guidance.

Threshold	BLEU1	BLEU2	BLEU3	BLEU4	METEOR	$ROUGE_L$	CIDEr
0.4	0.548	0.491	0.445	0.403	0.304	0.511	1.557
0.5	0.578	0.517	0.470	0.426	0.317	0.523	1.579
0.6	0.571	0.511	0.464	0.421	0.314	0.520	1.576
Soft guidance	0.638	0.573	0.519	0.469	0.399	0.636	1.615

4.7 Lightweight EIE-all and More Experiments

In addition to the aforementioned experiments, the appendix contains supplementary experimental results and showcases, including a lightweight version of **EIE-all**. This variant utilized expert guidance during the training phase to enhance the knowledge learning, whilst omitting the expert soft guidance branch during inference. In addition, the performance of a model with a stronger and recently proposed expert classifier was also investigated, indicating that model performance can be further improved with better expert guidance. For more details, please refer to the appendix.

5 Limitations

Albeit exceeding the performance of SOTA, our proposed model warrants further improvement in its clinical accuracy. In addition, we will bestow our model the ability to predict the location of abnormalities in future studies.

6 Conclusion

A transformer-based framework was proposed to address the clinically critical yet largely understudied task of follow-up chest X-ray summary generation. Two mechanisms, namely expert soft guidance and masked entity modeling loss, were proposed to bestow expert insight on abnormality recognition for improving the overall model performance. Extensive experiments demonstrated that the performance of our model is competitive with or exceeds the state-of-the-art.

A Appendix

A.1 Lightweight EIE-all

In this section, we present a streamlined version of EIE-all, denoted as **EIE-light**, which relied on expert guidance exclusively during the training phase, excluding the expert soft guidance branch during inference while maintaining commendable performance. To achieve this objective, we propose a random dropout training strategy. Specifically, we assigned an enabling probability β to the expert soft guidance. During a random training iteration, we drew a sample from a uniform distribution between 0 and 1, yielding a number r. If $r < \beta$, the guidance from the expert classifier was replaced with zero vectors of the same dimension. Otherwise, the guidance was enabled and put into the injection layer. The results of **EIE-light** are demonstrated in Table S. 1. Comparison with State-of-the-Art (SOTA): EIE-all surpasses existing SOTA by a substantial margin across all metrics. Furthermore, each proposed mechanism notably enhances the performance of the base model. The combination of both mechanisms leads to further improvement.† Results were reported from EKAID. ‡ Results were reproduced using the official codes., and the results of the hyperparameter sensitivity study of β are illustrated in Figure Fig. S. 1.

Notably, **EIE-light** outperformed state-of-the-art methods by a considerable margin, achieving gains of 1.1%, 2.1%, 2.9%, and 3.4% in BLEU, and 4.3% and 2.9% in METEOR and ROUGE_L, respectively. Furthermore, there was a notable 0.538 improvement in CIDEr, representing a 52.39% increase. It is worth mentioning that **EIE-light** exhibited superior performance compared to **EIE-mem** and comparable performance with **EIE-esg**, indicating that even without guidance during inference, the guidance during training significantly enhances the performance of **EIE-base** and **EIE-mem**. Additionally, as depicted in Fig. S.1., there was minimal fluctuation with changes in β, maintaining results consistently higher than 1.30, with the optimum performance achieved at $\beta = 0.6$.

A.2 Expert Soft Guidance with More Observations

As stated in the main paper, we adopted PCAM [15], which was the top1 solution of CheXpert [14] classification, as our expert classifier. However, PCAM can only produce the predictions of the 5 most common diseases contained in the CheXpert test set, resulting in suboptimal guidance. Therefore, it is intuitive that the performance would be further improved if the expert classifier could provide guidance on more observations. A recent work [16] is capable of offering the classification results for all 14 observations of CheXpert, denoted as **Expert-14**, which was then incorporated into our model and was assessed accordingly. When the expert classifier of **EIE-esg** was substituted by **Expert-14**, we denote it as **EIE-esg-14**. When the classifier of **EIE-all** was replaced with **Expert-14**, it is denoted as **EIE-all-14**.

Table S.1. Comparison with State-of-the-Art (SOTA): EIE-all surpasses existing SOTA by a substantial margin across all metrics. Furthermore, each proposed mechanism notably enhances the performance of the base model. The combination of both mechanisms leads to further improvement.† Results were reported from EKAID. ‡ Results were reproduced using the official codes.

Method	BLEU1	BLEU2	BLEU3	BLEU4	METEOR	$ROUGE_L$	CIDEr
MCCformers†	0.214	0.190	0.170	0.153	0.319	0.340	0.000
IDCPCL†	0.614	0.541	0.474	0.414	0.303	0.582	0.703
EKAID	0.628	0.553	0.491	0.434	0.339	0.577	1.027
EIE-base	0.574	0.510	0.456	0.407	0.416	0.590	0.968
EIE-mem	0.614	0.547	0.492	0.441	0.405	0.616	1.421
EIE-esg	0.638	0.573	0.519	0.469	0.399	0.636	1.615
EIE-all	0.646	0.583	0.528	0.477	0.401	0.653	1.698
EIE-light	0.639	0.574	0.520	0.468	0.382	0.606	1.565
EIE-esg-14	0.656	0.591	0.537	0.487	0.408	0.655	1.747
EIE-all-14	0.670	0.604	0.549	0.498	0.391	0.639	1.780

All the results are demonstrated in Table S.1. Comparison with State-of-the-Art (SOTA): EIE-all surpasses existing SOTA by a substantial margin across all metrics. Furthermore, each proposed mechanism notably enhances the performance of the base model. The combination of both mechanisms leads to further improvement.† Results were reported from EKAID. ‡ Results were reproduced using the official codes. We can see that model performance improved further when more guidance was available. Specifically, **EIE-esg-14** brought more performance gain on all metrics than **EIE-esg**: 1.8%, 0.9%, 1.9% and 0.132 on BLEU, METEOR, $ROUGE_L$ and CIDEr, respectively. Together with the guidance from all 14 observations and MEM, **EIE-all-14** consistently achieved further improvements in performance. Compared with **EIE-all**, **EIE-all-14** gained 2.4%, 2.1%, 2.1% and 2.1% in BLEU, 0.4% and 0.082 in $ROUGE_L$ and CIDEr respectively. These results suggest that more expert guidance could further boost the overall performance of our proposed model. It is worth noting that although PCAM can only predict the presence of the most common abnormalities on chest X-ray, it can significantly improve the overall performance. This could be due to the fact that PCAM can not only offer the presence probability of these abnormalities, but also the presence probability of other related abnormalities.

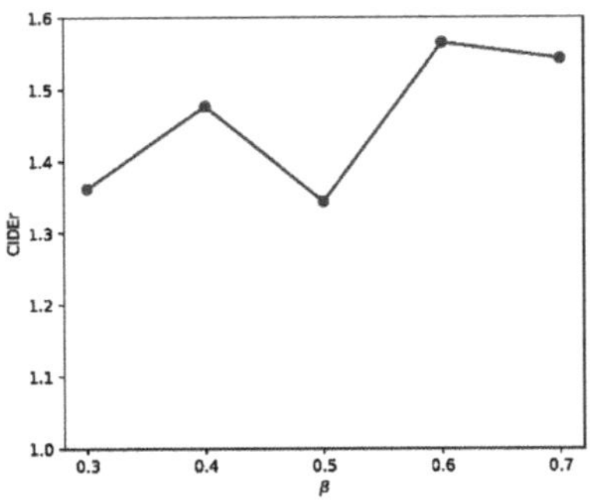

Fig. S.1. Hyperparameter β sensitivity study of **EIE-light**

A.3 Entity Selection

To enrich expert insight into abnormality prediction and complement **EIE-esg**, we have selected relevant terms associated with diseases from CheXpert, namely atelectasis, edema, pneumothorax, cardiomegaly, consolidation, cardiac silhouette, fracture, lung opacity, pleural effusion, and pneumonia.

A.4 Metrics

Following [1], we have incorporated clinical efficacy metrics, Acc5 and Acc14, to showcase the effectiveness of our model. Specifically, Acc5 represents the micro-averaged accuracy across the 5 most common observations: atelectasis, cardiomegaly, consolidation, edema, and pleural effusion. Acc14 is the example-based averaged accuracy across all 14 observations in CheXpert, encompassing pneumonia, fracture, consolidation, enlarged cardiomediastinum, no finding, pleural other, cardiomegaly, pneumothorax, atelectasis, support devices, edema, pleural effusion, lung lesion and lung opacity. We employed the code from [1] to calculate these clinical efficacy metrics, utilizing CheXbert [17] labeler to determine the presence of the most common observations. It is noteworthy that, following [1], instances labeled as *negative, uncertain, or no mention* were all treated as negative results.

References

1. Tanida, T., Müller, P., Kaissis, G., Rueckert, D.: Interactive and explainable region-guided radiology report generation. In: Proceedings of the IEEE/CVF Conference on Computer Vision and Pattern Recognition, pp. 7433–7442. IEEE, Vancouver (2023)
2. Miura, Y., Zhang, Y., Tsai, E.B., Langlotz, C.P., Jurafsky, D.: Improving factual completeness and consistency of image-to-text radiology report generation, In: Conference of the North-American-Chapter of the Association-for-Computational-Linguistics - Human Language Technologies, pp. 5288–5304. ACL (2021)
3. Hu, X., Gu, L., An, Q., Zhang, M., Liu, L., et al.: Expert knowledge-aware image difference graph representation learning for difference-aware medical visual question answering. In: Proceedings of the 29th ACM SIGKDD Conference on Knowledge Discovery and Data Mining, pp. 4156–4165. ACM (2023)

4. Qiu, Y., Yamamoto, S., Nakashima, K., Suzuki, R., Iwata, K., Kataoka, H., Satoh, Y.: Describing and localizing multiple changes with transformers. In: Proceedings of the IEEE/CVF International Conference on Computer Vision, pp. 1971–1980. IEEE, New York (2021)
5. Yao, L., Wang, W., Jin, Q.: Image difference captioning with pre-training and contrastive learning. In: Proceedings of the AAAI Conference on Artificial Intelligence, pp. 3108–3116. AAAI, Vancouver (2022)
6. Johnson, A.E., et al.: MIMIC-CXR, a de-identified publicly available database of chest radiographs with free-text reports. Sci. Data **6**(1), 317 (2019)
7. Jhamtani, H., Berg-Kirkpatrick, T.: Learning to describe differences between Pairs of similar images. In: 2018 Conference on Empirical Methods in Natural Language Processing, pp. 4024–4034. ACL, Brussels (2018)
8. He, K., Zhang, X., Ren, S., Sun, J.: Deep residual learning for image recognition. In: Proceedings of the IEEE Conference on Computer Vision and Pattern Recognition, pp. 770–778. IEEE, Seattle (2016)
9. Devlin, J., Chang, M.W., Lee, K., Toutanova, K.: Bert: Pre-training of deep bidirectional transformers for language understanding. arXiv preprint arXiv:1810.04805 (2018)
10. Papineni, K., Roukos, S., Ward, T., Zhu, W.J.: Bleu: a method for automatic evaluation of machine translation. In: Proceedings of the 40th Annual Meeting of the Association for Computational Linguistics, pp. 311–318. ACL, Univ Penn (2002)
11. Denkowski, M., Lavie, A.: Meteor universal: language specific translation evaluation for any target language. In: Proceedings of the Ninth Workshop on Statistical Machine Translation, pp. 376–380. ACL, Baltimore (2014)
12. ROUGE, L.C.: A package for automatic evaluation of summaries. In: Proceedings of Workshop on Text Summarization of ACL, Spain, vol. 5 (2004)
13. Vedantam, R., Lawrence Zitnick, C., Parikh, D.: Cider: Consensus-based image description evaluation. In: Proceedings of the IEEE Conference on Computer Vision and Pattern Recognition, pp. 4566–4575. IEEE, Boston (2015)
14. Irvin, J., et al.: CheXpert: a large chest radiograph dataset with uncertainty labels and expert comparison. In: Proceedings of the AAAI Conference on Artificial Intelligence, pp. 590–597. AAAI, Honolulu (2019)
15. Ye, W., Yao, J., Xue, H., Li, Y.: Weakly supervised lesion localization with probabilistic-cam pooling. arXiv preprint arXiv:2005.14480 (2020)
16. Pellegrini, C., Özsoy, E., Busam, B., Navab, N., Keicher, M.: Radialog: a large vision-language model for radiology report generation and conversational assistance. arXiv preprint arXiv: 2311.18681 (2023)
17. Smit, A., Jain, S., Rajpurkar, P., Pareek, A., Ng, A.Y., Lungren, M.P.: CheXbert: combining automatic labelers and expert annotations for accurate radiology report labeling using bert. arXiv preprint arXiv:2004.09167 (2020)

Integrating Multimodal Patient Data into Attention-Based Graph Networks for Disease Risk Prediction

Xiayuan Huang(✉)

Yale University, New Haven, CT 06510, USA
xiayuan.huang@yale.edu

Abstract. In the era of digital medicine, electronic health records (EHRs) encompass a vast array of patient data from diverse sources. Harnessing this readily available information is crucial for personalized medicine and predictive healthcare. This study explores the potential of integrating multimodal data into an attention-based graph network for enhancing disease risk prediction. Leveraging the intrinsic structure of medical data, we propose a novel method called MiGAT, which fuses information from various modalities, including clinical events and sequencing data. By applying graph attention networks (GAT), our model effectively captures the intricate inter-relationships among patients, particularly emphasizing genetic similarity. This enables our model to assess and prioritize disease risks based on the genetic closeness of patients. Results indicate a substantial improvement in predictive accuracy compared to baseline models, reaffirming the potential of our multimodal integration strategy. Our findings underscore the importance of synergizing diverse EHR data sources through advanced computational models for better-informed clinical decision-making and more effective patient care.

Keywords: EHR · Machine Learning · Disease Risk Modeling · Multimodal Data Integration · Health Informatics

1 Introduction

The modern healthcare landscape is replete with data sources, and among the most influential developments in recent years is the advent of multimodal patient data (Zhao, Wang, Yu, & Shang, 2022; Li, Ma, & Gao, 2022). The intersection of digital health and advanced biomedical research has given rise to a data-rich environment, where Electronic Health Records (EHRs) and genetic data stand out as two pivotal pillars. EHRs, representing the digitized version of a patient's medical history, encapsulate structured information like diagnoses, procedures, and medications. On the other hand, genetic data provides a deeper dive into a patient's genetic information, revealing the intricacies of their DNA sequencing and potential susceptibilities to diseases. Both data sources, in isolation, have empowered a myriad of predictive models that harness deep learning techniques, with tools like Recurrent Neural Networks (RNNs) (Choi E., et al., 2016) becoming pivotal for diagnosis prediction based on EHRs.

J. Finkelstein et al. (Eds.): AIME 2024, LNAI 14845, pp. 194–203, 2024.
https://doi.org/10.1007/978-3-031-66535-6_22

Yet, a crucial gap remains (Che, Wang, Zhou, & Ma, 2023). Despite the strides in modeling within each data type, there is a conspicuous absence of unified models that seamlessly integrate both EHR and genetic data for enhanced disease risk prediction. Such integration promises a synergistic approach where the observed clinical manifestations from EHRs are enriched by the underlying genetic propensities, creating a comprehensive patient profile that could dramatically enhance the accuracy and precision of predictive models.

A primary limitation in modern medical diagnostics and predictive modeling is the notable absence of family medical history (Huang, et al., 2018; Huang, et al., 2021; Polubriaginof, et al., 2018). This lineage of health records offers invaluable insights, serving as a barometer for potential hereditary conditions and genetic predispositions. Familial patterns, passed down through generations, can highlight susceptibilities or resilience to certain ailments, making them instrumental for comprehensive health risk assessments. Unfortunately, many contemporary predictive models bypass this treasure trove of information, leading to potential gaps in accuracy and foresight.

As the medical world leans more into the embrace of machine learning, particularly deep neural networks, it faces the conundrum of their interpretability. These sophisticated models, such as RNN, with their multilayered architectures (Ma, et al., 2018), are undeniably potent in their analytical capabilities. However, they often function as enigmatic 'black boxes' (Guidotti, et al., 2018), producing results without a clear, traceable rationale. Such opacity can be a roadblock in a field where transparency is paramount, especially when clinical decisions hinge on these predictions. For deep learning to be fully integrated into medical practice, the bridge of understanding between its computations and outcomes must be strengthened.

Motivated by these observations, we introduce the Multimodal Attention-based Graph Network prediction model (MiGAT), designed to address the limitations. MiGAT stands as a paradigm shift in predictive modeling, adeptly merging key inputs like diagnosis codes with an inventively built genetic similarity graph. This synthesis ensures a holistic approach, enhancing the predictive prowess by harnessing both historical medical data and genetic correlations, paving the way for more accurate and comprehensive disease risk predictions.

The core contributions of our work can be encapsulated as follows:

- We develop an innovative attention-based graph network framework (Veličković, et al., 2018) tailored for disease risk prediction, a pivotal arena within health informatics. Notably, our approach marks the pioneering effort to integrate clinical features with genetic data within an attention-based graph network predictive model for risk prediction tasks. The ingeniously built genetic similarity graph (Wang, et al., 2021) inherent to MiGAT is adept at encapsulating the profound impact of patients who share significant genetic ties. By emphasizing these genetic connections, the model is better equipped to discern patterns and relationships, ensuring that the influence of hereditary factors on disease susceptibility is not overshadowed or overlooked.
- Our model incorporates an attention-centric mechanism to gauge the interplay between a patient's diagnosis codes and their clinical features. Furthermore, it discerns the influence of clinical features (Gui, Ge, & Hu, 2019) by considering the genetically closely related patients by genetic similarity.

- Through empirical studies, our MiGAT model showcases superior performance against baseline methods using a real-world multimodal patient dataset sourced from All of Us Research Program (Ramirez, et al., 2022). We also uncover key attributes of the proposed model that potentially account for its enhanced performance. Our conclusions are drawn from experiments that encompass a quantitative assessment of the significance levels of clinical features and a qualitative examination rooted in case study analyses.

2 Problem Statement

2.1 Data

The NIH's *ALL of Us* Research Program (*ALL of Us*) is a longitudinal cohort study aimed at advancing precision medicine and improving human health through partnering with one million or more diverse participants across the United States. In this study, we extracted 31,625 patients with EHR and short-read whole genome sequencing data. The detailed summary of this dataset is described in Table 1.

Table 1. Summary of dataset.

Number of participants	*31.625*			
Number of visits	*85,287*			
Average number of visits per patient	*2.7*			
Number of unique diagnosis codes	*9,087*			
Race	White	Black and African American	Asian	other
----	52.8%	20.8%	3.1%	23.3%
Age	18–44	45–64	> 65	
White	25.6%	31.7%	42.7%	
Black and African American	29.3%	48.2%	22.5%	
Asian	51.2%	30.0%	18.8%	

EHR data: for each patient, the clinical record can be viewed as a sequence of visits $V_1 \ldots V_T$, where each visit record V_t contains diagnosis information x_t. The number of visits for each patient can be varied. The diagnosis information $\boldsymbol{x_t} \in \{0, 1\}^{|\mathcal{D}|}$ is a multi-hot binary vector, where $|\mathcal{D}|$ is the number of unique diagnosis codes, and $\mathcal{D} = \{d_1, d_2, \ldots d_{|\mathcal{D}|}\}$ $x_{t,i} = 1$ indicates that the patient was diagnosed with disease d_i in the t-th visit; otherwise 0.

Genetic data: for each patient, short-read whole genome sequencing (WGS) was measured. We use "Plink" (Purcell, Neale, Todd-Brown, & Thomas, n.d.) to estimate the genetic similarity. By setting a reasonable threshold for genetic similarity, we can construct a connected graph $\mathcal{G}$ for genetically related patients.

2.2 Task

Disease Risk Prediction Task. Based on the above notations, we define our task as follows. Given the patient's diagnosis information x_t, and the patient genetic similarity graph $\mathcal{G}$, we can construct a graph neural network with attention mechanism. The goal of this task is to predict disease risks of the next visit denoted as $\hat{Y}_{t+1}$.

3 Method

Figure 1 provides a comprehensive overview of the proposed MiGAT framework, comprising two primary segments:

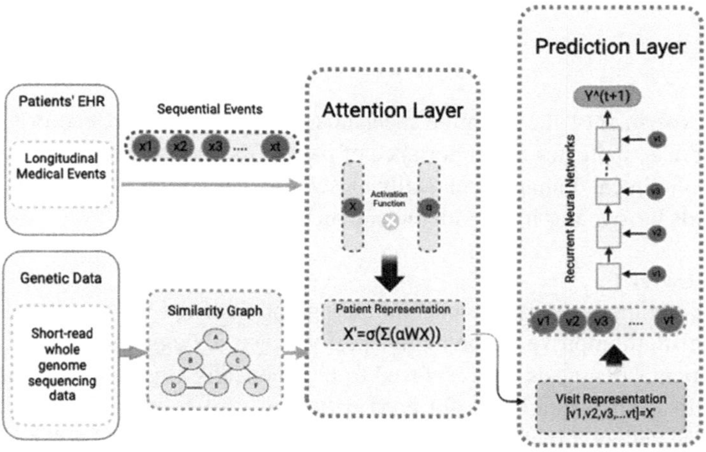

Fig. 1. Framework of MiGAT. MiGAT consists of two parallel processes: 1) Construction of genetic similarity graph: the patient representation X' aggregates health information from genetically related individuals (denoted as A, B, C... in the Similarity Graph); and 2) Sequential medical events X captures each patient's diseases information. The final representation v_t merges both patient's EHR information and individual's EHR information who is connected to the patient via genetic similarity graph to represent a patient's disease status for each visit. Finally, a series of visits is fed into a neural network model for the risk prediction task.

Attention Layer
In building the similarity graph of patients, we use short-read whole genome sequencing data to calculate the pairwise genetic similarity. Patients are linked by edges, indicating significant genetic relatedness surpassing a specified threshold. Each connection can be viewed as resembling a first-degree relative, even though they may not belong to the same family. Each node in the graph represents a patient and node feature contains a patient's diagnosis codes vector, denoted as $\boldsymbol{x_t} \in \{0, 1\}^{|\mathcal{D}|}$, $|\mathcal{D}|$ is the total number of diagnosis codes in the EHR data. We apply attention mechanism to infuse connected patient's disease status information into the patient's disease status we are looking at. We formulate

patient's final disease representation as a linear combination attention coefficient α_{ij}, and the disease features x_j corresponding to them with a non-linear activation function $\sigma(\cdot)$, as following:

$$x_{i'} = \sigma\left(\sum_{j \in \mathcal{N}_i} \alpha_{ij} W x_j\right),$$

where $x_{i'}$ denotes the final representation of patient i, $\mathcal{N}_i$ contains the indices the connected patients of patient i, W is the weight matrix which is applied to each node for a shared linear transformation. The attention coefficient α_{ij} of genetic similarity graph is calculated by the following softmax function:

$$\alpha_{ij} = softmax(e_{ij}) = \frac{\exp(e_{ij})}{\sum_{k \in \mathcal{N}_i} \exp(e_{ik})}, \quad \text{where}$$

$$e_{ij} = \alpha(Wx_i, Wx_j).$$

In our experiments, the attention mechanism is a single-layer feedforward neural network. Here e_{ij} indicates the importance of patient j's disease features to patient i. In the context of clinical domain, genetically closely related patient's current and previous health records have major impact on the patient's disease risk.

Prediction Layer

We train the attention mechanism on similarity graph with a predictive model via attention mechanism to improve the predictive performance. By aggregating final representation of patient's diagnosis status infused from genetically closely related patients, we obtain the final representation v_t for each patient at visit t, indicating all the clinical events for one patient. Finally, the visit representation v_t for patient i will be used as an input to pass to a predictive model for predicting the target label Y_t using recurrent neural networks. In this study, we choose LSTM as the option for experiments. The prediction task is to predict the disease risk of next visit Y_{t+1} at timestep $t+1$ given all the previous visit history up to the current timestep $v_1, v_2, \ldots, v_t$.

Loss Function

We compute the cross entropy loss between the ground truth Y_t and the predicted $\widehat{Y_t}$ using:

$$\mathcal{L} = -\frac{1}{T-1} \sum_{t=1}^{T-1} (y_t^{\mathrm{T}} \log(\hat{y}_t) + (1 - y_t)^{\mathrm{T}} \log(1 - \hat{y}_t))$$

4 Experiments

We conduct experiments on All of Us dataset. In our study, the task of predicting the onset of a specific disease during a patient's subsequent clinical visit is framed as a binary classification challenge. To set this in motion, we pinpoint individuals from the EHR data based on their inaugural documented occurrence of the target disease, designating them as potential case patients for each disease model. A comprehensive total of 202 predictive models were crafted, each tailored to a distinct cancer code, to rigorously test the efficacy and resilience of MiGAT across a wide range of disease forecasts. Taking care to preserve the continuity of patient data, we omitted those whose first documented onset of disease coincided with their first two clinical consultations. We establish a control group for each model that consists of patients, each with the same length of clinical visit history as the corresponding case patient, but without the disease in record, matching the same sex and similar age using R package 'MatchIt' (Ho, Imai, King, & Stuart, 2011). This methodology ensures a balanced representation of prior EHR data for prognosticating ensuing clinical episodes in both the case and control factions.

We select three baseline models that epitomize three distinct categories of predictive models. XGBoost (Chen & Guestrin, 2016) stands as a testament to the prowess of conventional machine learning techniques. On the other hand, RNN represents the recurrent deep neural networks, known for capturing sequential patterns in data. Lastly, GRAM (Choi E., Bahadori, Song, Stewart, & Sun, 2017) is emblematic of models leveraging the attention mechanism, highlighting its capacity to discern and prioritize pertinent features in prediction tasks.

5 Results

Cancer Risk Prediction

Table 2 offers a detailed performance comparison between our novel MiGAT model and three established baseline models. What immediately stands out is the superior predictive prowess of MiGAT, as evidenced by its impressive 6.3% surge in average AUCs over the top-performing baseline. This pronounced enhancement accentuates the significant advantages of integrating multimodal patient data into the diagnostic prediction framework. Figure 2 shows a Kernel Density Estimation (KDE) of AUC distribution of 202 cancer diagnosis codes. MiGAT displays performance advances on overall AUC distribution.

What truly sets MiGAT apart is its intrinsic capability to factor in genetic similarities amongst patients. This feature is paramount in sharpening the accuracy of disease risk predictions, especially when predicting genetically influenced diseases. A closer look at the results pertaining to the 'Neoplasms' diagnosis codes further consolidates this point, with MiGAT markedly outpacing all other contenders.

Table 2. Average performance on cancer risk prediction with 95% confidence interval.

Methods	AUC
XGBoost	0.723 ± 0.009
RNN	0.721 ± 0.007
GRAM	0.779 ± 0.007
MiGAT	$\mathbf{0.828 \pm 0.004}$

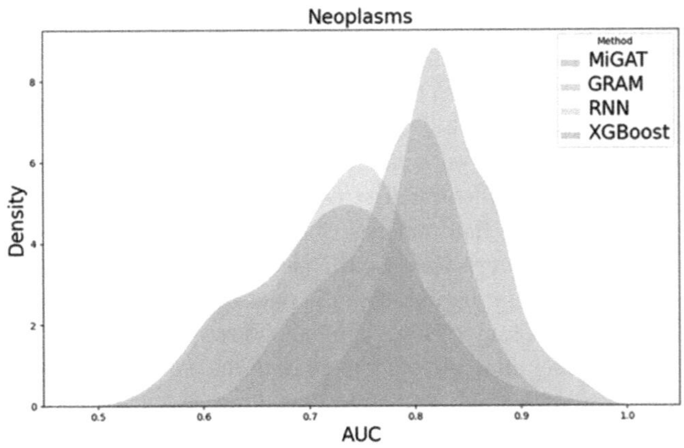

Fig. 2. Kernel Density Estimation (KDE) of AUC distribution of MiGAT and other three baseline models, showing that MiGAT outperforms all other methods across entire diseases.

Case Study

The findings in Table 3 present a comparative analysis of the AUC for three critical diseases: heart failure, lung cancer, and chronic kidney disease (CKD). Notably, our proposed model consistently secures a superior position compared to other benchmark models. This consistent edge underscores the effectiveness and advancements of our model in the realm of disease risk prediction.

Two major factors appear to drive this heightened performance: the incorporation of the genetic similarity graph and the attention mechanism. By leveraging the genetic similarity graph, our model is uniquely positioned to capture underlying genetic predispositions, which traditional models might overlook. This provides a more holistic view of the patient, enabling predictions that account for hereditary risks, and thus potentially enhancing the model's accuracy.

Furthermore, the adoption of the attention mechanism allows our model to dynamically weigh the significance of different patients. This means it can discern which clinical features are more relevant for a given prediction task.

We further analyze the impact of clinical feature on prediction outcomes. Feature importance values grant us insight into which variables play a pivotal role in distinguishing between case and control patients for a particular model. However, their insight remains circumscribed, failing to elucidate interactions among features. Additionally, as these values are consistently positive, they do not reveal whether a feature's presence amplifies or mitigates risk. In our analysis, we delineate the top 10 standout features for three popular diseases: heart failure, lung cancer and CKD It's noteworthy that we adjusted for variables age and sex. Table 4 shows the attention-based important features of our proposed MiGAT model.

The examination of the top-10 ranked features yields some noteworthy findings, emphasizing their relevance to the onset of specific diseases. This observation is not only based on our analysis, but it is also grounded in existing medical literature, which further validates the efficacy and credibility of our predictive model.

Taking the instance of essential hypertension, it stands prominently among the top-10 features. What is compelling about this feature is its established linkage with heart failure and CKD. Numerous scientific studies and clinical observations have identified essential hypertension as a pivotal risk factor for heart failure and CKD.

In essence, these results do not only vouch for the model's current utility but also emphasize its potential as a tool for future medical investigations. The interplay between data-driven predictions and empirical medical knowledge can pave the way for more informed healthcare decisions and strategies

Table 3. Model performance (AUC) for Heart Failure, Lung Cancer, and CKD.

Methods	Heart Failure	Lung Cancer	CKD
XGBoost	0.71	0.61	0.69
RNN	0.76	0.67	0.76
GRAM	0.82	0.72	0.82
MiGAT	**0.84**	**0.76**	**0.86**

Table 4. Feature importance ranking of MiGAT for Heart Failure, Lung Cancer, and CKD.

Rank	Heart Failure	Lung Cancer	CKD
1	Essential hypertension ICD-9 401	Nondependent tobacco Use disorder ICD-9 305.1	Essential hypertension ICD-9 401
2	Hyperlipidemia ICD-9 272.4	Nondependent abuse of drugs ICD-9 305	Diabetes mellitus without mention of complication ICD-9 250.0
3	Chest pain ICD-9 786.5	Chronic airway obstruction ICD-9 496	Disorders resulting from impaired renal function ICD-9 588
4	Cough ICD-9 786.2	Other diseases of lung ICD-9 518	Acute renal failure ICD-9 584
5	Abdominal pain ICD-9 789.0	Swelling, mass, or lump in chest ICD-9 786.6	Diabetes with renal manifestations ICD-9 250.4
6	Low back pain ICD-9 724.2	Pulmonary collapse ICD-9 518.0	Malignant neoplasm of kidney ICD-9 189.0
7	Chronic pain ICD-9 338.4	Other diseases of lung ICD-9 518.8	Unspecified essential hypertension ICD-9 401.9
8	Obesity ICD-9 278.0	Dyspnea ICD-9 786.0	Liver disease ICD-9 570

(continued)

Table 4. (*continued*)

Rank	Heart Failure	Lung Cancer	CKD
9	Gastroesophageal reflux disease without esophagitis ICD-9 530.81	Chest pain ICD-9 786.5	Anemia ICD-9 285.9
10	Dyspnea ICD-9 786.0	Acute respiratory failure ICD-9 518.81	Obesity ICD-9 278.0

6 Conclusions

This study illustrates the transformative potential of fusing multiple modalities of patient data within an attention-based graph neural network framework, specifically for predicting disease risk. The prowess of our MiGAT model, armed with its ability to adeptly process both EHR and pertinent genetic data, offers a promising avenue in the arena of medical diagnostics. Beyond its robust performance across diverse diagnoses, MiGAT's agility in discerning attention weights stands as a testament to its innovative design, providing valuable insights into the interplay of various disease indicators. Moreover, our experiments demonstrate that MiGAT employs attention weights, aiding in identifying the significance of features and uncovering correlations linked to diseases.

References

Che, L., Wang, J., Zhou, Y., Ma, F.: Multimodal federated learning: a survey. Sensors 6986 (2023)
Chen, T., Guestrin, C.: XGBoost: a scalable tree boosting system. In: Proceedings of the 22nd ACM SIGKDD International Conference on Knowledge Discovery and Data Mining (2016)
Choi, E., et al.: Multilayer representation learning for medical concepts. In: ACM SIGKDD International Conference on Knowledge Discovery and Data Mining (2016)
Choi, E., Bahadori, M., Song, L., Stewart, W.F., Sun, J.: GRAM: graph-based attention model for healthcare representation learning. In: SIGKDD (2017)
Gui, N., Ge, D., Hu, Z.: AFS: an attention-based mechanism for supervised feature selection. In: The Thirty-Third AAAI Conference on Artificial Intelligence (2019)
Guidotti, R., Monreale, A., Ruggieri, S., Turini, F., Giannotti, F., Pedreschi, D.: A survey of methods for explaining black box models. ACM Comput. Surv., 1–42 (2018)
Ho, D., Imai, K., King, G., Stuart, E.A.: MatchIt: nonparametric preprocessing for parametric causal inference. J. Stat. Softw., 1–28 (2011)
Huang, X., et al.: Applying family analyses to electronic health records to facilitate genetic research. Bioinformatics, 635–642 (2018)
Huang, X., et al.: E-Pedigrees: a large-scale automatic family pedigree prediction application. Bioinformatics, 3966–3968 (2021)
Li, R., Ma, F., Gao, J.: Integrating multimodal electronic health records for diagnosis prediction. In: AMIA *Annual Symposium* Proceedings (2022)
Ma, F., You, Q., Xiao, H., Chitta, R., Zhou, J., Gao, J.: KAME: knowledge-based attention model for diagnosis prediction in healthcare. In: CIKM (2018)
Polubriaginof, F.C., et al.: Disease heritability inferred from familial relationships reported in medical records. Cell, 1692–1704 (2018)

Purcell, S., Neale, B., Todd-Brown, K., Thomas, L.F. (n.d.). http://pngu.mgh.harvard.edu/purcell/plink/
Ramirez, A.H., et al.: The all of us research program: data quality, utility, and diversity. Cell (2022)
Veličković, P., Cucurull, G., Casanova, A., Romero, A., Liò, P., Bengio, Y.: Graph attention networks. ICLR (2018)
Wang, S., et al.: Estimating genetic similarity matrices using phylogenies. J. Comput. Biol., 587–600 (2021)
Zhao, X., Wang, X., Yu, F., Shang, J.: UniMed: multimodal multitask learning for medical predictions. In: BIBI 2022 (2022)

Integrative Analysis of Amyloid Imaging and Genetics Reveals Subtypes of Alzheimer Progression in Early Stage

Bing He[1], Neel Sangani[3], Ruiming Wu[2], Pradeep Varathan[1], Alice Patania[4], Shannon L. Risacher[3], Kwangsik Nho[3], Liana G. Apostolova[3], Andrew J. Saykin[3], Li Shen[2], and Jingwen Yan[1,3](✉)

[1] Indiana University Luddy School of Informatics, Computing and Engineering, Indianapolis, IN, USA
jingyan@iupui.edu
[2] Perelman School of Medicine, University of Pennsylvania, Philadelphia, PA, USA
[3] Indiana University School of Medicine, Indianapolis, IN, USA
[4] Department of Mathematics and Statistics, University of Vermont, Burlington, VT, USA

Abstract. Alzheimer's disease (AD) progresses along a continuum and begins many years before symptom onset. Amyloid beta and neurofibrillary tangles are two AD hallmarks that precede changes in cognitive performance and have been utilized in the current framework for classification and staging of AD patients. This framework, however, relies on the dichotomous classification of individual biomarkers (e.g., amyloid positive or negative) and therefore cannot capture the full spectrum of AD progression. In this study, we propose a multi-factorial pseudotime approach that integrates heterogeneous genotype and amyloid imaging data for continuous staging of AD. Our new staging score demonstrated strong capability in differentiating early-stage subjects with distinct progression rates, suggesting its potential in supplementing existing diagnostic framework for early risk stratification.

Keywords: Alzheimer's Disease · imaging genetics · early risk assessment

1 Introduction

Alzheimer's disease (AD) is an irreversible neurodegenerative disorder with a long prodromal phase. It is highly heritable with 58–79% heritability estimated from twin studies [1]. Amyloid-beta plaques (A) and neurofibrillary tangles (T) are two neuropathological hallmarks of AD, which are believed to precede structural neurodegenerative changes and cognitive decline in the brain. This hypothesis laid the foundation for the current AT classification framework, which has been used to assist with early detection, staging, and patient screening for clinical trials [2]. Particularly, amyloid PET amyloid plays a

J. Yan—Supported by NSF award 1942394 and NIH grant R21AG072101, U19AG074879, R01LM013463 and U01AG068057.

J. Finkelstein et al. (Eds.): AIME 2024, LNAI 14845, pp. 204–211, 2024.
https://doi.org/10.1007/978-3-031-66535-6_23

central role in this framework as one of the earliest pathological features and have been widely used to provide biological evidence. Although amyloid beta accumulation takes place on a continuum, amyloid-PET scans are commonly dichotomized as pathological (positive) versus normal (negative) in the current AT system [2]. Moreover, individuals with a high level of amyloid burden could remain cognitively intact, suggesting that amyloid alone is not sufficient for early risk assessment.

To address these problems, we propose a multi-factorial pseudotime approach to integrate the genotype and amyloid imaging data for continuous staging of AD patients. Our pseudotime approach integrates information from imaging and genetic data through the fusion of two modality-specific similarity networks rather than direct concatenation, which further avoided the heterogeneity problem. Subsequently, graph-based pseudotime analysis was performed to generate a risk (or staging) score for each subject indicating its risk of developing AD. This risk score showed strongest association with multiple cognitive performances when compared to age and conventional diagnostic biomarkers used in the current AT framework. More importantly, it also demonstrated capability in stratification of MCI subjects and in particular early MCI subjects. MCI subgroups with high/low risk scores showed significant difference in time to AD onset and declining rate in longitudinal cognitive performances, which is otherwise not observed between diagnostic groups. Taken together, these results suggested a great potential of our risk score in patient stratification for clinical trials.

2 Methods

2.1 Data

Genotype and amyloid-PET imaging data were downloaded from the Alzheimer's disease Neuroimaging Initiative (ADNI) database (http://adni.loni.usc.edu/). Amyloid imaging data has been pre-processed [3] and imputed genotype data has gone through quality control steps as shown in our previous studies [4]. All the regional Amyloid deposition was re-normalized using whole cerebellum as reference region. Finally, we have amyloid measurement in 68 cortical regions of interest (ROIs). For genotype data, we focus on 27 single nucleotide polymorphisms (SNPs) with significant association with AD [5]. Effect of age, sex and education on amyloid measures were regressed out with the weight derived from healthy controls. In total, we have 950 subjects with both genotype and amyloid imaging data, including 259 cognitive normal (CN), 298 early mild cognitive impairment (EMCI), 208 late MCI (LMCI) and 185 AD patients. Detailed demographic information of gender, age and years of education are in Appendix A Table 1.

2.2 Similarity Network Fusion

We first integrated genotype and amyloid imaging data through similarity network fusion [6]. Two distance matrices were generated for imaging data and genotype of 27 SNPs using Euclidean distance and genetic relationship matrix (GRM) respectively [7]. These matrices were transformed into similarity matrices using a scaled exponential similarity kernel, which further went through similarity network fusion, leading to one similarity network [6] (details in Appendix A).

2.3 Pseudotime Analysis

We performed pseudotime analysis using PHATE, which learns a two-dimensional trajectory embedding of the fused similarity network [8]. The relative position of each subject projected to the principal curve is denoted as pseudotime, ranging between 0 to 1, which can be interpreted as an estimation of disease risk. In this project, we extracted the principal curve and estimated the pseudotime of each subject using Slingshot [9]. Subjects with similar pseudotime risk are expected to have similar amyloid and genetic profile.

3 Results

3.1 Pseudotime Comparison Across Diagnostic Groups

Shown in (Fig. 1(a)) is the disease progression trajectory generated using fused similarity network with color indicating the scaled pseudotime from 0 to 1. We observed the significant difference of pseudotime distribution across diagnosis groups (Fig. 1(b)), where cognitive normal have the lowest pseudotime and AD patients have the highest. This result suggests the potential of multi-modal pseudotime to serve as the risk score for continuous staging of AD progression.

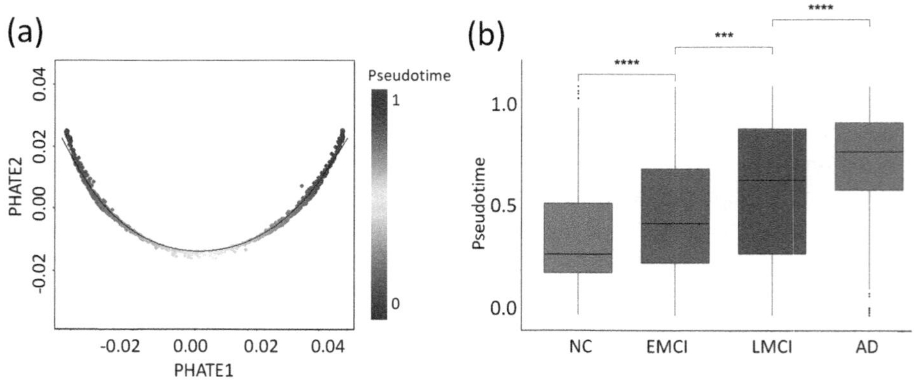

Fig. 1. (a) Inferred progression trajectory from PHATE. Each dot is a subject, colored by the pseudotime estimated using SlingShot. (b) Pseudotime risk distribution across diagnosis groups. Differences between diagnostic groups are all significant.

3.2 Longitudinal Progression Analysis of MCI Subgroups

It is worth mentioning that noticeable changes in clinical cognitive test won't appear until mid to late stages. It would be ideal if we can estimate the cognitive change in an earlier stage such that effective intervention could be applied. Toward this, we investigated whether estimated pseudotime is associated with longitudinal cognitive decline during

AD progression. In particular, we focus on MCI groups for the purpose of early detection. MCI subjects, including EMCIs and LMCIs, were divided into 2 subgroups: pseudotime ≥ 0.5 and <0.5.

Leveraging the cognitive test scores across follow-up visits, we examined the association of baseline pseudotime with longitudinal cognitive performance using linear mixed model. In Fig. 2, among 7 cognitive tests, ADAS11, ADAS13, MMSE and RAVLT immediate were found with significant different progression rate in two pseudotime subgroups. Cognitive performance in MCIs with higher pseudotime tend to decline much faster than in those with lower pseudotime. Such difference was however not observed between EMCIs and LMCIs defined using current AT framework, suggesting the great potential of fused pseudotime as risk score for candidate screening in clinical trials.

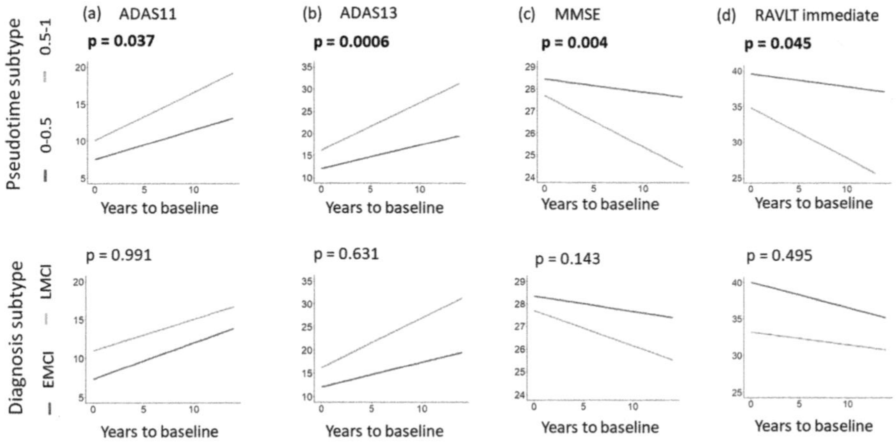

Fig. 2. Longitudinal progression of cognitive performance in MCI subgroups defined by pseudotime (top) and existing diagnosis (bottom).

3.3 Survival Analysis of MCI Subgroups

We performed the Kaplan-Meier survival analysis to examine whether MCI subgroups by fused pseudotime have different risk of developing AD. The event is defined as the conversion to AD and censorship as the MCI subject not converted up to the last visit. While EMCIs and LMCIs did demonstrate significant difference in survival risk ($p = 8.06e-12$), pseudotime showed substantially improved differentiation power with $p = 3.12e-22$. More importantly, even within EMCIs, pseudotime risk score still captures a subset with significantly higher risk of survival with $p = 1.62e-10$. Details are in Appendix B Fig. 2.

4 Conclusion

We proposed a novel computational approach to derive a AD risk score through integrative analysis of heterogeneous amyloid imaging and genotype data. Our risk score at baseline is strongly associated with longitudinal cognitive decline, in MCI groups and also in EMCI subsets. Considering that cognitive performance has been commonly used to assess the AD progression, but noticeable changes won't occur until late stage, our results suggest its great potential for early detection and progression subtyping.

Appendix A: Methods

Demographic Information

See Table 1.

Table 1. Demographic information of Amyloid participants

Subjects	CN	EMCI	LMCI	AD
Number	259	298	208	185
Gender(M/F)	126/133	172/126	123/85	111/74
Age(mean ± std)	75.52 ± 6.98	72.04 ± 7.32	74.47 ± 8.41	75.51 ± 7.99
Educ(mean ± std)	16.55 ± 2.61	16.13 ± 2.63	16.25 ± 2.81	15.84 ± 2.69

Similarity Network Fusion

First, we generated two subject similarity matrices using pre-processed amyloid data and genotype of 27 SNPs respectively. Suppose we have subjects $\{x_1, x_2, \ldots, x_n\}$, their similarity matrix can be denoted as $W^{(A)} \subset R^{n \times n}$. For the amyloid imaging data, Euclidean distance between subjects x_i and x_j was first calculated as $\rho_A(x_i, x_j)$. Subject similarity matrix for amyloid was obtained by transforming the distance matrix with a scaled exponential similarity kernel (Eq. 1). Here, $\varepsilon_{i,j}$ helps eliminate the scaling problem and μ is a hyper parameter set to 0.5 by default [6]. For genotype, genetic similarity matrix $W^{(G)} \subset R^{n \times n}$ was calculated using genetic relationship matrix (GRM), which is the covariance matrix calculated from the SNP information of the individuals, i.e., from the minor allele counts

$$W_{ij}^{A} = \exp(-\frac{\rho_A^2(x_i, x_j)}{\mu \varepsilon_{i,j}}) \quad (1)$$

Each similarity matrix was normalized to form **P**, representing the full kernel containing all information from one subject to all others. This normalization eliminates the scaling issue of self-similarity in diagonal entries and ensures that $\sum_j P(i, j) = 1$

(Eq. 2), guaranteeing that each subject's self-similarity is consistently higher than its similarity to other neighbors. In addition, another local affinity matrix **S** was generated to encode the relative similarity of each subject to their **K** most similar neighbors. As such, we obtained four matrices $P^{(A)}$, $S^{(A)}$, $P^{(G)}$ and $S^{(G)}$ for amyloid and genotype data respectively

$$P_{i,j} = \begin{cases} \frac{W_{ij}}{2\sum_{k\neq i} W_{i,k}}, & j \neq i \\ \frac{1}{2}, & j = i \end{cases}, \qquad S_{i,j} = \begin{cases} \frac{W_{ij}}{2\sum_{k\in N_i} W_{i,k}}, & j \in N_i \\ 0, & otherwise \end{cases} \tag{2}$$

Next, we performed the network fusion of two similarity matrices. This is an iterative process where both networks keep getting updated until convergence. Final fused network is expected to represent the subject relationships supported by both amyloid imaging and genotype data. Let $P^A_{t=0} = P^A$ and $P^G_{t=0} = P^G$ be the initial matrices when t = 0. The fusion process will iteratively update two similarity matrices, as shown in Eq. 3. Here, the alternating multiplication of the squared KNN similarity in two modalities essentially combines the local information across modalities. It helps reinforce the shared information and thereby achieves a balanced fusion of two modalities. Final fused network $P^{(F)}$ is simply the average of P^A_{t+1} and P^G_{t+1}

$$P^A_{t+1} = S_A \times P^G_t \times (S_A)^T, \qquad P^G_{t+1} = S_G \times P^A_t \times (S_G)^T \tag{3}$$

Appendix B

Association of Pseudotime with Cognitive Performances

We examined the correlation of estimated pseudotime with clinical cognition scores, including ADAS11, ADAS13, MMSE, RAVLT delayed recall, RAVLT immediate, trailmaking A, and trailmaking B. These cognitive performances have been commonly used in clinical routines to assess the progression of AD. When compared with pseudotime derived using amyloid data only, pseudotime from fused network showed much stronger correlation with all cognitive performances (Fig. 1), suggesting multi-factorial cause of cognitive decline in AD. It also outperformed amyloid composite SUVR, a driving factor of AD diagnosis in the current AT framework

Survival Risk Stratification in MCI and EMCI Groups

See Fig. 2.

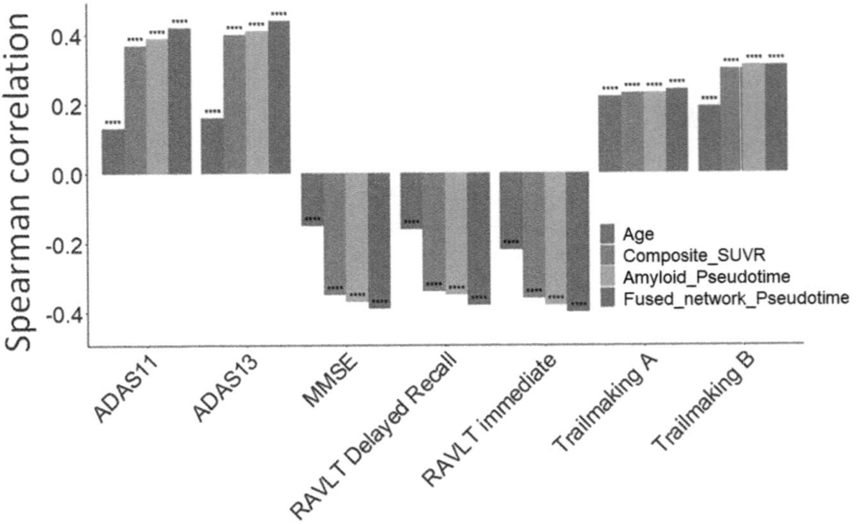

Fig. 1. Correlation of estimated pseudotime, amyloid composite SUVR and age with performance in clinical cognitive tests.

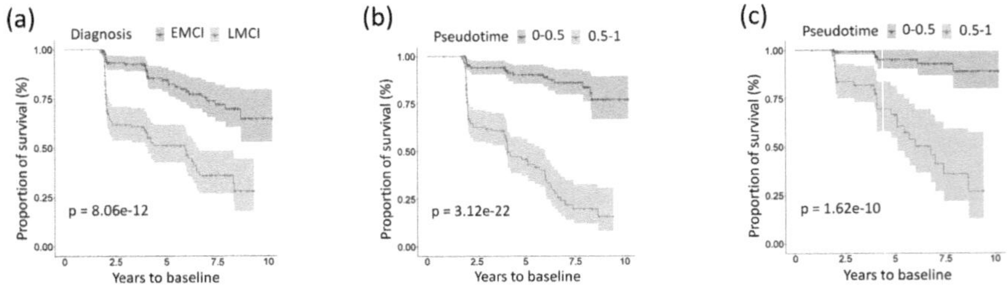

Fig. 2. Survival difference between (a) MCI diagnosis groups (EMCI vs LMCI), (b) MCI pseudotime subgroups (EMCI + LMCI), and (c) EMCI pseudotime subgroups.

References

1. Sims, R., Hill, M., Williams, J.: The multiplex model of the genetics of Alzheimer's disease. Nat. Neurosci. **23**(3), 311–322 (2020)
2. Counts, S.E., Ikonomovic, M.D., Mercado, N., Vega, I.E., Mufson, E.J.: Biomarkers for the early detection and progression of Alzheimer's disease. Neurotherapeutics **14**(1), 35–53 (2017)
3. Landau, S.M., et al.: Amyloid-β imaging with Pittsburgh compound b and florbetapir: comparing radiotracers and quantification methods. J. Nucl. Med. **54**(1), 70–77 (2013)
4. Nho, K., et al.: Genome-wide transcriptome analysis identifies novel dysregulated genes implicated in Alzheimer's pathology. Alzheimer's Dement. **16**(9), 1213–1223 (2020)
5. Desikan, R.S., et al.: Genetic assessment of age associated Alzheimer disease risk: development and validation of a polygenic hazard score. PLoS Med. **14**(3), e1002258 (2017)
6. Wang, B., et al.: Similarity network fusion for aggregating data types on a genomic scale. Nat. Methods **11**(3), 333–337 (2014)
7. Speed, D., Hemani, G., Johnson, M.R., Balding, D.J.: Improved heritability estimation from genome-wide SNPs. Am. J. Hum. Genet. **91**(6), 1011–1021 (2012)

8. Moon, K.R., et al.: Visualizing structure and transitions in high-dimensional biological data. Nat. Biotechnol. **37**(12), 1482–1492 (2019)
9. Street, K., et al.: Slingshot: cell lineage and pseudotime inference for single-cell transcriptomics. BMC Genomics **19**, 1–16 (2018)

Modular Quantitative Temporal Transformer for Biobank-Scale Unified Representations

Mátyás Antal[1(✉)], Márk Marosi[1(✉)], Tamás Nagy[1,2], András Millinghoffer[1], András Gézsi[1], Gabriella Juhász[2], and Péter Antal[1]

[1] Budapest University of Technology and Economics, Magyar Tudósok Körútja 2., 1117 Budapest, Hungary
matyas.antal@edu.bme.hu, marosi@mit.bme.hu
[2] Semmelweis University, Nagyvárad tér 4., 1089 Budapest, Hungary

Abstract. Transformers provide a novel approach for unifying large-scale biobank data spread across different modalities and omic domains. We introduce Modular Quantitative Temporal Transformer (MQTT), a modular architecture for multimodal data that offers a robust fusion mechanism for biobank data, which can integrate (1) systematically missing modality data, (2) quantitative data, and (3) longitudinal data. We apply the model for the fusion of personal, laboratory, diagnostic, clinical, and drug prescription data from the UK Biobank. We investigate MQTT's modular representations, focusing on multimorbidity and polypharmacy. We demonstrate its applicability by a novel stratification of major depressive disorder resulting in subtypes related to treatment-resistant depression with novel genes significant at the GWAS level.

Keywords: Transformer · pre-training · multimodal data fusion · patient stratification · UK Biobank · major depressive disorder · drug prescription

1 Introduction

Large-scale biobanks, such as the UK Biobank, provide an unprecedented amount of heterogeneous data to learn latent patient representations and stratify patients and subtype diseases [2]. However, multimodal data fusion with systematically missing modalities is still an open problem, especially the integration of time series data, such as the irregularly sampled drug prescription data [19].

Transformers have proven to be powerful candidates for multimodal data fusion and time series data analysis in biomedicine [26]. We focus on three open issues in the multimodal application of transformers: **autonomous treatment of modalities**, **mixed numeric-nominal data types**, and **longitudinal data**. This article introduces a novel transformer model, MQTT, to tackle

M. Antal and M. Marosi—These authors contributed equally to this work.

J. Finkelstein et al. (Eds.): AIME 2024, LNAI 14845, pp. 212–226, 2024.
https://doi.org/10.1007/978-3-031-66535-6_24

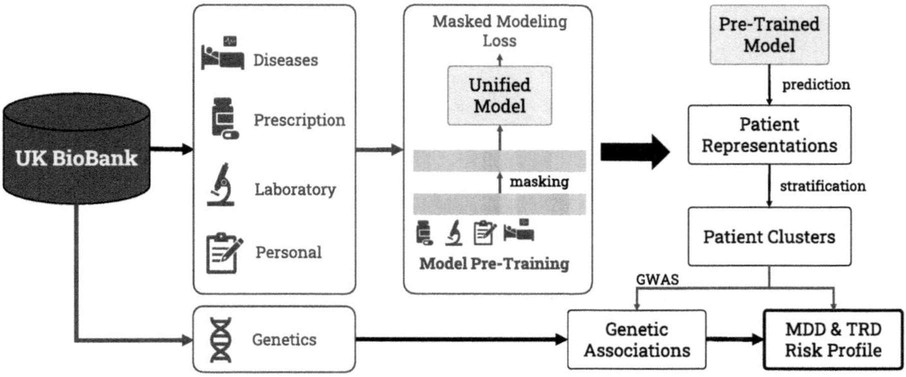

Fig. 1. Overview of Data and Methods The MQTT model integrates diverse data modalities, including personal, laboratory, prescription, and disease information, while genetic data is processed separately. Clustering and GWAS analyses are conducted on latent representations to investigate MDD subtypes and their genetic underpinnings.

these challenges: (1) the modular availability of data (i.e., systematically missing modalities), (2) real-valued measurements, and (3) longitudinal data, including precise quantitative temporal information about the occurrence of the events. MQTT can integrate demographic, metabolomic, diagnostic, clinical, and drug prescription data with varying levels of availability at biobank scale, which we demonstrate using data from the UK Biobank [2]. We evaluate the representation learning ability of MQTT in patient stratification focusing on multimorbidity (MM) [15], polypharmacy (PP) [10], and the subclassification of major depressive disorder (MDD) [7], primarily focusing on the subtype of treatment-resistant depression (TRD) [6]. Figure 1 provides an overview of the data sets, the MQTT model architecture, and the evaluation workflow of our study. Following convention in the field, we use the expressions multimodal and cross-domain interchangeably.

2 Earlier Works

The multimorbidity analysis of large-scale biobanks allows the identification of novel patient stratifications and shared molecular background of clusters of patients in the latent space of patient representations [13,21]. Latent representations based on the metabolomic data from the UK Biobank have also been shown to increase performance in downstream predictive tasks [1]. In addition to non-longitudinal data, drug prescription time series data is an additional substantial resource, but its use is frequently hindered by heterogeneous coding, limited availability, and its longitudinal nature [5,14,24,25].

Transformers were rapidly adopted for multimodal and temporal biomedical data [16,17,20,22,23]. Liu et al. [18] proposed the ARMOUR architecture, using modality-specific Transformer encoders for fusing textual data. Coincidentally, the encoding scheme for real-valued inputs proposed in our work has also been concurrently suggested by Golkar et al. [9] in their research.

3 Methodology

The Transformer architecture was adapted for learning from diverse modalities in multiple ways [26]. The proposed extensions aim to overcome the limitations that hinder the application of transformers to the data types typically encountered in health-related databases. The conventional Transformer design is primarily geared towards processing textual content, or more broadly, information that manifests as sequences or sets of distinct tokens. In our approach, we redefine tokens to represent either events or measurement values, and we extend the embedding layer to accommodate the integration of temporal and quantitative data. We introduce multiple encoders to distinguish between the impacts of different modalities and to mitigate the issue of missing modalities. Our model is pretrained with a Masked Language Modeling (MLM) objective, utilizing the initial element of every sequence for representation. Figure 2 shows the MQTT architecture, for an in-depth description of the model architecture, please refer to the Appendix A.

4 Experiments

The MQTT architecture can integrate personal information, laboratory parameters, disease occurrences, and drug prescription data. Computational and technical limitations currently exclude the uniform integration of genome-wide polymorphic or sequencing genetic data, so we use it to investigate and evaluate the learned representations. In the qualitative part of our experiments, we used dimensionality reduction methods. We investigated basic descriptors, such as sex, dominant diseases, and the continuity of the MM and PP scores in low-dimensional maps. In the quantitative experiments, we clustered the participants using the learned patient-level latent representations and performed GWAS analyses for these clusters (we performed clustering and GWAS for the participants after genetic quality control with a sample size of 330k).

Additionally, we performed multiple MDD case-only GWAS analyses to explore better the differential genetics of these MDD-filtered clusters. To support the exploration of MDD subtypes, we used a novel treatment-resistant depression (TRD) score based on the drug prescription data [4], which were used to characterize the MDD-filtered clusters, and we also performed a restricted MDD case-only GWAS for participants with existing TRD score.

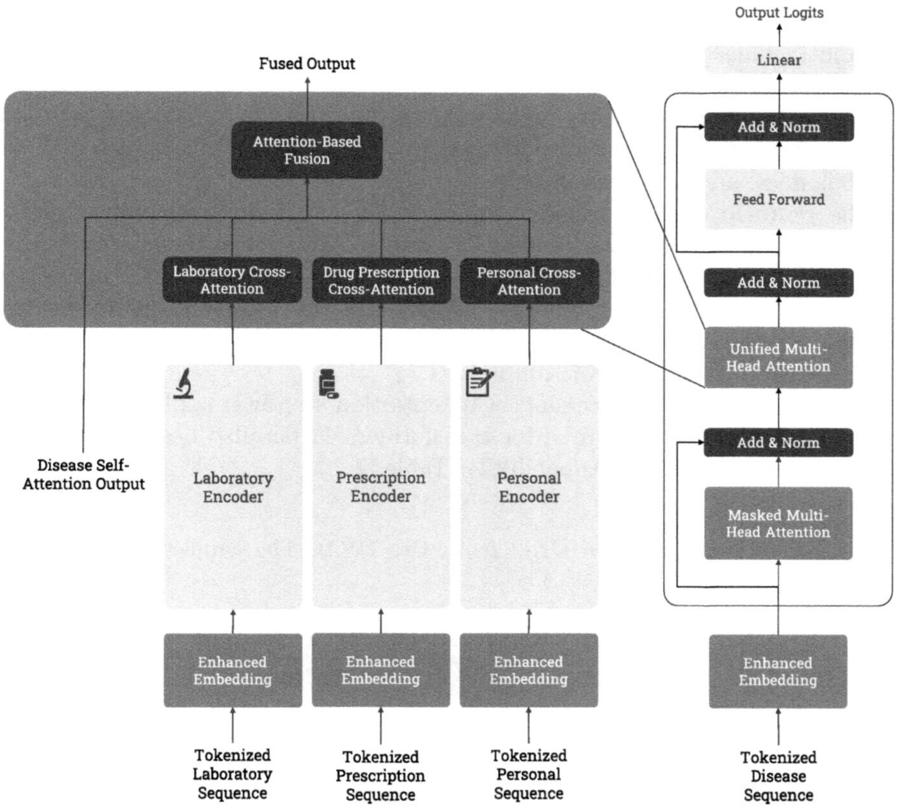

Fig. 2. Overview of the MQTT architecture The MQTT model integrates diverse data modalities, including personal, laboratory, prescription, and disease information. Our tokenized sequences also contain temporal and scalar data.

4.1 Data

We categorized the UKB data into the following four distinct modalities.

The **disease** modality is made from the first occurrence of 3-character ICD-10 codes (International Classification of Diseases, 10th Revision) for each patient. The total number of codes is 1127.

The **drug prescription** modality encompasses primary care prescription data sourced from 4 system providers. The data spans from 1990 to 2016, and each provider used a combination of coding systems (British National Formulation (BNF), read_v2, and dm+d). For 3 providers, free-text information was also available about the prescribed drug and quantity. As a straightforward mapping between the coding systems was not available, and to provide a unified pipeline, we extracted the chemblID [8] of the contained compounds from the free-text information using PRESNER [4]. This public name entity recognition pipeline was developed using the same UK Biobank prescription data. We removed compounds whose parent was contained in the same free-text identifier or had fewer than 10 occurrences. The final count of unique chemblID-s was 981.

The **personal** modality contains information recorded in the assessment centers and includes UKB fields containing basic descriptors and various lifestyle factors.

The **laboratory** modality contains data obtained from biological samples. This includes most UKB fields from blood count, blood biochemistry, NMR metabolomics, and urine assays.

The **genetic data** originally contained 805,161 genotyped and a further 96,254,167 imputed variants for 488,377 samples. We applied the quality control steps described in [3] to this dataset, and kept only biallelic single nucleotide polymorphisms (SNPs) for multiple GWAS (for the complete quality control pipeline, please refer to the Appendix C). Thus, the final dataset contained 6,277,868 variants for 334,257 samples.

After the initial preprocessing, a tokenization step was applied, converting the data to the format required for transformers. A detailed description of our approach can be found in Appendix B (Table 1).

Table 1. Overview of Data Used from the UKB. The samples of the different modalities are missing systematically.

Data Block	No. of Variables	No. of Samples
Diseases	1127	502,383
Prescriptions	981	201,879
Personal	41	502,383
Laboratory	316	–
NMR	249	121,724
Other	67	501,195
Genetic data	6,277,868	334,257

4.2 GWAS Calculations

Association relations between genetic variants and phenotypes were assessed using linear and logistic regression models, for continuous and binary variables, respectively. In both cases, gender, age, an indicator variable for the genotyping array used, and the coordinates corresponding to the top 10 principal component vectors of genotypic data were added to the models as covariates.

4.3 Clustering of the Latent Representations

We clustered all participants in the 312-dimensional space defined by their joint latent representations. We evaluated several clustering methods, including k-means, agglomerative hierarchical clustering, and HDBSCAN, each across a wide range of parameters. Finally, we used the k-means clustering algorithm with four clusters, as it showed the best coherence with the minimum cluster size surpassing 10% of all individuals and achieving the highest silhouette score.

In the case of GWAS analyses, the posterior log-odds of the cluster memberships were used as target variables, where the posterior probability of cluster membership for the ith participant and the jth cluster was the normalized exponentiated negative Euclidean distance between the point that corresponds to the ith participant and the jth cluster's cluster center in the space defined by the latent representations:

$$Pr_j(i) = \frac{exp(-||p_i - c_j||)}{\sum_k exp(-||p_i - c_k||)},$$

where p_i and c_j are the latent vectors of the ith participant and the center of the jth cluster, respectively.

5 Results

Following the pre-training phase using masked modeling loss, the model generates sequence representations. These are obtained as the first token of the output from each modality. Such representations are derived for all encoders (including personal data, laboratory results, and drug prescriptions) and the decoder pertaining to disease. The disease decoder uniquely offers a fused representation that integrates information from the other modalities.

To facilitate visualization and analysis of these complex, high-dimensional latent spaces, we employ Uniform Manifold Approximation and Projection (UMAP). UMAP serves to reduce the dimensionality, thus enabling a more manageable graphical depiction of individual and combined latent spaces. Color-coding these visual representations with various descriptors can give insights into the structure and characteristics of the learned representations. We show the latent spaces color-coded by multimorbidity (MM) in Fig. 3. For further plots please refer to our GitHub repository in Sect. 7.

5.1 Patient Stratification and Its Genetic Associations

To investigate the biological relevance of the participant-level latent representations, we first applied the k-means algorithm for clustering all the participants. This yielded four distinct clusters, as determined by various internal evaluation metrics. Notably, reflecting the central importance of MDD in the multimorbidity landscape. In the first two clusters, the risks of both MDD and TRD were significantly reduced. Conversely, in the third cluster, MDD and TRD risks were elevated. In the fourth cluster, however, only the risk of MDD was increased, with no elevation in TRD risk, please refer to the Appendix D for the exact data.

To investigate the genetic underpinnings of these clusters, we conducted a GWAS analysis on the UK Biobank cohort (N = 334,257), utilizing the posterior

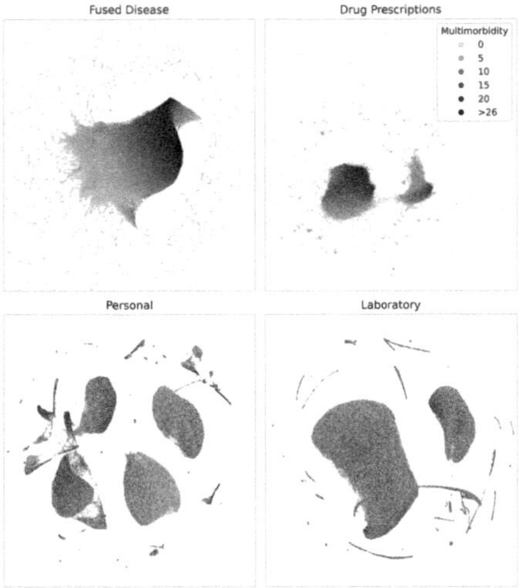

Fig. 3. Latent Representations of Pre-trained Models. Each plot shows a UMAP from the representations learned by the respective modality. The color code indicates the MM score of each patient, calculated by the total number of different ICD10 code onsets.

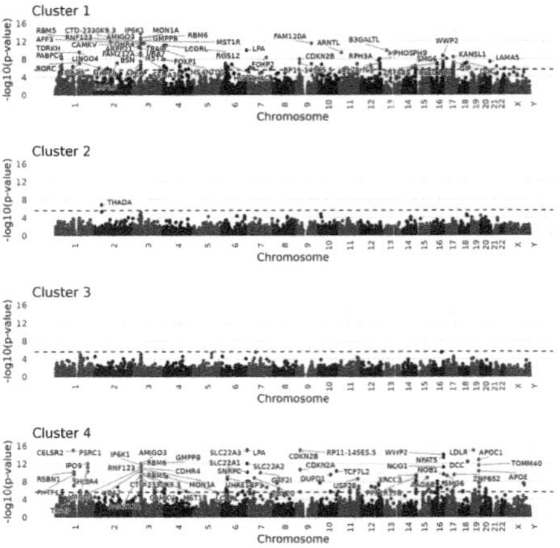

Fig. 4. Gene-level Manhattan plots of the four clusters. The top 50 genome-wide significant genes are indicated with labels, and the significance threshold is depicted with a dashed dark red line.

log-odds of the cluster memberships as target variables in the analysis. Based on the SNP-level summary statistics, we performed MAGMA gene-level analysis to identify putative significant genes. Figure 4 presents the gene-level Manhattan plots for each cluster. Clusters 1 and 4 exhibited the highest number of significant genes, with Cluster 1 identifying 149 genes and Cluster 4 revealing 169 genes. The significant genes in Cluster 1 are associated with various DNA damage response mechanisms, including depurination, depyrimidination, base-excision repair, DNA damage/telomere stress-induced senescence, and packaging of telomere ends. In contrast, the genes identified in Cluster 4 are implicated in diverse biological processes, such as cholesterol metabolism, spermidine transport, dopamine uptake, and toxin transmembrane transporter activity.

Next, we conducted a specific GWAS using only MDD cases within the four clusters, aiming to uncover the unique genetic characteristics of these MDD subtypes. However, the gene-level results from this case-only GWAS revealed only moderate significance. Specifically, it identified a single significant gene in each Cluster 1 and 2.

In addition, we conducted a focused GWAS limited to MDD cases with available TRD scores (N = 10,148), specifically considering only participants with drug prescription data to compute the TRD score as the target variable. This gene-level analysis identified a unique significant gene, *TMPRSS6* (Holm's adjusted p-value = 0.0007). The *TMPRSS6* gene encodes a protein called matriptase-2, which plays a crucial role in the body's iron regulation. Matriptase-2 is a part of the signaling pathway that controls the levels of the protein hepcidin, a key regulator of iron balance. Mutations in the *TMPRSS6* gene can lead to iron-refractory iron deficiency anemia due to reduced functional matriptase-2. Previous research has established a link between iron-deficiency anemia and depression [11,12]. Our results, however, suggest a link between *TMPRSS6* and the regulation of iron homeostasis in the context of treatment resistance in MDD.

6 Conclusion

Biobanks offer an unprecedented repertoire of multimodal data, including non-longitudinal and longitudinal data; however, modalities such as the metabolomics and medications data in the UK Biobank are frequently systematically missing. We proposed a novel modular transformer architecture, MQTT, which extends transformers to the novel domains of real-world biobanks. MQTT's modular architecture provides a simultaneous view of latent representations for the modalities and their balanced joint representation and it provides a unified solution for mixed/real-valued and temporal data. MQTT allows the exploration of conditions where the unified analysis of complex, quantitative, temporal drug-disease patterns is necessary, such as in treatment-resistant major depressive disorder. Our research illustrates the biological significance of the joint latent representations through patient stratification, revealing distinct risk profiles for MDD and TRD across four clusters using the UK Biobank. Further, our GWAS analysis elucidates a molecular background for these clusters, particularly highlighting the role of the *TMPRSS6* gene in iron homeostasis and its potential link to treatment resistance in MDD.

In conclusion, this work demonstrates that transformer architectures can be effectively adapted to manage the complex, heterogeneous data typical of real-world biobanks, addressing modular, quantitative, and temporal aspects. Through the development of the MQTT model, our study shows potential for improved patient stratification and disease subtyping by leveraging large-scale, multimodal data, especially utilizing drug prescription data at scale to explore conditions such as treatment-resistant depression and major depressive disorder. Notably, this adaptation has been achieved with a modest computational burden, suggesting that such advanced data integration and analysis can be more accessible to a wider range of research and clinical settings.

7 Code Availability

The code for this study is publicly available to facilitate replication of our results and further research. It includes preprocessing scripts for the UK Biobank data, the MQTT model implemented in PyTorch and Huggingface Transformers, and scripts for patient stratification, and visualization of latent representations. For access to the code, detailed documentation, and more plots please visit our public GitHub repository at https://github.com/Talma18/AIME-2024-UKB.

Acknowledgments. This study was supported by the Hungarian National Research, Development, and Innovation Office K 139330, K 143391 and PD 134449 grants; the European Union (EU) Joint Program on Neurodegenerative Disease (JPND) Grant SOLID JPND2021-650-233; the Hungarian Brain Research Program (grant: 2017-1.2.1-NKP-2017-00002) and the Hungarian Brain Research Program 3.0 (NAP2022-I-4/2022); the National Research, Development, and Innovation Fund of Hungary under Grant TKP2021-EGA-02 and TKP2021-EGA-25, the European Union project RRF-2.3.1-21-2022-00004 within the framework of the Artificial Intelligence National Laboratory.
This research has been conducted using the UK Biobank Resource under Application Number 1602.

A Model Architecture

A.1 Transformers

The core mechanisms of the proposed Transformer models are as follows.

Embedding of Input. In transformer models, sets of tokens are transformed into a continuous vector space via a trainable mapping, commonly referred to as embedding. As transformers do not inherently recognize sequence order, a positional encoding is incorporated for sequential input tokens. Additionally, each token is frequently assigned a specific type, for example, to indicate special tokens. The model learns a separate mapping that translates these token types into vectors. The final embedding for each token is then calculated as the sum of the components:

$$E_{\text{final}} = E_{\text{token}} + E_{\text{type}} + E_{\text{positional}} \tag{1}$$

Attention Mechanism. The attention mechanism in the model assigns varying degrees of importance to different tokens, either in relation to one another or based on another distinct token set. The attention function is defined by:

$$\text{Attention}(Q, K, V) = \text{softmax}\left(\frac{QK^T}{\sqrt{d_{\text{model}}}}\right) V \tag{2}$$

Here, Q (query), K (key), and V (value) are parameters derived from linear projections of the embedded tokens. When the inputs are derived from the same token set, the process is termed self-attention. Conversely, when Q comes from a different set of embeddings, it is called cross-attention.

Encoder-Decoder Architecture. The Transformer model employs an encoder-decoder architecture for handling various sequence-to-sequence tasks. The encoder maps an input sequence to a sequence of continuous representations. For each position in the input sequence, the encoder performs self-attention, weighing the importance of other elements in the sequence when encoding that particular element. The output of the encoder is a series of these representations, which are then passed to the decoder. For each element in the output sequence, the decoder uses self-attention to focus on different parts of the output sequence and cross-attention to focus on the output of the encoder, effectively integrating information from the input sequence.

Masked Language Modeling (MLM). Training the model to predict the original tokens in a partially masked input has become a commonly employed loss function.

$$\mathcal{L}_{\text{CCE}} = -\sum_{i \in \text{masked}} \sum_{c=1}^{C} y_{ic} \log \hat{y}_{ic} \tag{3}$$

where C represents the total number of classes (token types), y_{ic} is the ground truth label for the ith masked token and its class c, and $\hat{y}_{ic}$ is the predicted probability of the ith token being of class c.

A.2 Enhanced Embedding for Non-uniform Data

We use two extensions to the standard embedding techniques to encode non-uniformly sampled longitudinal data and sequences comprising mixed numeric-nominal data types. Firstly, we augment positional embeddings with date-specific embeddings. This addition enables the transformer to discern the sequence order and the temporal intervals between consecutive input tokens, capturing essential temporal information. Secondly, we employ a unique, learnable 'numeric' token to process numeric data.

Temporal Embeddings. Tailored to encode temporal information into continuous vector representations, this module adopts a traditional vocabulary-based embedding technique. It encodes temporal units-years, months, and days-into corresponding vector embeddings as different tokens. Using the dates as embeddings facilitates using time intervals and including absolute temporal information, like seasons. The final date embedding is obtained by adding the three embeddings.

$$E_{\text{date}} = E_{\text{year}} + E_{\text{month}} + E_{\text{day}} \tag{4}$$

Numeric Embeddings. This approach creates embeddings by encoding nominal and numerical tokens differently. Nominal tokens are processed using a conventional embedding layer, while numerical tokens undergo encoding via a linear projection.

$$E_{\text{mixed}} = \begin{cases} \text{EmbeddingLayer}(x_i) & \text{if } x_i \text{ is nominal} \\ \text{LinearLayer}(x_i) & \text{if } x_i \text{ is nominal} \end{cases} \tag{5}$$

where x_i is the ith token in the input.

The Final Embedding. The final embedding is obtained as the sum of the embeddings.

$$E_{\text{final}} = E_{\text{mixed}} + E_{\text{date}} + E_{\text{type}} + E_{\text{positional}} \tag{6}$$

A.3 Unified Cross-Attention Decoder

The Unified Cross-Attention Decoder is designed to integrate information from multiple encoders simultaneously. To this end, each layer contains a cross-attention layer for each encoder. It aggregates the output of each attention layer (including self-attention), via attention base fusion.

Attention Mechanism for Information Fusion. The process of attention-based information fusion begins with the outputs from attention layers, symbolized as $X_i \in \mathbb{R}^{B \times L \times H}$, where B stands for batch size, L for sequence length, and H for the dimension of embedding vectors. The index i corresponds to the ith attention layer output. These outputs are combined by considering the importance, or 'attention', given to each source over the sequence, batch, and hidden representation dimensions.

To form a comprehensive input $X \in \mathbb{R}^{N \times B \times L \times H}$, we stack the outputs from each attention layer, X_i. The quantity N denotes the number of data types or modalities being processed, equal to the number of encoder outputs plus one for the initial input modality.

The final fused output Y is calculated by the following equation:

$$Y = \sum_{i=1}^{N} \text{softmax}(XW)_i \odot X_i \tag{7}$$

where $W \in \mathbb{R}^{H \times H}$ represents a set of parameters the model can learn. In this equation, $\odot$ denotes element-wise multiplication, and $Y \in \mathbb{R}^{B \times L \times H}$ is output.

This fusion method effectively takes a weighted average of the outputs, considering both self-attention and cross-attention mechanisms. Analyzing these weights, we can gain insight into the contributions of each modality to the final output for a given sequence.

A.4 Loss Function

We use the MLM loss function with a single modification. Our model predicts not only the probabilities for the class of a masked token but also predicts a numeric value. This allows us to calculate both the usual categorical cross-entropy loss and mean squared error, respectively, for the masked nominal and numeric tokens.

$$\mathcal{L}_{\text{i}} = \begin{cases} -\sum_{c=1}^{C} y_{ic} \log \hat{y}_{ic} & \text{if } x_i \text{ is nominal} \\ (z_i - \hat{z}_i)^2 & \text{if } x_i \text{ is numerical} \end{cases} \tag{8}$$

where the i represents the index of the masked token, x. C represents the total number of classes (token types), y_c is the ground truth label for class c, $\hat{y}_c$ is the predicted probability of class c, z is the ground truth numeric value and $\hat{z}$ is the predicted numeric value.

The total loss of a sequence is the sum of the losses for each token:

$$\mathcal{L}_{\text{sequence}} = \sum_{i \in \text{masked}} \mathcal{L}_{\text{i}} \tag{9}$$

This loss can be applied to both the encoders and the decoder. To help the different modalities develop useful representations, even if the learning signal through the cross-attention is weak, we apply the masked token prediction loss to all encoders and the decoder.

$$\mathcal{L}_{\text{total}} = \mathcal{L}_{\text{decoder}} + \sum_{\text{encoders}} \mathcal{L}_{\text{encoder}} \tag{10}$$

B Tokenization

Tokenization in the context of Transformers means transforming the input sequence into a sequence of tokens. The set of all possible tokens is called the vocabulary of the model. In our case, we transform each patient into five sequences.

- A mixed numerical and nominal token sequence encoding events and measurements.
- The sequence of token types. We use 4 types of tokens: special tokens, event tokens, categorical measurement tokens, and numerical measurement tokens.
- 3 sequences encoding the year, month, and day of each event. Measurement results and special tokens receive a special no-date token, which is mapped to the 0 embedding vector.

All encoders and the decoder use the same vocabulary for the embedding layer, but these parameters are not shared. We now describe how each modality's sequences are obtained from a patient's information. For a detailed example of step-by-step tokenization, please refer to our GitHub repository in Sect. 7.

- **Disease** sequences are obtained by first sorting the ICD10 onset codes by date and replacing all codes with smaller than 1:50000 prevalence with a rare disease code. The tokens of the sequence are the ICD10 codes, and the date for each token is the onset date of the code.
- **Prescription** sequences are obtained in a similar fashion. Still, no chemblIDs were removed at this stage since an occurrence-based filter step had already been performed during data collection.
- **Personal** and **laboratory** information are both handled in a field-wise manner. Each field is assigned two tokens. The first token encodes the fieldID, while the second encodes the field value. The fieldID tokens are typed as event tokens, while the value tokens can be categorical or numerical measurements. Special numeric values (e.g. -3: Do not know) are handled as categorical measurements, and all numeric information is standardized. Date information is only added to event tokens. Otherwise, the no-date token is used. These sequences are also sorted by the date of the measurements.

C The Complete Quality Control Pipeline for Genetic Data

The complete quality control pipeline for genetic data consisted of the following steps.

- Remove variants with a minor allele frequency less than 0.01.
- Remove variants and samples in an iterative manner so that for the resulting dataset missingness rates calculated both for samples and for variants will be less than 0.01.
- Remove variants that deviate from the Hardy–Weinberg equilibrium (HWE)with a p-value less than 1×10^{-5}.
- Remove samples so that in the resulting dataset, no two samples will have a kinship value greater than 0.044, calculated by King software.
- Remove samples for which the sex reported does not conform to the one inferred from genetic data.
- Remove samples that are outliers regarding their inbreeding coefficient (heterozygosity) using a threshold of 3 standard deviations.

Table 2 summarizes how the dimensions of genetic data changed after each quality control step. As can be seen, most variants were eliminated because of their low minor allele frequency, whereas most of the samples removed were eliminated because of kinship relations.

Table 2. Genetic QC. The tables show the number of variables and samples after each quality control step.

QC step	No. of Variables	No. of Samples
Initial input	96,254,167	488,377
Minor allele frequency	8,203,835	408,998
Missingness	6,345,444	408,267
HWE	6,277,868	408,267
Kinship	6,277,868	335,805
Sex check	6,277,868	335,645
Heterozigosity	6,277,868	334,257

D Risk of MDD and TRD in the Four Global Clusters

See Table 3

Table 3. Risk of MDD and TRD in the four global clusters. One-tailed Fisher's exact test was used to assess the risk of the disease in the corresponding cluster. P values are adjusted using Holm's correction method.

Cluster	Disease	OR	95% CI	Adj. P value
1	MDD	0.20	0.19–0.21	< 0.001
1	TRD	0.59	0.46–0.75	< 0.001
2	MDD	0.63	0.61–0.64	< 0.001
2	TRD	0.41	0.34–0.50	< 0.001
3	MDD	2.03	1.99–2.06	< 0.001
3	TRD	1.37	1.25–1.50	< 0.001
4	MDD	2.10	2.06–2.15	< 0.001
4	TRD	1.01	0.92–1.12	0.780

References

1. Buergel, T., et al.: Metabolomic profiles predict individual multidisease outcomes. Nat. Med. **28**(11), 2309–2320 (2022)
2. Bycroft, C., et al.: The UK biobank resource with deep phenotyping and genomic data. Nature **562**(7726), 203–209 (2018)
3. Coleman, J.R., Euesden, J., Patel, H., Folarin, A.A., Newhouse, S., Breen, G.: Quality control, imputation and analysis of genome-wide genotyping data from the illumina humancoreexome microarray. Brief. Funct. Genomics **15**(4), 298–304 (2016)
4. Colón-Ruiz, C., Fitzgerald, T.W., Segura-Bedmar, I., Birney, E., Herrero-Zazo, M.: Automated extraction and classification of drug prescriptions in electronic health records: introducing the presner pipeline. medRxiv 2023–10 (2023)
5. Darke, P., Cassidy, S., Catt, M., Taylor, R., Missier, P., Bacardit, J.: Curating a longitudinal research resource using linked primary care EHR data-a UK biobank case study. J. Am. Med. Inform. Assoc. **29**(3), 546–552 (2022)
6. Fabbri, C., et al.: Genetic and clinical characteristics of treatment-resistant depression using primary care records in two UK cohorts. Mol. Psychiatry **26**(7), 3363–3373 (2021)
7. Flint, J.: The genetic basis of major depressive disorder. Molecular Psychiatry 1–12 (2023)
8. Gaulton, A., et al.: ChEMBL: a large-scale bioactivity database for drug discovery. Nucleic Acids Res. **40**(D1), D1100–D1107 (2012)
9. Golkar, S., et al.: XVAL: a continuous number encoding for large language models. arXiv preprint arXiv:2310.02989 (2023)
10. Guthrie, B., Makubate, B., Hernandez-Santiago, V., Dreischulte, T.: The rising tide of polypharmacy and drug-drug interactions: population database analysis 1995–2010. BMC Med. **13**(1), 1–10 (2015)
11. Hidese, S., Saito, K., Asano, S., Kunugi, H.: Association between iron-deficiency anemia and depression: a web-based Japanese investigation. Psychiatry Clin. Neurosci. **72**, 513–521 (2018)

12. Hosseini, S.R., Zabihi, A., Ebrahimi, S.H., Amiri, S.R.J., Kheirkhah, F., Bijani, A.: The prevalence of anemia and its association with depressive symptoms among older adults in north of Iran. J. Res. Health Sci. **18**, e00431 (2018)
13. Jensen, A.B., et al.: Temporal disease trajectories condensed from population-wide registry data covering 6.2 million patients. Nature Commun. **5**(1), 4022 (2014)
14. Kiiskinen, T., et al.: Genetic predictors of lifelong medication-use patterns in cardiometabolic diseases. Nat. Med. **29**(1), 209–218 (2023)
15. Koné Pefoyo, A.J., et al.: The increasing burden and complexity of multimorbidity. BMC Public Health **15**(1), 1–11 (2015)
16. Li, Y., et al.: BEHRT: transformer for electronic health records. Sci. Rep. **10**(1), 7155 (2020)
17. Li, Y., et al.: Hi-BEHRT: hierarchical transformer-based model for accurate prediction of clinical events using multimodal longitudinal electronic health records. IEEE J. Biomed. Health Inform. **27**(2), 1106–1117 (2022)
18. Liu, J., Capurro, D., Nguyen, A., Verspoor, K.: Attention-based multimodal fusion with contrast for robust clinical prediction in the face of missing modalities. J. Biomed. Inform. **145**, 104466 (2023)
19. McInnes, G., Lavertu, A., Sangkuhl, K., Klein, T.E., Whirl-Carrillo, M., Altman, R.B.: Pharmacogenetics at scale: an analysis of the UK biobank. Clin. Pharmacol. Therapeut. **109**(6), 1528–1537 (2021)
20. Meng, Y., Speier, W., Ong, M.K., Arnold, C.W.: Bidirectional representation learning from transformers using multimodal electronic health record data to predict depression. IEEE J. Biomed. Health Inform. **25**(8), 3121–3129 (2021)
21. Prasad, B., Bjourson, A.J., Shukla, P.: Data-driven patient stratification of UK biobank cohort suggests five endotypes of multimorbidity. Briefings Bioinform. **23**(6), bbac410 (2022)
22. Rasmy, L., Xiang, Y., Xie, Z., Tao, C., Zhi, D.: Med-BERT: pretrained contextualized embeddings on large-scale structured electronic health records for disease prediction. NPJ Digital Med. **4**(1), 86 (2021)
23. Rupp, M., Peter, O., Pattipaka, T.: ExBEHRT: extended transformer for electronic health records. In: ICLR 2023 Workshop on Trustworthy Machine Learning for Healthcare (2023)
24. Schwarz, K., Banos, D.T., Rathmes, G., Krauthammer, M.: Drug administration clusters in the UK biobank. arXiv:2207.08665 (2022)
25. Wu, Y., et al.: Genome-wide association study of medication-use and associated disease in the UK biobank. Nat. Commun. **10**(1), 1891 (2019)
26. Xu, P., Zhu, X., Clifton, D.A.: Multimodal learning with transformers: a survey. IEEE Trans. Pattern Anal. Mach. Intell. (2023)

Multimodal Fusion of Echocardiography and Electronic Health Records for the Detection of Cardiac Amyloidosis

Zishun Feng[1(✉)], Joseph A. Sivak[2], and Ashok K. Krishnamurthy[1,3]

[1] Department of Computer Science, University of North Carolina at Chapel Hill, Chapel Hill, USA
fzs@cs.unc.edu
[2] Division of Cardiology, University of North Carolina at Chapel Hill, Chapel Hill, USA
[3] Renaissance Computing Institute, University of North Carolina at Chapel Hill, Chapel Hill, USA

Abstract. Cardiac amyloidosis, a rare and highly morbid condition, presents significant challenges for detection through echocardiography. Recently, there has been a surge in proposing machine-learning algorithms to identify cardiac amyloidosis, with the majority being imaging-based deep-learning approaches that require extensive data. In this study, we introduce a novel transformer-based multimodal fusion algorithm that leverages information from both imaging and electronic health records. Specifically, our approach utilizes echocardiography videos from both the parasternal long-axis (PLAX) view and the apical 4-chamber (A4C) view along with patients' demographic data, laboratory tests, and cardiac metrics to predict the probability of cardiac amyloidosis. We evaluated our method using 5-fold cross-validation on a dataset comprising 41 patients and achieved an Area Under the Receiver Operating Characteristic curve (AUROC) of 0.94. The experimental results demonstrate that our approach can achieve competitive results with a significantly smaller dataset compared to prior imaging-based methods that required data from thousands of patients. This underscores the potential of leveraging multimodal data to enhance diagnostic accuracy in the identification of complex cardiac conditions such as cardiac amyloidosis.

Keywords: Deep Learning · Multimodal Learning · Cardiac Amyloidosis · Echocardiography · Electronic Health Record

1 Introduction

Cardiac amyloidosis is a highly morbid condition and a significant cause of heart failure. It is most effectively treated when recognized early. While there are several echocardiographic features suggestive of cardiac amyloidosis, these often go unrecognized in many outpatient clinics. Since cardiac amyloidosis is rare, these

J. Finkelstein et al. (Eds.): AIME 2024, LNAI 14845, pp. 227–237, 2024.
https://doi.org/10.1007/978-3-031-66535-6_25

echocardiographic features might not be checked for every patient. Further, diagnosing cardiac amyloidosis from echocardiography requires significant expertise, but patients with symptoms may not present to expert readers. Typically, suspected cardiac amyloidosis is confirmed by cardiac magnetic resonance (CMR) which is not done routinely as it is an expensive and time-consuming procedure. As a result, there is significant value for methods that can flag patients with suspected cardiac amyloidosis from echocardiography and other routinely collected data such as Electronic Health Records (EHR).

In clinical practice, the diagnosis of cardiac amyloidosis requires information from multiple aspects, including lab tests, biopsy, and imaging tests. The confirmation of cardiac amyloidosis relies on biopsy, cardiac magnetic resonance, or nuclear imaging, which are costly and risky. On the other hand, there are lab tests that are fast and low-cost and can provide rich information for the diagnosis. For example, blood and urine tests can detect abnormal proteins that signify amyloidosis. Furthermore, amyloidosis may also affect organ functions like the liver, kidney, and thyroid. Therefore, integrating information from lab tests and other EHR has great potential for the early detection of cardiac amyloidosis.

In this study, we explore different ways of combining echocardiograms and EHR to detect cardiac amyloidosis early. Specifically, we train several models and evaluate their performance against decisions made by experts. The models we consider include: (a) an EHR-only model; (b) a model based on the echocardiogram PLAX view only; (c) a model based on the echocardiogram A4C view only; (d) a model that combines both the PLAX and A4C views; and finally (e) a multi-modality model that combines the PLAX, A4C view echocardiograms and EHR. To accomplish this, we introduce a transformer-based intermediate fusion model that can achieve an AUROC score of 0.94 on a small dataset with 41 patients (17 in the case group, 24 in the control group).

2 Related Works

2.1 Cardiac Amyloidosis Detection

Cardiac amyloidosis detection with machine learning is a relatively new approach that has not been studied much. In [14], the authors collected PLAX echocardiograms from 852 unique patients (81 in the case group, 771 in the control group) for cardiac amyloidosis detection. The authors trained a VGG network on single-frame echocardiograms and achieved an AUROC score of 0.87. [4] built an AI-based cardiac amyloidosis detection system with both electrocardiograms (ECG) and echocardiograms. The system first used an ECG model to select patients with a high risk of amyloidosis. And a trained echocardiography model was applied to give the final prediction. The echocardiography network was trained on a large dataset with 2828 patients (410 in the case group, 2418 in the control group). The authors trained a 3D CNN model on echocardiography videos and achieved an AUROC score of 0.96. However, the AUROC score of the proposed method in this work will possibly be improved with more training data.

2.2 Multimodal Learning with Medical Data

There are numerous modalities in medical data, such as imaging, lab tests, prescriptions, clinical notes, and genomics and [1,7] review multimodal AI methods in healthcare. Some examples include [13] that combined macroscopic and dermoscopic images, and patients' demographic data to classify skin lesions and detect melanoma. The authors used a 2-D convolutional neural network to extract features from images. Then, extracted features were combined with the metadata and sent to fully connected layers for the final prediction. This work is an example of intermediate fusion. [12] developed a multimodal ensemble method to fuse medical reports (text data) and EHR (structured data) with imaging data to predict ICD (International Classification of Diseases) codes for diseases.

3 Method

In this work, we explore cardiac amyloidosis detection performance of different modalities: 1) with electronic health records only, 2) with imaging data only, and 3) with both EHR and imaging data. For the imaging data-only models, we explore the single-view models with the PLAX view, the A4C view, and a double-view model that combines these two views. For the multimodal classification using both echocardiogram and EHR data, we explore the late fusion and intermediate fusion strategies.

3.1 Electronic Health Record Classification Model

The EHR features contain demographic data and vital signs, cardiac metrics, and lab test values. Given the limited size of the training data (41 patients) and total feature number (166 features), we use early fusion to combine these three features and a logistic regression model to predict the probability of cardiac amyloidosis (Fig. 1 a).

3.2 Echocardiography Video Classification Model

Single-View Model. There are two different views of echocardiograms: the PLAX view and the A4C view. We use the spatiotemporal model described in [2], which has shown the ability to extract spatiotemporal features from echocardiograms for prediction. The model consists of a convolutional encoder, a bidirectional LSTM with attention, and a feed-forward classifier. The echocardiographic frames are first sent to the convolutional encoder to extract spatial features. The features are then fed into the LSTM with an attention layer to extract temporal features. Finally, the feed-forward classifier predicts the probability of cardiac amyloidosis (Fig. 1 b).

Double-View Model. The double-view model combines two single-view models with the late fusion strategy, which has been shown as an effective strategy in video analysis [11]. Two single-view models first extract spatiotemporal features from the two views. Then, the two features are concatenated to form the joint feature, which is sent to a feed-forward classifier for the final prediction (Fig. 1 c).

3.3 Mutlimodal Classification Model

Late Fusion. The late fusion strategy is one of the most popular fusion strategies in medical applications [3,5,7]. Late fusion methods first extract one feature from each modality and concatenate these features for the final prediction. Similar to the double-view model, the late fusion multimodal model extracts spatiotemporal features from the PLAX and A4C echocardiograms and aggregates these features with the EHR features. The combined feature is then fed to a fully connected layer-based classifier for prediction (Fig. 1 d).

Intermediate Fusion. We introduce a transformer-based intermediate fusion model. The transformer-based model has shown the capability to learn multimodal features in many previous works [6,8,9]. The model first maps EHR and imaging data to 512-dimensional features. Specifically, feed-forward layers are used for EHR embeddings; pre-trained convolutional neural networks are

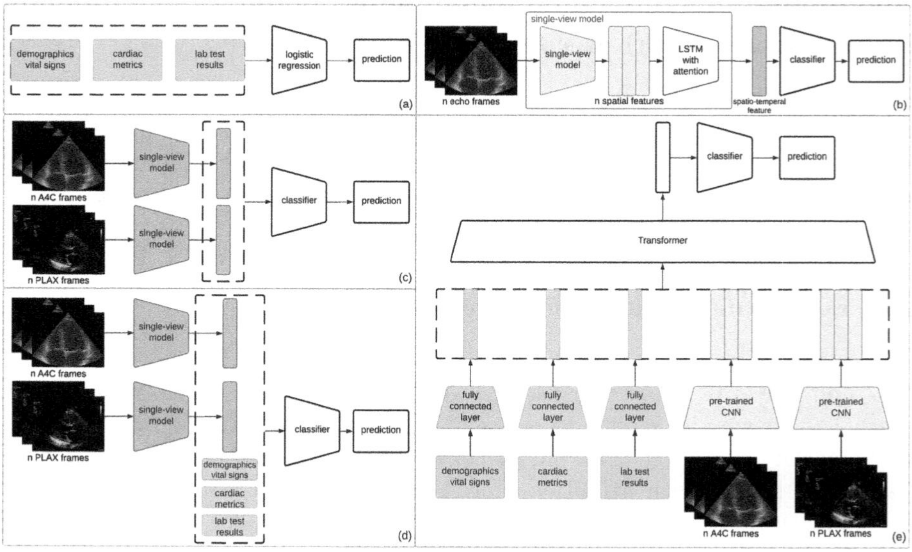

Fig. 1. Architecture of the intermediate fusion model for multimodal classification. (a) EHR only model. (b) Single-view model. (c) Double-view model with late fusion. (d) Multimodal model with late fusion. (e) Multimodal model with intermediate fusion based on transformer. Components in the dashed box are concatenated.

used for echocardiogram embeddings. Then, the EHR and imaging data embeddings are fed into a transformer encoder. The transformer layer extracts the joint features, which are sent to fully connected layers to predict the probability of cardiac amyloidosis. Figure 1 (e) shows the architecture of the intermediate fusion model.

3.4 Training Settings

Since the task is a binary classification task, we chose the binary cross-entropy as the loss function for all the above-mentioned models. All the models were trained using the Adam algorithm with a learning rate of 0.0001 for 20 epochs.

4 Data

4.1 Selected Cohort

Patients from UNC hospitals and clinics in the Carolina Data Warehouse for Health (CDWH)[1] between the ages of 18–99 with a diagnosis of amyloidosis (ICD-10: E85 and ICD-9 277.3) who underwent an echocardiography procedure within a 6-month time period prior to a cardiac magnetic resonance (CMR) procedure were included. All CMRs were between 01/01/2014 to 01/01/2022. The patients were labeled as positive or negative for cardiac amyloidosis based on the CMR, which is considered the gold standard for detecting the disease.

Table 1. Demographics and vital signs of patients in the dataset.

Characteristic	All (n = 41)	Positive (n = 17)	Negative (n = 24)
Age, mean(std), y	66.1(12.0)	71.7(11.3)	62.0(10.8)
Sex: number(percentage)			
Female	14 (34.1%)	5(29.4%)	9(37.5%)
Male	27 (65.9%)	12(70.6%)	15(62.5%)
Race:number(percentage)			
Black or African American	20 (48.8%)	8(47.1%)	12(50%)
White	21 (51.2%)	9(52.9%)	12(50%)
Blood pressure: mean(std), mmHg			
Diastolic	72.5 (15.4)	68.1(10.2)	72.5(17.5)
Systolic	121.6 (23.9)	112.5(17.6)	128.3(25.6)
Weight: mean(std), kg	88.9(19.8)	84.1(18.5)	92.3(19.9)
Height: mean(std), m	1.76(0.1)	1.75(0.1)	1.77(0.1)
BMI: mean(std)	28.7(5.7)	27.5(4.7)	29.6(6.1)

[1] https://tracs.unc.edu/index.php/services/informatics-and-data-science/cdw-h.

From the EHR we collected demographic data, vital signs, and lab tests from one year before to one year after the diagnosis date (CMR-based diagnosis date). We extracted cardiac metrics from echocardiography imaging reports. After patient selection, the final dataset contains 41 patients with 17 positive and 24 negative samples. Table 1 shows the patients' characteristics.

4.2 Data Preprocessing

All the EHR data and imaging data contain protected health information (PHI). To protect patients' identifiers, de-identification is always the first step. We first stored all data in a protected environment and removed all the PHI in the data before the development.

The raw electronic health records are usually sparse due to missing values, which cannot be used for prediction directly. Therefore, preprocessing is a necessary step for EHR data. Similarly, the echocardiograms are saved in DICOM (Digital Imaging and Communications in Medicine) files containing static frames and videos from different views. Thus, we need to select the proper echocardiographic videos for development and prediction.

Preprocessing of Electronic Health Record. The electronic health records used in this study were extracted from CDWH in the PCORnet[2] data model. In this study, we used demographic data, vital signs, and laboratory test results to form the EHR features. The demographic data and vital signs were extracted as the values during the event when the patient underwent the echocardiography procedure. Since the demographic data and vital signs are the basic information that is routinely collected, there are no missing values for these variables and no data imputation was applied. There were 414 different lab tests in the dataset with a large number having very low coverage across patients. We assume that these low-coverage lab tests are not related to the target disease, and only selected the lab tests that happened within 3 months before and 1 month after the echocardiography procedure and had at least a 10 percent coverage across patients. All the missing values were filled with the mean value across all patients for whom the test was available. This resulted in 149 lab tests being included in the dataset for training and testing.

Preprocessing of Echocardiograms. The DICOM file for each patient contains echocardiograms from different videos with different settings. To extract the PLAX and A4C views, we used a pre-trained VGG network [10] to identify these two views of echocardiograms in the DICOM file. One of the authors visually checked and confirmed that the correct echocardiogram views were selected by the VGG classification. Further, only the videos with at least 30 frames were selected to ensure that an entire cardiac cycle was included.

[2] The National Patient-Centered Clinical Research Network. https://pcornet.org/.

Several cardiac metrics are also included in the DICOM files. The echocardiogram technician manually labeled these cardiac metrics during the procedure and saved the frame with measurements as static images in the DICOM files. To form the cardiac metrics feature vector from the DICOM files we extracted the measurements, including wall thickness, chamber diameter, chamber volumes, stroke volume, ejection fraction, fractional shortening, and sphericity index.

5 Experiments and Results

5.1 Experiment Setting

In this study, we evaluated multimodal learning from two aspects: 1) modality types and fusion strategy and 2) the contribution of each modality. Limited by the dataset size, we used 5-fold cross-validation for all experiments, and the average score of all 5 folds was reported as the final score. As the task is binary classification, we used accuracy, sensitivity, specificity, and area under the receiver operating characteristic (AUROC) to evaluate the prediction performance.

5.2 The Impact of Multimodal Learning

We first evaluated the prediction performance with 1) data from different modalities and 2) different fusion strategies. For the modalities, we tested models with 1) EHR data only, 2) the PLAX view echocardiograms only, 3) the A4C echocardiograms only, 4) the fusion of PLAX and A4C echocardiograms, and 5) the fusion of all three data sources. For the fusion strategies, we tested intermediate fusion and late fusion.

Table 2. The prediction performance with different modalities and fusion strategies.

	Method	Data Modality			Accuracy	Sensitivity	Specificity	AUROC
		EHR	PLAX	A4C				
1	Logistic regression	✓			0.778	0.900	0.720	0.769
2	CNN+LSTM+Attn		✓		0.851	0.897	0.818	0.810
3	CNN+LSTM+Attn			✓	0.854	0.810	0.885	0.772
4	Late fusion		✓	✓	0.883	0.893	0.875	0.863
5	Late fusion	✓	✓	✓	0.912	0.894	0.925	0.895
6	Intermediate fusion	✓	✓	✓	0.927	0.918	0.933	0.941

The experimental results are shown in Table 2. From the table, the single modality data (EHR, PLAX, and A4C echocardiograms) can provide good prediction performance. The EHR feature can provide an acceptable predictive power but with a high false positive rate. The imaging data contains more helpful information than the EHR data, and the two echocardiogram views can provide information from different aspects. Thus, combining the two views of echocardiograms can aggregate the information for the two aspects. Line 4 in the table proves that claim; the late fusion of PLAX and A4C views of echocardiograms gives better accuracy and much higher AUROC. Finally, fusing the EHR and imaging data can achieve the highest prediction performance. Moreover, the intermediate fusion can combine the information in an earlier stage and achieve a higher performance than the late fusion method, which aggregates the information from each modality before the final classifier.

5.3 Ablation of EHR Components

From Table 2, we know that adding the EHR feature to the imaging data can achieve better performance with both intermediate fusion and late fusion. The EHR data consists of three components: 1) demographics and vital signs, 2) cardiac metrics, and 3) lab test results. To investigate the impact of each component in the EHR data, we conducted an ablation study, and the results are shown in Table 3

Table 3. The ablation study of EHR components. Each component is removed from the EHR data to evaluate the impact.

Method	EHR components			Accuracy	Sensitivity	Specificity	AUROC
	Demo&Vital	metrics	Labs				
Intermediate		✓	✓	0.912	0.882	0.933	0.900
Intermediate	✓		✓	0.899	0.906	0.892	0.885
Intermediate	✓	✓		0.878	0.859	0.892	0.855
Intermediate	✓	✓	✓	0.927	0.918	0.933	0.941

From Table 3, we can find that dropping the lab test values influenced the performance the most. And the demographic data and vital signs contribute the least to the prediction model.

5.4 Contribution to Performance

To further study the contribution of each modality, we calculated the transformer model's attention weights of each modality. We drew pie charts to show the importance of each modality and each component in the EHR data. Moreover, we also computed the weights of the EHR feature mapping layer to show the contribution of each entry in the EHR data.

From Fig. 2, we can find that the imaging data are much more important than the EHR data. The contribution of two echocardiogram views is almost the same. For the EHR data, the lab test result is still the most significant component. Although lab tests are 90% (149/166) of EHR feature uses, they only contribute 83% to the EHR performance in prediction, which indicates the proposed model

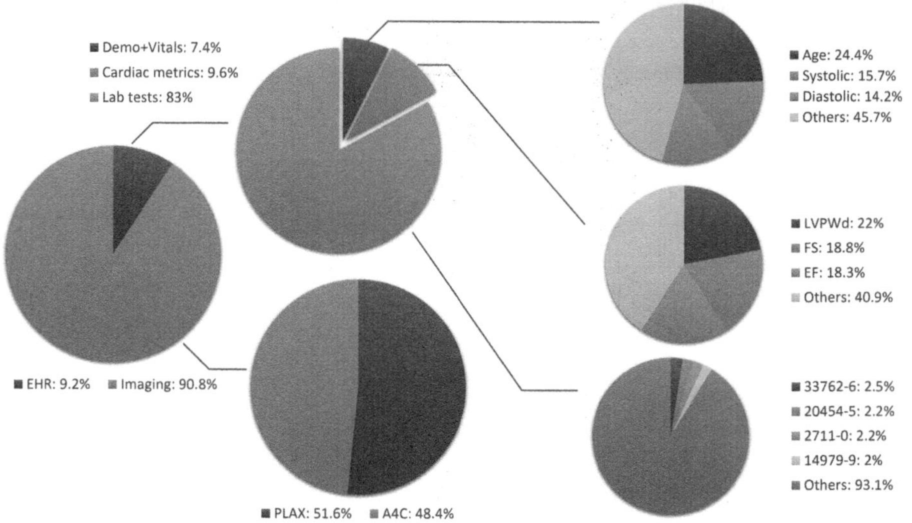

Fig. 2. Pie charts of feature importance. Left: overall importance. Middle: the importance of EHR and imaging. Right: detailed importance of each entry in the EHR data. The LOINC codes of lab tests are shown in the lower-right corner.

is able to detect non-related lab tests and find other useful non-lab entries. We also find that some lab tests have an importance higher than 2% (the average is 0.6%), such as B-type Natriuretic Peptide (BNP) and proteins in urine, which can be considered valuable indicators for cardiac amyloidosis analysis. The top three entries for the cardiac metrics are the wall thickness and indices such as ejection fraction and fractional shortening. For the demographics and vital signs, patients' age and blood pressure contribute the most.

5.5 Comparison with Related Studies

We also compared the proposed multimodal learning-based cardiac amyloidosis detection method with the related studies. Here, we list the method, data type, and prediction performance.

Table 4 shows that the multimodal learning method is able to achieve competitive results even with a small amount of training data. Moreover, by comparing [14] and the proposed method, utilizing data that contains more information is more helpful than enlarging the size of the dataset. Given more training data, the proposed method will possibly achieve a better performance.

Table 4. Comparison with other cardiac amyloidosis detection works.

	Method	Data Type	Number of Patients (Case/Control)	AUROC
[14]	VGG16	single PLAX view echocardiogram frame	852 (81/771)	0.87
[4]	3D CNN	A4C view echocardiogram video	2828 (410/2418)	0.96
proposed	multimodal learning	EHR PLAX, A4C videos	41 (17/24)	0.94

6 Conclusion

In this work, we explored the potential of multimodal learning for cardiac amyloidosis detection. We proposed a transformer-based intermediate fusion method for fusing electronic health records and two views of echocardiograms. The model can select valuable identifiers for cardiac amyloidosis diagnosis. Additionally, the model can achieve competitive prediction performance with a small amount of training data.

Acknowledgments. This work was partially funded by NSF Grant 1633295 BIGDATA: F: Collaborative Research: From Visual Data to Visual Understanding.

References

1. Acosta, J.N., Falcone, G.J., Rajpurkar, P., Topol, E.J.: Multimodal biomedical AI. Nat. Med. **28**(9), 1773–1784 (2022)
2. Feng, Z., Sivak, J.A., Krishnamurthy, A.K.: Two-stream attention spatio-temporal network for classification of echocardiography videos. In: 2021 IEEE 18th International Symposium on Biomedical Imaging (ISBI), pp. 1461–1465. IEEE (2021)
3. Feng, Z., Sivak, J.A., Krishnamurthy, A.K.: Improving echocardiography segmentation by polar transformation. In: Camara, O., et al. (eds.) International Workshop on Statistical Atlases and Computational Models of the Heart, pp. 133–142. Springer (2022). https://doi.org/10.1007/978-3-031-23443-9_13
4. Goto, S., et al.: Artificial intelligence-enabled fully automated detection of cardiac amyloidosis using electrocardiograms and echocardiograms **12**(1), 2726 (2021)
5. Huang, S.C., et al.: Multimodal fusion with deep neural networks for leveraging CT imaging and electronic health record: a case-study in pulmonary embolism detection. Sci. Rep. **10**(1), 22147 (2020)
6. Khan, S., Naseer, M., Hayat, M., Zamir, S.W., Khan, F.S., Shah, M.: Transformers in vision: a survey. ACM Comput. Surv. (CSUR) **54**(10s), 1–41 (2022)
7. Kline, A., et al.: Multimodal machine learning in precision health: a scoping review. NPJ Digit. Med. **5**(1), 171 (2022)
8. Li, L.H., et al.: VisualBERT: A simple and performant baseline for vision and language. arXiv:1908.03557 (2019)
9. Lu, J., et al.: ViLBERT: pretraining task-agnostic visiolinguistic representations for vision-and-language tasks. In: NeurIPS (2019)

10. Simonyan, K., Zisserman, A.: Very deep convolutional networks for large-scale image recognition. arXiv preprint arXiv:1409.1556 (2014)
11. Snoek, C.G., Worring, M., Smeulders, A.W.: Early versus late fusion in semantic video analysis. In: Proceedings of the 13th Annual ACM International Conference on Multimedia, pp. 399–402 (2005)
12. Xu, K., et al.: Multimodal machine learning for automated ICD coding. In: Machine Learning for Healthcare Conference, pp. 197–215. PMLR (2019)
13. Yap, J., Yolland, W., Tschandl, P.: Multimodal skin lesion classification using deep learning. Exp. Dermatol. **27**(11), 1261–1267 (2018)
14. Zhang, J., et al.: Fully automated echocardiogram interpretation in clinical practice: feasibility and diagnostic accuracy. Circulation **138**(16), 1623–1635 (2018)

Multi-view k-Nearest Neighbor Graph Contrastive Learning on Multi-modal Biomedical Data

Yilin Zhang[1], Shuo Chen[2], Rozalina G. Mccoy[2], Chixiang Chen[2], and Yuzhou Chen[3](✉)

[1] Department of Mathematics, University of Maryland, College Park, USA
yzhang26@umd.edu
[2] School of Medicine, University of Maryland, College Park, USA
{shuochen,Rozalina.McCoy,chixiang.chen}@som.umaryland.edu
[3] Department of Computer and Information Sciences, Temple University, Philadelphia, USA
yuzhou.chen@temple.edu

Abstract. Machine learning (ML) and deep learning (DL) on biomedical datasets is an important and ubiquitous task with applications from drug discovery to brain image segmentation to diagnosis and classification tasks in medicine. The primary challenges in this domain are (i) finding a way to encode biomedical data so that it can be easily exploited by ML and DL models and (ii) capacities of ML/DL models largely come from high-quality annotated labels, which are expensive to collect. To learn promising representation of biomedical data and handle unlabeled data, we propose Multi-View k-Nearest Neighbor Graph Contrastive Learning (MV-KGCL) in this paper. Specifically, MV-KGCL extends the graph neural network-based contrastive learning, which is powerful in graph representation learning, but is degraded when structural information is not clear enough or there is non-graph observation. MV-KGCL tackles these direct causes of the problem of graph neural network-based contrastive learning models. That is, we extract various k-nearest neighbor (kNN) graphs from multiple views of entity features, and then conduct contrastive learning over these kNN graphs in the local and the global views. Experiments on two datasets demonstrate that MV-KGCL outperforms the state-of-the-art methods due to better evaluation metrics of clustering, capturing meaningful patterns and heterogeneity within the data, and enabling more interpretable and clinically meaningful clusters that are predictive to cognition conditions.

Keywords: Unsupervised Learning · UK Biobank · Graph Neural Networks

1 Introduction

Nowadays, real-world data including the UK Biobank (UKB) [20] offers a comprehensive collection of genetic, clinical, lifestyle, and imaging data, making it an ideal resource for exploring various health-related questions. To handle high-dimensional and multi-modal databases, machine learning-based clustering algorithms have gained prominence as they can effectively identify clusters or subgroups within the data

J. Finkelstein et al. (Eds.): AIME 2024, LNAI 14845, pp. 238–248, 2024.
https://doi.org/10.1007/978-3-031-66535-6_26

without requiring prior assumptions. Such techniques including auto-detection [7, 17] are especially useful in biomedical research since conditional annotation and disease diagnosis often involve significant labor and substantial costs. For example, traditional approaches to detect Alzheimer's disease (AD) biomarkers involve laborious and time-consuming manual annotation and disease diagnosis processes, such as manual region of interest delineation/visual rating scales applied to neuroimaging data [18]. These processes require substantial expertise, training, and personnel resources, leading to significant labor and financial costs. Another related example could be found is that current providers have limited tools to effectively measure cognition over time and have to rely on traditional methods designed to only detect severe impairments. Comprehensive neuropsychological evaluations are costly, and screenings and task assessments are limited in accuracy and may not allow for tracking subtle changes over time and provide insights into different cognitive domains [6]. Thus, there is an urgent need to develop artificial intelligence (AI) tools capable of directly learning and extracting information from diverse markers with multi-modality (e.g., neuroimaging data, blood chemistry, metabolite markers, etc.). These tools can help alleviate the labor-intensive and costly nature of annotation and diagnosis in biomedical research.

In this paper, we introduce a new graph neural network (GNN)-based self-supervised learning model, i.e., Multi-View k-Nearest Neighbor Graph Contrastive Learning (MV-KGCL). To the best of our knowledge, MV-KGCL is the first kind in literature that creatively uses knowledge of multi-view visual representation learning [23] to identify unique sub-clusters of subjects by making use of multi-modal biomedical markers. The resulting clusters capture meaningful patterns and heterogeneity within the data and enable interpretable and clinically meaningful subgroups that are predictive to health conditions. Specifically, our goal is to develop and apply the proposed model to classify unlabeled participants by only utilizing diverse markers with multi-modality from each subject as input, without requiring prior knowledge of ground-truth labels. At the heart of our MV-KGCL is the capability of encode the topological and graph structural information from multiple views concealed in the participant biomedical features. Our proposed model not only considers multi-view properties of participant biomedical features from both local and global perspectives but conducts both local- and global-level self-supervised learning (here we consider the contrastive learning), and hence provides more critical hidden information better explaining complex subject-level features with multi-modality. To see the unsupervised clustering performance, we apply and evaluate our method on the real-world UKB database including dataset1 (multi-modal neuroimaging data) and dataset2 (blood chemistry and metabolites). Experiments show that our method consistently achieves the best clustering performance among all the state-of-the-art models and leads to more clinically interpretable clustering that is predictive to subject's blood pressure condition and cognition function, respectively. These validate the effectiveness and generalization of the proposed model.

Our contributions are summarized as follows:

- We introduce a novel model for sub-clusters of participants identification from GNN-based self-supervised learning. To the best of our knowledge, that is the first self-supervised geometric deep learning model with unique data application based upon UKB.

- We utilize various k-nearest neighbor graph construction techniques to generate different informative graph structures, and develop both local-global and local-local contrastive learning schemes to improve contrastive GNN training.
- We evaluate the proposed model on two real-world datasets with four widely used evaluation metrics and show that our model outperforms all other models consistently.

2 Related Work

Traditional Statistical and Machine Learning Methods. Clustering is a fundamental task in data analysis, essential for grouping similar data points based on their intrinsic characteristics. Clustering algorithms can be categorized into various types, including partitioning, hierarchical, density-based, and model-based algorithms. Popular methods such as K-Means, hierarchical clustering, Gaussian mixture models, and random forest [2,14] have been utilized for this purpose. However, these traditional clustering methods may encounter limitations when dealing with high-dimensional and multi-modal data.

Graph Representation Learning and Graph Contrastive Learning. Inspired by the success of convolutional neural networks (CNN) on image-based tasks, graph neural networks (GNNs) have emerged as a powerful tool for graph representation learning. Based on the spectral graph theory, Bruna *et al.* [4] introduces a graph-based convolution in Fourier domain. ChebNet [5] integrates spectral graph convolution with Chebyshev polynomials. In addition, Graph Convolutional Networks (GCNs) [11] simplifies the graph convolution with a localized first-order approximation. More recently, there have been proposed various approaches based on accumulation of the graph information from a wider neighborhood, using diffusion aggregation and random walks [1,12]. Recently, contrastive learning (CL) has emerged as a successful method for unsupervised graph representation learning. Graph contrastive learning is a self-supervised learning approach to learn an encoder (e.g., GNNs without the final classifier) for extracting embeddings from the unlabeled input data. For instance, GCC [19] proposes a pretraining framework based on local-local CL which constructs multiple graph views by sampling subgraphs based on random walks. For global-local CL, the works [8,24] follow the InfoMax principle [16] to maximize the Mutual Information (MI) between the representation of local features and global features. Moreover, another graph contrastive learning mode, i.e., global-global CL [25] studies the relationships between the global context representations of different samples, which performs better on graph-level tasks. Different from these methods, we propose a novel Multi-View k-Nearest Neighbor Graph Contrastive Learning with the two-stream multi-view contrasting mode (i.e., based on both global-local CL and local-local CL) that not only captures crucial topological and geometrical information but enhances the latent graph representation learning.

3 Methodology

Consider a graph structure $\mathcal{G} = \{\mathcal{V}, \mathcal{E}, \boldsymbol{A}, \boldsymbol{X}\}$, where $\mathcal{V}$ is a node set with cardinality $|\mathcal{V}|$ of N, and $\mathcal{E} \subseteq \mathcal{V} \times \mathcal{V}$ is an edge set. The $\boldsymbol{A} \in \mathbb{R}^{N \times N}$ with entries $\{a_{uv}\}_{1 \leq u,v \leq N}$

is the symmetric adjacency matrix of $\mathcal{G}$, that is, $a_{uv} \neq 0$ for any $e_{uv} \in \mathcal{E}$ and $a_{uv} = 0$, otherwise. $\boldsymbol{X} = [\boldsymbol{x}_1^\top, \boldsymbol{x}_2^\top, \ldots, \boldsymbol{x}_N^\top] \in \mathbb{R}^{N \times d}$ is the node feature matrix where d is the number of features and $\boldsymbol{x}_i$ denotes the node features of the node u_i. In this study, we aim to use informative markers to identify sub-clusters of participants within the studied cohort that could potentially serve as predictors of downstream health conditions, such as cognition decline. For instance, the input features could be multi-modal brain imaging measures to recover disparate profiles predictive to cognition conditions. See more details about applications in Section of Datasets.

3.1 k-Nearest Neighbor Graphs

First, in order to capture multi-view graph structural information of nodes in both topology and feature spaces, we build a k-nearest neighbor (kNN) graph, i.e., $\mathcal{G}_{\text{kNN}} = (\mathcal{V}, \mathcal{E}_{\text{kNN}}, \boldsymbol{X})$, where $\mathcal{E}_{\text{kNN}}$ is the set of edges of the kNN graph. In our study, we first define the similarity matrix $\boldsymbol{S}_{\text{kNN}} \in \mathbb{R}^{N \times N}$ among N nodes and we consider five different methods as follows *(i) Cosine Similarity*: It uses the cosine value of the angle between two vectors to measure the similarity, i.e., $\boldsymbol{S}_{uv} = \frac{\boldsymbol{x}_u \cdot \boldsymbol{x}_v}{|\boldsymbol{x}_u||\boldsymbol{x}_v|}$; *(ii) Heat Kernel*: It is based on the idea of the heat equation, a partial differential equation that describes how heat propagates over time t, i.e., $\boldsymbol{S}_{uv} = \exp(-||\boldsymbol{x}_u - \boldsymbol{x}_v||^2/t)$. As such, given M similarity matrices (i.e., views) $\{\boldsymbol{S}_i\}_{i=1}^M$, we can extract M kNN graphs, i.e., $\{\mathcal{G}_{\text{kNN}}^i\}_{i=1}^M$ by setting a pre-defined threshold ϵ. That is, there is an edge between nodes u and v if $\boldsymbol{S}_{uv} > \epsilon$, otherwise, there is no edge between them. Therefore, considering multi-view graph representation learning, we can extract specific information encoded in topology, feature, and different structural spaces.

3.2 Multi-view Contrastive Graph Neural Network

Multi-view Global-Local Contrastive Learning Loss. The proposed AM-GCL leverages a self-supervised learning task which conducts contrastive learning between the original networks and corrupted networks by maximizing mutual information.

Following the Deep Graph Infomax (DGI) [24], the self-supervised auxiliary task focuses on maximizing the mutual information between a local patch (i.e., node-level representation) and its global context (i.e., graph-level representation). The fundamental motivation behind this design is to encourage the node representations to encapsulate the global information gleaned from the entirety of the graph. Specifically, for i-th kNN graph $\mathcal{G}_{\text{kNN}}^i$, we first perform stochastic augmentation to generate multiple graph views from it. For instance, we use an augmentation function such as topology augmentation and feature augmentation $\mathcal{T}_i$ to generate a perturbed graph, i.e., $\tilde{\mathcal{G}}_{\text{kNN}}^i = \mathcal{T}_i(\mathcal{G}_{\text{kNN}}^i) = \mathcal{T}_i(\tilde{\boldsymbol{S}}_i, \tilde{\boldsymbol{X}}_i)$. Then we can obtain node embeddings for the perturbed graph by applying a GNN encoder $f(\cdot)$ over $\tilde{\mathcal{G}}_{\text{kNN}}^i$. After that, to summarize the obtained local patches into the global context, we employ a readout function $\Gamma(\cdot)$ (e.g., mean or sum pooling) over the node embedding, as the following equation shows.

$$\begin{aligned} \boldsymbol{Z}_i^{(\ell)} &= f(\mathcal{G}_{\text{kNN}}^i) = \sigma(\boldsymbol{S}_i \boldsymbol{H}_i^{(\ell)} \boldsymbol{W}_i^{(\ell)}), \quad \boldsymbol{Q}_i = \Gamma(\boldsymbol{Z}_i) \\ \tilde{\boldsymbol{Z}}_i^{(\ell)} &= f(\tilde{\mathcal{G}}_{\text{kNN}}^i) = \sigma(\tilde{\boldsymbol{S}}_i \tilde{\boldsymbol{H}}_i^{(\ell)} \tilde{\boldsymbol{W}}_i^{(\ell)}), \quad \tilde{\boldsymbol{Q}}_i = \Gamma(\tilde{\boldsymbol{Z}}_i). \end{aligned} \tag{1}$$

Then, we utilize a noise-contrastive type objective with a standard binary cross-entropy (BCE) loss [24] (i.e., a global-local contrastive learning mode) to maximize mutual information which are defined as

$$\mathcal{L}_{\text{G-L}}^{i} = -\frac{1}{|\mathcal{V}| + |\tilde{\mathcal{V}}|}(\sum_{u \in \mathcal{V}} \mathbb{E}_{\mathcal{G}_{\text{kNN}}^{i}}[\log(\Xi(\boldsymbol{z}_{i,u}, \boldsymbol{Q}_i))] + \sum_{v \in \tilde{\mathcal{V}}} \mathbb{E}_{\tilde{\mathcal{G}}_{\text{kNN}}^{i}}[\log(1 - \Xi(\tilde{\boldsymbol{z}}_{i,u}, \boldsymbol{Q}_i)]),$$

where Ξ denotes a discriminator to provide probability scores for sampled pairs. The objective is to minimize $\mathcal{L}_{\text{G-L}}^{i}$, which is equivalent to maximize the mutual information. As a result, we can obtain M types of losses $\{\mathcal{L}_{\text{G-L}}^{i}\}_{i=1}^{M}$ for M different kNN graphs.

Multi-view k-Hop Neighborhood Contrastive Loss. Armed with the weighted k-hop neighborhood, we propose the multi-view k-hop neighborhood contrastive loss. That is, for each node u, its k-hop neighbors are regarded as the positive samples, while the remaining nodes are sampled as the negative ones. The loss encourages positive samples to be closer to the target node and pushes negative samples away from the target one in terms of feature distance. Formally, the multi-view k-hop neighborhood contrastive loss for the node u can be formulated as

$$\mathcal{L}_{\text{L-L}}^{j} = -\frac{1}{|\mathcal{V}|}\sum_{u \in \mathcal{V}} \log(\frac{\sum_{v}^{N} \delta_{j,uv} \exp(\boldsymbol{z}_{j,u} \cdot \boldsymbol{z}_{j,v}/\tau)}{\sum_{v}^{N} \exp(\boldsymbol{z}_{j,u} \cdot \boldsymbol{z}_{j,v}/\tau)}), \tag{2}$$

where $\delta_{j,uv}$ denotes the similarity between connected nodes u and v in the j-th kNN graph $\mathcal{G}_j$. Similarly, we can have M types of multi-view k-Hop neighborhood contrastive losses $\{\mathcal{L}_{\text{L-L}}^{j}\}_{j=1}^{M}$ for M kNN graphs. Finally, our objective function conflates the above two main types of losses as

$$\mathcal{L} = \sum_{i=1}^{M} \alpha_i \mathcal{L}_{\text{G-L}}^{i} + \sum_{j=1}^{M} \beta_j \mathcal{L}_{\text{L-L}}^{j}, \tag{3}$$

where $\{\alpha_i\}_{i=1}^{M}$ and $\{\beta_j\}_{j=1}^{M}$ are tunable parameters for controlling intensities and important weights for different losses.

4 Experiments

Datasets. We apply the proposed method to the UK Biobank (UKB) data[1]. The UKB data are originally collected from over $500,000$ participants in the UK. For each participant, data on genetics, clinical characteristics, metabolites, health, and lifestyle are acquired. The study design of UKB enables investigations into a wide range of conditions, including those that are challenging or even impossible to study retrospectively, such as dementia and rapidly fatal conditions like pancreatic or lung cancer [3]. Our study focuses on a cohort with multi-modal brain imaging variables, including white matter (WM) fiber track integrity (39 measures) and gray matter (GM) cortical thickness (100 measures), nuclear magnetic resonance (NMR)-metabolomics (249

[1] https://www.ukbiobank.ac.uk.

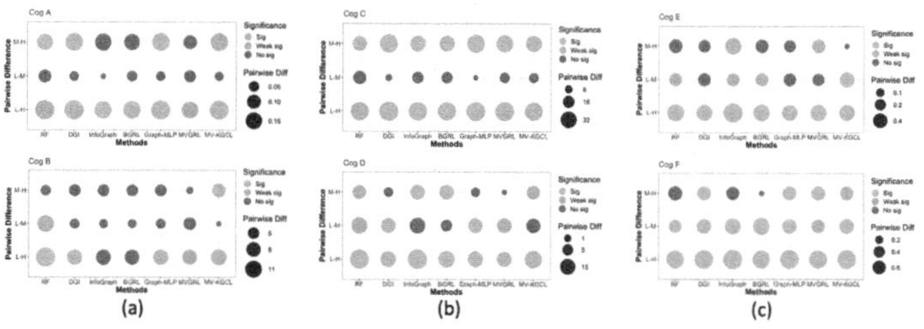

Fig. 1. Visualization of statistical analyses for 7 methods on (a) Cog A and Cog B, (b) Cog C and Cog D, and (c) Cog E and Cog F. Colors represent significance level based on p-value, and circle sizes represent pair-wise difference in terms of cognition function.

measures), and blood biochemistry (BB, 28 measures). Within the full cohort (2589 participants) and the reduced cohort (1756 participants), we summarize the demographic information, including age, gender (male/female), education (college/university attainment), household income (above or below median income), and body mass index (see Table 1). Note that the reduced cohort has the full record of multi-modal data as well as complete cognition measures. It can be seen that these two cohorts are very similar in terms of demographics.

Table 1. Descriptive Characteristics for two studied cohort.

n (%) or mean (±SD)	Full Cohort	Reduced cohort
Age	54.2 (±7.6)	53.8 (±7.5)
BMI	26.5 (±4.4)	26.4 (±4.3)
Education		
College or University degree	1444 (56)	1015 (58)
Others	1145 (44)	741 (42)
Average Household Income		
≥$31000	1735 (67)	1203 (69)
<$31000	854 (33)	553 (31)
Sex		
Female	1081 (42)	731 (42)
Male	1508 (58)	1025 (58)

For WM fiber tracts, we extract and calculate fractional anisotropy (FA) values from diffusion MRI data processed using the widely adopted ENIGMA-Diffusion Tensor Imaging (DTI) Phenotypes analysis pipeline [13,15,26]. In addition to these markers describing metabolism and physiological changes in the brain, each participant's cognition function is evaluated using 9 metrics. These metrics are obtained from multiple cognition tests, including assessments of the number of incorrect matches per round, maximum digits remembered correctly, duration to complete the alphanumeric path, etc. In this study, our aim is to identify sub-clusters of participants within the studied cohort that could potentially serve as predictors of health conditions, such as cognition decline. Specifically, we utilize condition-specific markers, including brain-imaging measures (WM and GM) for assessing cognition function and metabolite-related measures (NMR and BB) for examining chronic conditions. Additionally, we employ 9 metrics of cognition function as benchmarks to assess the clinical relevance and interpretability of the identified clusters. To obtain the labels that can be used as the "ground truth", we first apply factor analysis (details in 'Statistical Analyses') to extract the shared latent factor as the general physical measure from the physical features including *Pulse rate*, *Pulse wave Arterial Stiffness index*, *Average diastolic blood pressure* and *Average systolic blood pressure*. Based on the ascending order of this general physical measure, we then divide the full cohort into three equal-sized clusters and assigned a label to each cluster. These labels would be treated as the "ground truth" to evaluate the accuracy of the clusters learned from our proposed model and other methods. Similarly, for the reduced cohort we extract the shared latent factor as the general metric from 9 metrics of cognition functions and then divide the reduced cohort into three equal-sized clusters, whose labels would be used as the "ground true labels".

Table 2. Unsupervised evaluation metrics comparison with three clusters for blood pressure and cognition.

Model	Blood Pressure			Cognition		
	SIL	CHI	DBI	SIL	CHI	DBI
K-Means	0.184	918.926	1.420	0.326	1641.562	1.000
RF	0.164	878.914	1.526	0.325	1576.675	0.992
DGI	0.425	2631.105	0.871	0.646	6024.280	0.524
InfoGraph	0.515	4882.502	0.656	0.581	6405.657	0.538
BGRL	0.442	2542.489	0.908	0.614	6081.163	0.550
Graph-MLP	0.463	3566.955	0.812	0.638	5946.063	0.530
MVGRL	0.490	4196.663	0.717	0.574	4129.812	0.712
MV-KGCL (ours)	**0.747**	**20411.886**	**0.409**	**0.947**	**11183.681**	**0.451**

Baselines and Experiments Setup. We compare the proposed MV-KGCL with two types of state-of-the-art methods, including (i) machine learning approaches: (1) K-Means and (2) Random Forest (RF) [9], and (ii) GNN-based contrastive learning models: (1) Deep Graph Infomax (DGI) [24], (2) InfoGraph [21], (3) Bootstrapped Graph Latents (BGRL) [22], (4) Multilayer-Perceptron-Based Framework (Graph-MLP) [10], and (5) Contrastive Multi-View Representation Learning Framework (MVGRL) [8]. We implement our proposed MV-KGCL with Python framework on one NVIDIA RTX A5000 GPU with 24 GiB RAM. For all datasets, the MV-KGCL is trained by the Adam optimizer with the proposed multi-view contrastive loss (see Eq. 3). The tuning of MV-KGCL on each dataset is done via an extensive grid search for learning rate, number of the hidden layer, number of hidden units of a hidden layer, dropout

Table 3. Accuracy comparison with three clusters for blood pressure and cognition.

Dataset	K-Means	RF	DGI	InfoGraph	BGRL	Graph-MLP	MVGRL	MV-KGCL (ours)
Blood Pressure	32.61%	33.08%	34.26%	32.00%	35.05%	34.77%	33.90%	**36.83%**
Cognition	36.04%	33.31%	36.84%	37.07%	34.22%	37.13%	36.62%	**45.73%**

Table 4. Number of group pairs that are significantly different from each other.

	Cog A	Cog B	Cog C	Cog D	Cog E	Cog F
RF	1	0	1	0	0	1
DGI	1	0	1	1	0	2
InfoGraph	1	0	1	1	0	1
BGRL	1	0	1	1	0	1
Graph-MLP	1	0	1	1	0	2
MVGRL	0	0	1	1	1	2
MV-KGCL (ours)	**1**	**2**	**2**	**1**	**2**	**2**

probability, and strength of the L_2 regularization. For the kNN graph construction, we conduct a grid search for the threshold ϵ which is performed over the search space $\{0.3, 0.5, 0.8, 0.9\}$. In addition, for unsupervised learning evaluation metrics, we use the silhouette score (SIL; the higher the better), Calinski-Harabasz Index (CHI; the higher the better), and Davies-Bouldin Index (DBI; the smaller the better). We train the models with up to 500 epochs to ensure full convergence. The data and source code is available at www.dropbox.com/scl/fo/fuzhnn2nha52xc9k0tdqi/h?dl=0&rlkey=egfoid848ztgdb3zv78pv4odk.

Statistical Analyses *Factor Analysis:* To obtain a comprehensive description of cognitive function from 9 domains and assessments, we recover the latent and shared factors that may partially explain the joint variation among all cognition indices. To this end, we apply factor analysis to recover one latent factor, which would be interpreted as the general index of cognition function for the downstream statistical analysis. *Screening:* We conduct a screening step to select features that are most related to the general cognition index we acquire from the previous factor analysis. Using a linear model, we regress the general cognition index on each feature and selected features with p-value less than 0.1. *Regression:* We are interested in exploring and assessing whether the learned clusters based on multi-modal neuroimaging data were predictive to downstream cognitive functions. To this end, we apply ANOVA by imposing linear structures to compare the mean of all 9 cognition indices, respectively, among different clusters and evaluate how significant the difference is between clusters in terms of the p-value. Figures and tables would be found in the Results section.

Results. Tables 2 and 3 show the comparison of our proposed MV-KGCL model with the state-of-the-art baselines for participant unsupervised clustering tasks with and without ground truth labels on two datasets respectively. The results indicate that our MV-KGCL consistently achieves the best performance on all datasets compared to all baselines. As shown in Table 2, we observe that our model outperforms all the baselines across all three evaluation metrics. For instance, on high blood pressure, we find that (i) compared to machine learning approaches (i.e., K-Means and RF), MV-KGCL yields more than 355.49% relative improvement to RF in SIL and (ii) MV-KGCL improves the performance over the best GNN-based contrastive learning baseline (i.e., InfoGraph) with a relative gain of 45.05% in SIL. On cognition, similar to blood pressure, we can conclude that the proposed multi-view contrastive learning framework is effective for the neural networks to learn the intrinsic participant representation from topology space and feature space. That is, our MV-KGCL obtains relative gains of 190.49% and 46.59% over the K-Means and DGI (i.e., the best machine learning and GNN-based contrastive learning baselines) respectively. Furthermore, comparing with DGI and InfoGraph, we can learn that there does exist a structural difference between multiple views of the given participant dataset and performing conventional contrastive learning on one specific graph does not necessarily capture important topology information. The comparison results on datasets having ground truth information are presented in Table 3. As shown, our proposed MV-KGCL model outperforms current state-of-the-art baselines by a significant margin, i.e., (i) reaching relative gains of 9.50% on average compared with 7 baselines on blood pressure, and (ii) with a relative gain up to 37.29% on cognition. Moreover, our MV-KGCL consistently outperforms MVGRL on all datasets, indicating the effectiveness of the proposed two-stream multi-view contrastive learning mechanism in MV-KGCL, because it can learn participant information from multiple views both locally and globally. Table 4 provides a comprehensive evaluation of the predictive performance of different methods using six selected indices of cognition function. These indices encompass Cog A: Number of incorrect matches in a round, Cog B: Duration to complete the numeric path (trail #1), Cog C: Duration to complete the numeric path (trail #2), Cog D: Mean time to correctly identify matches, Cog E: Number of puzzles correct, and Cog F: Number of symbol digit matches made correctly. In determining the predictive capabilities of the recovered clusters, we consider the clustering to be predictive if the corresponding index shows a significant difference among any pair of clusters. As depicted in the table, our proposed MV-KGCL method exhibits superior predictive performance when compared to the competing methods for the downstream cognition indices. Specifically, MV-KGCL outperforms all other methods in predicting the indices of Cog B, Cog C, and Cog E by successfully distinguishing two pairs of clusters, while other methods either distinguish only one pair or fail to distinguish any pair at all. More information about the pairwise difference among three clusters learned from 7 methods can be found in Fig. 1, where we use "L", "M" and "H" to label the cluster with low, medium and high estimated measure of cognition function indices. As shown in these sub-figures, our MV-KGCL model in general leads to more (weak) significance with relatively larger group difference than other methods, meaning it achieves better ability to differentiate cognition function indices among clusters.

5 Conclusion

Topology, structural information, and features are pivotal for improving graph representation learning. While recent studies on graph representation learning mainly focus on modeling local/global structures of one pre-defined graph, fewer efforts have been made to study multiple views of the graph and integrating both local and global modeling into a graph representation learning framework. Therefore, in this paper, we propose a novel contrastive learning model, i.e., Multi-View k-Nearest Neighbor Graph Contrastive Learning (MV-KGCL) for participant-level analysis of real-world biomedical databases. Not limited to the realm of clustering, the proposed framework of dimension reduction can be readily extended to other fields, including supervised prediction. For example, the application of MV-KGCL to Medicare Claims data allows us to utilize high-dimensional patient-specific characteristics and facility-level factors as input features, thereby enhancing the accuracy of predicting post-fracture recovery for beneficiaries. Additionally, the cluster labels obtained or the predicted outcomes can be subjected to further analysis to explore the potential influence of environmental factors or lifestyles on the observed variability among clusters or outcomes. To achieve a deeper understanding and facilitate the translation of results into clinical decision-making and practice, advanced causal inference models such as propensity score weighting and regression calibration can be incorporated into the MV-KGCL framework. These approaches offer valuable avenues for investigation and should be pursued in future research endeavors.

References

1. Abu-El-Haija, S., et al.: MixHop: higher-order graph convolutional architectures via sparsified neighborhood mixing. In: ICML (2019)
2. Afanador, N.L., Smolinska, A., Tran, T.N., Blanchet, L.: Unsupervised random forest: a tutorial with case studies. J. Chemometr. **30**(5), 232–241 (2016)
3. Allen, N.E., Sudlow, C., Peakman, T., Collins, R., Biobank, UK: UK biobank data: come and get it (2014)
4. Bruna, J., Zaremba, W., Szlam, A., LeCun, Y.: Spectral networks and locally connected networks on graphs. In: ICLR (2014)
5. Defferrard, M., Bresson, X., Vandergheynst, P.: Convolutional neural networks on graphs with fast localized spectral filtering. In: NeurIPS, vol. 29 (2016)
6. Diaz-Orueta, U., Blanco-Campal, A., Lamar, M., Libon, D.J., Burke, T.: Marrying past and present neuropsychology: is the future of the process-based approach technology-based? Front. Psychol. **11**, 361 (2020)
7. Esteva, A., et al.: Dermatologist-level classification of skin cancer with deep neural networks. Nature **542**(7639), 115–118 (2017)
8. Hassani, K., Khasahmadi, A.H.: Contrastive multi-view representation learning on graphs. In: ICML, pp. 4116–4126 (2020)
9. Ho, T.K.: Random decision forests. In: ICDAR, vol. 1, pp. 278–282. IEEE (1995)
10. Hu, Y., You, H., Wang, Z., Wang, Z., Zhou, E., Gao, Y.: Graph-MLP: node classification without message passing in graph. arXiv preprint arXiv:2106.04051 (2021)
11. Kipf, T.N., Welling, M.: Semi-supervised classification with graph convolutional networks. In: ICLR (2017)

12. Klicpera, J., Bojchevski, A., Günnemann, S.: Predict then propagate: graph neural networks meet personalized PageRank. In: ICLR (2019)
13. Kochunov, P., et al.: Heritability of fractional anisotropy in human white matter: a comparison of human connectome project and ENIGMA-DTI data. Neuroimage **111**, 300–311 (2015)
14. Köhn, H.F., Hubert, L.J.: Hierarchical Cluster Analysis. Wiley StatsRef: Statistics Reference Online, pp. 1–13 (2014)
15. Lee, H., Chen, C., Kochunov, P., Elliot Hong, L., Chen, S.: Modeling multivariate age-related imaging variables with dependencies. Stat. Med. **41**(22), 4484–4500 (2022)
16. Linsker, R.: Self-organization in a perceptual network. Computer **21**(3), 105–117 (1988)
17. McKinney, S.M., et al.: International evaluation of an AI system for breast cancer screening. Nature **577**(7788), 89–94 (2020)
18. van Oostveen, W.M., de Lange, E.C.: Imaging techniques in Alzheimer's disease: a review of applications in early diagnosis and longitudinal monitoring. Int. J. Mol. Sci. **22**(4), 2110 (2021)
19. Qiu, J., et al.: GCC: graph contrastive coding for graph neural network pre-training. In: SIGKDD, pp. 1150–1160 (2020)
20. Sudlow, C., et al.: UK biobank: an open access resource for identifying the causes of a wide range of complex diseases of middle and old age. PLoS Med. **12**(3), e1001779 (2015)
21. Sun, F.Y., Hoffman, J., Verma, V., Tang, J.: InfoGraph: unsupervised and semi-supervised graph-level representation learning via mutual information maximization. In: ICLR (2019)
22. Thakoor, S., Tallec, C., Azar, M.G., Munos, R., Veličković, P., Valko, M.: Bootstrapped representation learning on graphs. In: ICLR 2021 Workshop on Geometrical and Topological Representation Learning (2021)
23. Tian, Y., Krishnan, D., Isola, P.: Contrastive multiview coding. In: Vedaldi, A., Bischof, H., Brox, T., Frahm, J.-M. (eds.) ECCV 2020. LNCS, vol. 12356, pp. 776–794. Springer, Cham (2020). https://doi.org/10.1007/978-3-030-58621-8_45
24. Velickovic, P., Fedus, W., Hamilton, W.L., Liò, P., Bengio, Y., Hjelm, R.D.: Deep graph infomax. In: International Conference on Learning Representations, vol. 2(3), p. 4 (2019)
25. You, Y., Chen, T., Sui, Y., Chen, T., Wang, Z., Shen, Y.: Graph contrastive learning with augmentations. In: NeurIPS, vol. 33, pp. 5812–5823 (2020)
26. Zhao, B., et al.: Common genetic variation influencing human white matter microstructure. Science **372**(6548), eabf3736 (2021)

Quasi-Orthogonal ECG-Frank XYZ Transformation with Energy-Based Models and Clinical Text

Lorenzo Simone(✉), Davide Bacciu, and Vincenzo Gervasi

Department of Computer Science, University of Pisa, Pisa 56122, Italy
lorenzo.simone@di.unipi.it

Abstract. The transformation of 12-Lead electrocardiograms to 3D vectorcardiograms, along with its reverse process, offer numerous advantages for computer visualization, signal transmission and analysis. Recent literature has shown increasing interest in this structured representation, due to its effectiveness in various cardiac evaluations and machine learning-based arrhythmia prediction. Current transformation techniques utilize fixed matrices, often retrieved through regression methods which fail to correlate with patient's physical characteristics or ongoing diseases. In this paper, we propose the first quasi-orthogonal transformation handling multi-modal input (12-lead ECG and clinical annotations) through a conditional energy-based model. Within our novel probabilistic formulation, the model proposes multiple transformation coefficients without relying on a single fixed approximation to better highlight relationships between latent factors and structured output. The evaluation of our approach, conducted with a nested cross validation on PTB Diagnostic dataset, showcased improved reconstruction precision across various cardiac conditions compared to state-of-the-art techniques (Kors, Dower, and QSLV).

Keywords: generative models in healthcare · energy-based models · electrocardiography · natural language processing · signal analysis

1 Introduction

The leading approach for early detection of potentially life-threatening conditions relies on electrocardiography, which monitors the heart's electrical activity through electrocardiograms (ECGs). The interpretation of these recordings is notably challenging, as abnormalities may be localized within specific leads of the standard 12-lead ECG or occur sparsely throughout the recording duration.

From an application domain's perspective, our contribution is centered on improving state of the art methodologies to connect this representation with vectorcardiography, being an integral part of broader electrocardiography field. A vectorcardiogram (VCG) presents an alternative three-dimensional visalization, emphasizing the QRS, P, and T complexes within a distinct three-dimensional

J. Finkelstein et al. (Eds.): AIME 2024, LNAI 14845, pp. 249–253, 2024.
https://doi.org/10.1007/978-3-031-66535-6_27

orthogonal format. Its peculiarity lies in its derivation of vector loops, obtained from the successive instantaneous mean electric forces originating from the heart throughout the cardiac cycle. The sequence, magnitude, and spatial direction of the instantaneous mean vectors are influenced by the order of depolarization and repolarization of the atria and ventricles [1]. Recent literature confirms the effectiveness of this structured representation [8], especially in assessing electrically inactive areas, intraventricular conduction disorders, ventricular preexcitation, axis deviation patterns, evaluating specific aspects of Brugada syndrome, and in machine-learning based arrhythmia prediction approaches [9].

To date, state-of-the-art methods to recover a vectorcardiogram from an electrocardiogram and vice versa, rely on signal multiplication with fixed transformation matrices. Dower et al. [2] utilized a model-based approach to derive these matrices, minimizing the error between physically measured and transformed VCGs. Subsequently, Kors et al. [5] conducted a comparative analysis between two quasi-orthogonal lead systems and a novel regression matrix with improved reconstruction fidelity. More recently, Guillem et al. [4] introduced a regression-based transformation matrix specifically optimized for QRS complex reconstruction, referred to as QLSV. Although these methods have served as gold standard for years due to their ease of application and simplicity, the growing interest in vectorcardiography applications and the increased demand in precision highlight the necessity for a tailored approach. With these objectives in mind, we devised the first personalized method that integrates patient information throughout the transformation process. The proposed approach guides the sampling process towards optimal coefficients with a generative model, which leverages an embedded representation of clinical annotations.

Our paper contributions can be summarized as follows: (1) we introduce the first methodology for a personalized vectorcardiography transformation leveraging textual clinical summaries in a multi-modal input setting; (2) we introduce a novel probabilistic approach using a conditional energy-based model which, in contrast to classical regression methods, predicts a set of candidate coefficients for a transformation matrix; (3) we achieved state-of-the-art results in terms of precision and fidelity of reconstruction across various heart conditions, including myocardial infarction, atrial fibrillation, bradycardia, tachycardia, and right bundle branch block.

2 Proposed Methodology

The target problem consists in retrieving a Frank-XYZ vectorcardiogram from a 12-lead ECG, also characterized as $\mathbf{E} = \{\mathbf{e}_i\}_{i=1}^{T}$, where $\mathbf{e}_i \in \mathbb{R}^{12}$ for T time steps. Thus, the objective can be easily defined as finding the coefficients of the optimal transformation matrix $\mathbf{M}^* \in \mathbb{R}^{3\times 12}$, in order to obtain an associated 3D vectorcardiogram $\mathbf{V} = \{(\mathbf{v}_{ix}, \mathbf{v}_{iy}, \mathbf{v}_{iz})\}_{i=1}^{T}$. It is commonly framed as a least squares optimization problem within the coefficient space $\mathbf{M}^* = \arg\min_M \left\|\mathbf{E}\mathbf{M}^T - \mathbf{V}\right\|_2^2$.

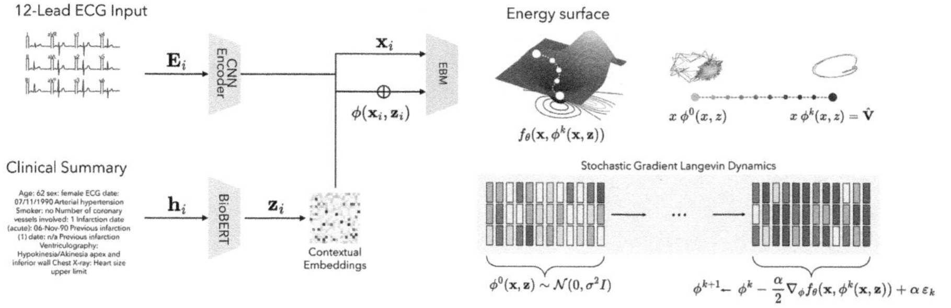

Fig. 1. Workflow of the proposed methodology for energy-based inference and sampling through MCMC Langevyn dynamics.

Rather than retrieving a general regression matrix, we learn a likely multi-modal distribution informed by clinical annotations through Energy-based models (EBMs [6]). We build on a training dataset containing pairs of ECG and ground truth recorded vectorcardiograms, $\mathcal{D} = \{(\mathbf{E}_i, \mathbf{V}_i)\}_{i=1}^n$, along with clinical annotations $\mathbf{h}_i$. Each input signal $\mathbf{E}_i$ is processed by a Convolutional Neural Network (CNN) for sequential data, yielding an intermediate representation $\mathbf{x}_i$. On the other hand clinical annotations $\mathbf{h}_i$ are encoded into contextual embeddings $\mathbf{z}_i : \mathbb{R}^{l \times d}$ using BioBERT [7], a well-known large language model for biomedical text. Then, we reframe the classical unconditional EBM framework extending it to learn the conditional distribution $p(\phi(\mathbf{x}, \mathbf{z})|\mathbf{x}; \theta)$ parametherized by θ, for a candidate matrix $\phi(\mathbf{x}, \mathbf{z})$ as:

$$p(\phi(\mathbf{x}, \mathbf{z})|\mathbf{x}; \theta) = \frac{\exp(-f_\theta(\mathbf{x}, \phi(\mathbf{x}, \mathbf{z})))}{\int \exp(-f_\theta(\mathbf{x}, \phi(\mathbf{x}, \mathbf{z})))d\phi(\mathbf{x}, \mathbf{z})} = \frac{\exp(-f_\theta(\mathbf{x}, \phi(\mathbf{x}, \mathbf{z})))}{Z_\theta} \quad (1)$$

In this context, $f_\theta(\mathbf{x}_i, \phi(\mathbf{x}_i, \mathbf{z}_i)) : \mathcal{X} \times \mathcal{M} \rightarrow \mathbb{R}$ represents an energy function, which, given a transformation matrix $\phi(\mathbf{x}_i, \mathbf{z}_i)$ and the associated input $\mathbf{x}_i$ returns a scalar value indicating the quality of the reconstruction $\hat{\mathbf{V}}_i = \mathbf{x}_i \phi(\mathbf{x}_i, \mathbf{z}_i)$. As a training objective, we aim to maximize the log-likelihood of the conditional probability given its gradient:

$$\nabla_\theta - \log p(\phi(\mathbf{x}, \mathbf{z})|\mathbf{x}; \theta) = \mathbb{E}_{\hat{\phi} \sim p_\theta}[-\nabla_\theta f_\theta(\mathbf{x}, \hat{\phi}(\mathbf{x}, \mathbf{z}))] - \nabla_\theta f_\theta(\mathbf{x}, \phi(\mathbf{x}, \mathbf{z})). \quad (2)$$

We approximate the expectation over the EBM distribution $\hat{\phi} \sim p_\theta$ via Markov Chain Monte Carlo (MCMC) Langevin dynamics from a prior isotropic Gaussian $\phi^0 \sim \mathcal{N}(0, \sigma^2 I)$, that is:

$$\phi^{k+1} = \phi^k - \frac{\alpha^2}{2} \nabla_\phi f_\theta(\mathbf{x}, \phi^k(\mathbf{x}, \mathbf{z})) + \alpha \varepsilon_k, \quad \text{where} \quad \varepsilon_k \sim \mathcal{N}(0, 1) \quad (3)$$

As a result, the sampling process iteratively utilizes the encoded multivariate signal x_i and a candidate transformation matrix, initialized with ϕ^0. This defines a trajectory $\{\phi^0, \phi^1, \ldots, \phi^k\}$ over the energy surface, pointing towards stable basins corresponding to transformation matrices maximizing the quality of the reconstruction according to the energy function f_θ (see Fig. 1).

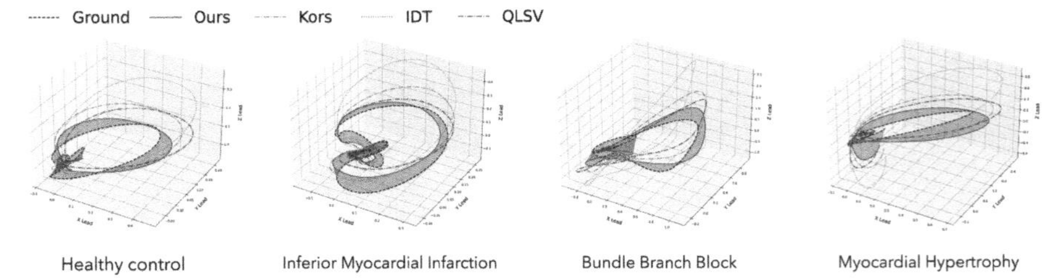

Fig. 2. Visual comparison of vectorcardiography reconstructions for healthy controls and patients suffering from various clinical conditions.

3 Experimental Results

We performed a nested cross validation procedure ($k = 5$), to validate our approach on the PTB Diagnostic Dataset [3], the most extensive collection of electrocardiography records paired with vectorcardiography ground truths (GT). For a precise assessment of reconstructed VCGs, we utilized key metrics from time series and point-cloud analysis, including Root Mean Squared Error (RMSE), correlation coefficient, Chebyshev distance, and Dynamic Time Warping (DTW). We report the averaged test set results with standard deviations in Table 1 divided in sections referring to invidual Frank-XYZ axis. Our method achieves state-of-the-art performances, notwithstanding the additional challenge of accounting for variability in clinical annotations across unseen samples. The precision of standard approaches, especially along Y and Z-lead, aligns with medical literature concerns regarding the usage of a universal coefficient matrix that neglects any correlation with patient characteristics. In Fig. 2, we compare our reconstructions for patients suffering from a variety of conditions, including myocardial infarction, bundle branch block, and myocardial hypertrophy, including distances from the ground truth within the blue shadowed region.

Table 1. Test set comparison with state-of-the-art transformation methods, including distance and correlation metrics within each Frank-XYZ lead.

	RMSE (↓)	Correlation (↑)	Chebyshev (↓)	DTW (↓)
$\text{Kors}_{[x]}$	0.0337 ± 0.01	$\mathbf{0.8511 \pm 0.01}$	0.1160 ± 0.02	24.03 ± 4.02
$\text{Dower}_{[x]}$	0.0678 ± 0.01	0.8068 ± 0.01	0.2421 ± 0.05	48.80 ± 11.42
$\text{QLSV}_{[x]}$	0.0482 ± 0.01	0.8390 ± 0.02	0.1705 ± 0.03	37.13 ± 8.92
Ours$^{[x]}$	$\mathbf{0.0312 \pm 0.00}$	$\underline{0.8487 \pm 0.01}$	$\mathbf{0.1052 \pm 0.01}$	$\mathbf{23.19 \pm 4.16}$
$\text{Kors}_{[y]}$	0.0403 ± 0.00	$\mathbf{0.8031 \pm 0.01}$	0.1273 ± 0.02	32.12 ± 4.34
$\text{Dower}_{[y]}$	0.0465 ± 0.01	0.7464 ± 0.03	0.1506 ± 0.03	36.82 ± 4.59
$\text{QLSV}_{[y]}$	0.0497 ± 0.01	0.7813 ± 0.03	0.1642 ± 0.02	43.78 ± 4.41
Ours$^{[y]}$	$\mathbf{0.0336 \pm 0.00}$	$\underline{0.8019 \pm 0.01}$	$\mathbf{0.1108 \pm 0.02}$	$\mathbf{26.43 \pm 2.62}$
$\text{Kors}_{[z]}$	0.0577 ± 0.01	0.7995 ± 0.01	0.2150 ± 0.03	40.61 ± 3.80
$\text{Dower}_{[z]}$	0.0974 ± 0.01	0.7391 ± 0.04	0.3695 ± 0.04	72.00 ± 3.07
$\text{QLSV}_{[z]}$	0.0576 ± 0.01	0.7753 ± 0.02	0.2070 ± 0.03	40.21 ± 3.68
Ours$^{[z]}$	$\mathbf{0.0471 \pm 0.01}$	$\mathbf{0.8015 \pm 0.01}$	$\mathbf{0.1774 \pm 0.03}$	$\mathbf{31.56 \pm 2.97}$

4 Conclusions

In this study, we presented a novel approach for personalized electrocardiography to vectorcardiography transformation, validated across various cardiac diseases. We achieved state-of-the-art reconstruction quality by reformulating the classical regression approach via conditional energy-based models, informed by clinical summaries provided by clinicians at hospital admission. The empirical findings unveiled the limitations of classical regression approaches, being simple approximations, which fail to account for patient-specific information during the transformation process. Despite the inherent challenges due to the limited availability of datasets containing both ECG and clinical annotations, and to the complexity of our model compared to traditional fixed transforms, our conditional generative approach sets the stage for future personalized medicine applications. The detailed clinical summaries used for training reveal a wealth of information, from patients' physical attributes to medical conditions and treatments. A generative model can leverage textual data to grasp patterns and relationships with multivariate signals. This enables the conditioning of transformations based on variations in clinical summaries, offering a means to use text prompts to assess the effects of specific treatments, ultimately facilitating personalized medicine applications.

References

1. Burch, G., Abildskov, J., Cronvich, J.: Vectorcardiography. Circulation **8**(4), 605–613 (1953)
2. Dower, G.E., Machado, H.B., Osborne, J.: On deriving the electrocardiogram from vectorcardiographic leads. Clin. Cardiol. **3**(2), 87–95 (1980)
3. Goldberger, A., et al.: Physiobank, physiotoolkit, and physionet: components of a new research resource for complex physiologic signals. Circulation **101**(23), e215–e220 (2000)
4. Guillem, M., Sahakian, A.V., Swiryn, S.: Derivation of orthogonal leads from the 12-lead ECG. accuracy of a single transform for the derivation of atrial and ventricular waves. In: 2006 Computers in Cardiology, pp. 249–252. IEEE (2006)
5. Kors, J., Van Herpen, G., Sittig, A., Van Bemmel, J.: Reconstruction of the frank vectorcardiogram from standard electrocardiographic leads: diagnostic comparison of different methods. Eur. Heart J. **11**(12), 1083–1092 (1990)
6. LeCun, Y., Chopra, S., Hadsell, R., Ranzato, M., Huang, F.: A tutorial on energy-based learning. Predicting Structured Data **1**(0) (2006)
7. Lee, J., et al.: BioBERT: a pre-trained biomedical language representation model for biomedical text mining. Bioinformatics **36**(4), 1234–1240 (2020)
8. Riera, A.R.P., et al.: Significance of vectorcardiogram in the cardiological diagnosis of the 21st century. Clin. Cardiol. **30**(7), 319 (2007)
9. Willems, J.L., Lesaffre, E., Pardaens, J.: Comparison of the classification ability of the electrocardiogram and vectorcardiogram. Am. J. Cardiol. **59**(1), 119–124 (1987)

Explainable AI

Do You Trust Your Model Explanations? An Analysis of XAI Performance Under Dataset Shift

Lorenzo Peracchio, Giovanna Nicora(✉), Tommaso Mario Buonocore, Riccardo Bellazzi, and Enea Parimbelli

Department of Electrical, Computer and Biomedical Engineering, University of Pavia, Pavia, Italy
lorenzo.peracchio01@universitadipavia.it, {giovanna.nicora, riccardo.bellazzi,enea.parimbelli}@unipv.it

Abstract. Machine learning (ML) models are increasingly deployed in many critical settings such as medicine, where ensuring their trustworthiness is a priority. The inherent complexity and opacity of many classifiers, as well as the potential unreliability of their predictions, are contributing factors that hinder the use of Artificial Intelligence systems in clinical practice. In response, Explainability methods (XAI), such as AraucanaXAI, provide a human-interpretable explanation of the reasoning process behind the predicted outputs. However, the concordance between the prediction of the classifier and the one produced by the surrogate explainer model (i.e. *fidelity*) is not always guaranteed, especially if the sample is near the decision boundary or in a region of the feature space far from the training distribution (dataset shift), where the ML predictions may be unreliable. Reliability methods assess potential ML misclassification errors: we exploit a method able to assess reliability based on the local performance of the classifier and the distance of the sample from the training set, in order to identify unreliable predictions. We then check the robustness of AraucanaXAI on unreliable predictions on a simulated scenario of Myelodysplastic Syndromes risk assessment. We show that *fidelity* considerably drops when AraucanaXAI is used to explain predictions flagged as unreliable. Drawing from this, we provide guidance for XAI hyperparameters tuning. With this work, we study the interplay between reliability and explainability, crucial requirements for Trustworthy AI. Code is available at https://github.com/bmi-labmedinfo/araucana-xai/tree/master/notebooks.

Keywords: Artificial Intelligence Explainability · Machine Learning Reliability · Synthetic data

1 Introduction

As Artificial Intelligence (AI) and Machine Learning (ML) are showing their potential in supporting clinical decisions, concerns are rising regarding their transparency. The classification process behind many popular and high-performing algorithms, such as deep

J. Finkelstein et al. (Eds.): AIME 2024, LNAI 14845, pp. 257–266, 2024.
https://doi.org/10.1007/978-3-031-66535-6_28

networks and ensembles, is often perceived as obscure, thus preventing their applicability in clinical practice. To unveil the "black box" reasoning process, Explainability methods (XAI) have been developed throughout the years; these approaches are providing users with local (on single instances) or global (on a cohort) explanations, in various forms. For instance, perturbation-based approaches, such as LIME and AraucanaXAI [1], train a local "white box" classifier in the neighborhood of the instance whose prediction needs to be explained, thus locally substituting the more complex model with a ML explainable by the design, such as a regression or decision tree. Evaluation of XAI performance is critical, and different metrics have been proposed [2]. Among others, *fidelity* measures the degree of agreement in the predicted classification between the complex model and the local model. As the local model is expected to explain the behavior of the black box model, the assumption is that the local model and the global one predict the same class for a specific instance (*fidelity*). However, past studies highlighted that in some cases, for instance near the decision boundaries, the global and the local model disagree about the predicted class [1, 2]. Benchmarking different XAI methods is also essential: most studies evaluating XAI are comparing different methods to assess explanations agreement. In [3], authors compared popular XAI approaches on a clinical problem, finding high agreement among the XAI methods (greater than 80%) in indicating top features contributions. Yet, another study indicates higher level of disagreement, and calls for the development of strategies to deal with explanations discordance [4].

Along with lack of transparency, a major barrier for the clinical adoption of AI is poor generalization ability under dataset shift [5] and under-represented populations. As dataset shift occurs over time, the performance of the ML decreases [6], making the ML predicted classifications unreliable. We have previously developed different approaches to assess pointwise reliability of ML predictions [7–9]. Briefly, we flag a ML prediction on a single instance as reliable if 1) the instance comes from the same training distribution (i.e., no dataset shift has occurred) and 2) the model performs well on similar instances.

While explainability and reliability of predictions are recognized as essential aspects to fulfil in order to achieve trustworthy AI systems in healthcare, their interplay has not been studied yet. In this work, we aim at understanding the performance of XAI on unreliable predictions. In particular, we want to assess how robust is the perturbation-based method AraucanaXAI to dataset shift and unreliability issues. To do so, we will address a problem of Myelodysplastic Syndromes (MDS) risk assessment, whose related prognostic scoring system provides risk scores in a simple and deterministic way, ensuring transparency in the risk labelling process.

2 Materials and Methods

The goal of this work is to measure the performance of the AraucanaXAI [1] method on unreliable regions of the features set, i.e. regions where the ML may be wrong at a higher rate due to dataset shift or poor local fitting of the model. We simulated a clinical dataset under dataset shift, and we evaluated XAI performance on reliable and unreliable predictions. To evaluate the XAI method performance, we compute *fidelity* as percentage of cases in which the XAI proxy model and the complex "black box" model agree on predictions.

2.1 Dataset

We generated a simulated dataset based on the Revised International Prognostic Scoring System (IPSS-R) [10] for MDS risk assessment. MDS consist of a heterogeneous spectrum of myeloid clonal hemopathies leading to dysplasia and ineffective hematopoiesis in the bone marrow [11]; to assess the prognosis of primary untreated MDS adult patients, the standard IPSS proposed in 1997 has been refined in 2012 through a multivariate analysis of survival time and time to Acute Myeloid Leukemia transformation. The resulting prognostic categorization model IPSS-R considers five main clinical features: hemoglobin, absolute neutrophil count (ANC), platelets, bone marrow blasts (BMB), and cytogenic category. Each of these features contributes to the total risk score based on its value, as reported in Table 1.

Table 1. IPSS-R prognostic score values

Prognostic Score	0	0.5	1	1.5	2	3	4
Cytogenetic category	Very Good		Good		Intermediate	Poor	Very Poor
BMB (%)	<=2		>2% < 5%		>5% < 10%	>10%	
Hemoglobin (g/dL)	>=10		>=8 < 10	<8			
Platelets (x 10^9/L)	>=100	>=50 < 100	<50				
ANC (x 10^9/L)	>=0.8	<0.8					

The sum of the prognostic scores associated with each clinical feature gives the final risk score, which identifies the five risk categories illustrated in Table 2.

Table 2. IPSS-R prognostic risk categories/scores

Risk Category	Risk Score
Very Low	<=1.5
Low	>1.5 & <3
Intermediate	>=3 & <4.5
High	>=4.5 & <6
Very High	>6

This scoring system allows us to have a deterministic model that can be used to generate the simulated dataset: each sample (i.e. simulated patient) is characterized by the clinical features considered for the computation of the IPSS-R (see Table 1), and is labelled in a risk category based on the risk score resulting from the sum of the prognostic scores associated with its features' values (see Tables 1 and 2). Starting from this, we defined two high-level risk categories: samples whose risk score is equal or lower than 4.5 are labelled as "Low Risk", and samples whose risk score is higher than 4.5 are labelled as "High Risk". To generate the training samples, since we want to simulate a scenario of dataset shift in which the training set has only Independent and Identically Distributed (IID) samples while the test set used for evaluation has also Out-of-Distribution (OOD) samples, we relied on the usual values' ranges of the clinical features reported in the IPSS-R calculator of the MDS Foundation. In particular, we generated a completely IID dataset of 1000 patients by extracting values in the usual ranges, and a dataset of 700 both IID and OOD patients by extracting values in wider ranges (see Table 3 for details), with the constrain of allowing only one feature per sample to possibly be out of the usual range. This produced 347 completely IID samples and 353 samples with a feature out of its usual range. Extracting from the extended ranges may produce values that are not feasible in reality and, in addition, each feature's value, both IID and OOD, is randomly extracted from its range independently from the other features; altogether, this may possibly lead to generate samples that are not clinically reasonable. However, the goal of this procedure is to build a dataset to consider IID and a dataset of also OOD samples, which will be used to test the hypothesis proposed in this work. The completely IID dataset is then split into a training set and a validation set, used to train the classifier and to implement and tune the hyperparameters of the reliability method employed [7], which learns to identify as reliable the IID samples. The other dataset is used for evaluation (we will refer to it as "test set") and will be exploited to drive conclusions about the influence of dataset shift on the AraucanaXAI explanations. Additionally, we investigate the effect of AraucanaXAI hyperparameters' tuning, based on its performance on the reliable and unreliable samples.

Table 3. IPSS-R clinical features usual and extended ranges

Features	Usual Ranges	Extended ranges
Cytogenetic category	-	-
BMB (%)	[0–30]	[0–61]
Hemoglobin (g/dL)	[4–20]	[4–41]
Platelets (x 10^9/L)	[0–2000]	[0–4000]
ANC (x 10^9/L)	[0–15]	[0–31]

2.2 Experimental Setup

We designed an experiment to evaluate how the performances of AraucanaXAI change with the change of its hyperparameters, when used to explain reliable and unreliable predictions of a classifier. To this end, we relied on *fidelity:* this metric measures the concordance of the prediction between the XAI model and the classifier. In other words, *fidelity* estimates if the explanations accurately describe the classifier's behaviour in the entire feature space [12].

The first step consisted in training a classifier on the IID training set; we implemented a Support Vector Machine with a Radial Basis Function (RBF) kernel, a model that can be very complex and opaque, thus not inherently explainable [13]. The same training set was also used to implement the reliability method [7], which was then used to assess the reliability of the test set, in order to divide it in a reliable subset and an unreliable subset; the IID test samples should be flagged as reliable, while the OOD test samples should be detected as unreliable due to their distance from the training population.

Once the reliable and unreliable samples of the test set were determined, we applied AraucanaXAI to get the explanation of their classifier's prediction: in particular, we tested how the *fidelity* of the reliable and unreliable subsets changes with the change of the parameters' values. Given a single instance whose prediction needs to be explained, AraucanaXAI employs a decision tree as the "explainer" model, and trains it on a local neighbourhood of the instance. In this local area, it is possible to augment the local training set through oversampling [1]. The three principal parameters of AraucanaXAI to be tuned are the local neighbourhood size, the oversampling size, and the maximum depth of the tree. If we let the tree to grow unpruned, it guarantees a very high *fidelity*; yet, although the explanation is coherent to the original prediction, a tree characterized by excessive complexity may be difficult to navigate, thus the explainer may not be simple and highly interpretable. Hence, to produce clear explanations, we developed surrogate trees with a maximum depth of 2. Thus, the remaining parameters we considered for the experiments were the neighbourhood size and the oversampling size, which we varied within the established ranges of values reported in Table 4, and we computed the *fidelity* of the reliable and unreliable subsets with a tree implemented with each parameters' combination. The source code for the experimental setup is available at https://github.com/bmi-labmedinfo/araucana-xai/tree/master/notebooks.

Table 4. Tested parameters values of AraucanaXAI.

Parameter	Tested Values
Neighbourhood size	[0.05, 0.07, 0.10, 0.12, 0.15, 0.17, 0.20, 0.22, 0.25, 0.27, 0.30, 0.32, 0.35, 0.37, 0.40, 0.42, 0.45, 0.47, 0.50, 0.52]
Oversampling size	[30, 50, 70, 100, 130, 150, 170, 200, 230, 250, 270, 300, 330, 350, 370, 400, 420, 450, 470, 500]

3 Results

3.1 Reliability Analysis

The application of the reliability method on the test samples identified 358/700 (~51.14%) reliable predictions and 342/700 (~48.86%) unreliable predictions. It is worthwhile noticing that we designed the test set to have a 50% chance to generate each sample with an OOD feature, which is in line with the percentages of reliability we found. 336 out of the 342 unreliable samples and 17 out of the 358 reliable samples exhibit OOD features; however, for the reliable samples, the values of these features lie close to their usual range, indicating their reliability. Indeed, we designed our reliability method not to flag as unreliable those values that slightly deviate from the training data.

As illustrated in Fig. 1, we tested how the performances of the classifier (SVM) change when applied on the entire test set (in blue), and when applied only on its reliable (in green) and unreliable (in red) subsets: the SVM performs better overall when it is applied within its intended use, i.e. on samples from the same distribution as the training set (those flagged as reliable by the reliability method).

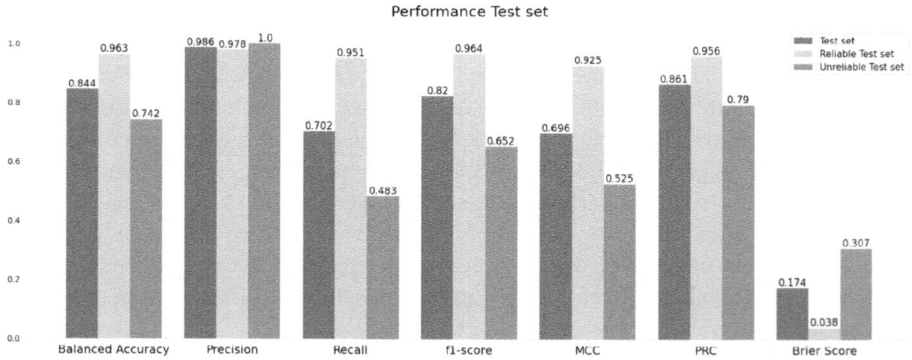

Fig. 1. Performances of the SVM on the total test set (in blue), on its reliable subset (in green), and on its unreliable subset (in red), as identified by the reliability method. (Color figure online)

3.2 AraucanaXAI Analysis

Figure 2 shows how the *fidelity* of the explanations obtained by AraucanaXAI on the reliable and unreliable subsets changes with the increase of the neighborhood size (as proportion of the training set), when no oversampling is performed. Notably, the *fidelity* is very high when AraucanaXAI is used to explain reliable predictions (average 0.933), while it considerably drops when dealing with unreliable ones (average 0.792); moreover, it is more stable to changes in neighborhood size (0.027 vs 0.077 of standard deviation). These results show that AraucanaXAI is more robust when it is used to explain reliable predictions, while its performances get worse when it is used on OOD samples, whose classifier's predictions cannot be trusted.

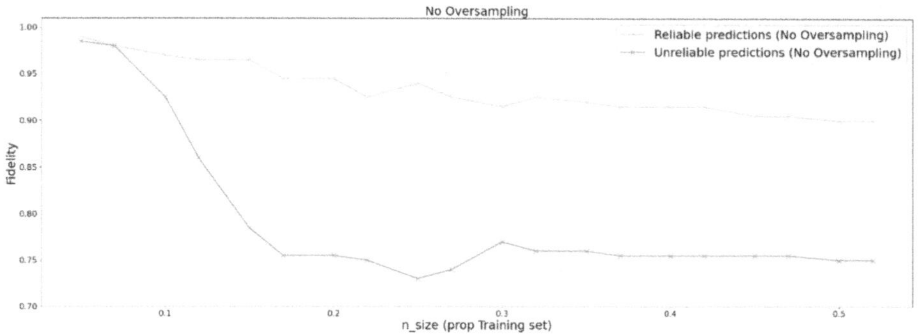

Fig. 2. *Fidelity* of the AraucanaXAI explanations of the reliable (in light green) and unreliable (in light red) predictions with different values of neighbourhood size (number of training samples computed as proportion of the training size). (Color figure online)

Then, we repeated the experiment including the oversampling step of AraucanaXAI, consisting of a random oversampling from a uniform distribution, performed to generate additional local training samples. The *fidelity* of the explanations of the reliable and unreliable predictions, computed with different combinations of oversampling size and neighbourhood size, is illustrated in Fig. 3, which proves that AraucanaXAI has lower performance when used to explain unreliable predictions, regardless of the chosen hyperparameters: in general, explanations of predictions produced out of the intended use of a classifier are less accurate than their reliable counterparts.

Focusing on the unreliable subset, performing oversampling leads to an improvement of *fidelity*: this provides guidance for hyperparameters tuning of AraucanaXAI, suggesting that the oversampling step is necessary when dealing with unreliable predictions to get more robust explanations.

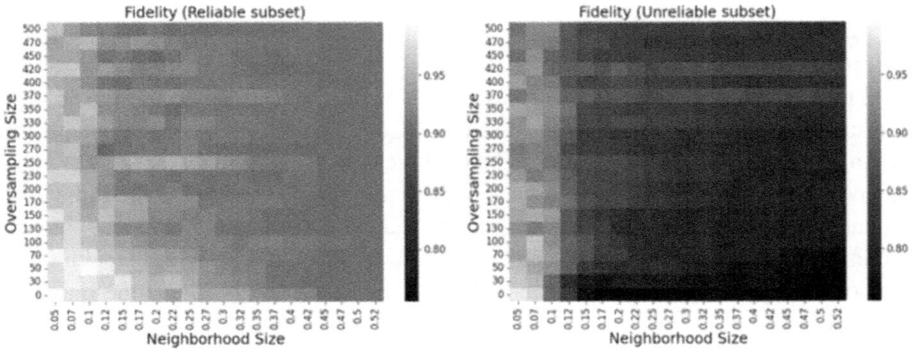

Fig. 3. *Fidelity* of AraucanaXAI on the reliable (left) and unreliable (right) subsets of the test set, with different combinations of neighbourhood size and oversampling size. The lower row of the heatmap refers to the non-oversampling case of Fig. 2.

Figure 4 gives a portrait of how *fidelity* changes when AraucanaXAI includes the oversampling of 370 samples, with respect to the non-oversampling experiment of Fig. 2;

remarkably, the unreliable predictions' *fidelity* significantly improves (Wilcoxon test's p-value: 2.38e-05); Table 5 reports the mean and standard deviation of the *fidelity* achieved with and without performing oversampling, for both the reliable and unreliable cases.

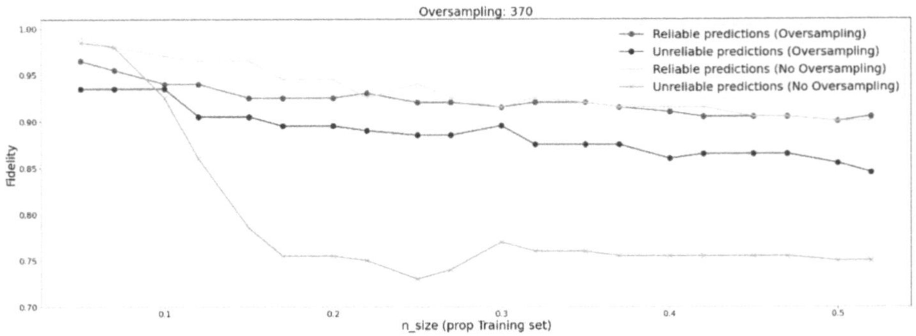

Fig. 4. "Oversampling-fidelity" of the AraucanaXAI explanations of the reliable (in green) and unreliable (in red) predictions with different values of neighbourhood size, with respect to the non-oversampling case (in light green and light red). (Color figure online)

Table 5. Average value and standard deviation of *fidelity* of the AraucanaXAI explanations computed with and without oversampling, of the reliable and unreliable predictions.

Oversampling	Reliable	Unreliable
Yes	0.909 (0.016)	0.872 (0.025)
No	0.933 (0.027)	0.792 (0.077)

4 Discussion

AI in medicine has shown promising results in several applications, from clinical decision support to biological knowledge discovery. Consequently, an increasing number of AI tools are implemented and translated into clinical practice: as of October 2023, more than 600 AI-based medical devices were approved by the FDA [14]. At the same time, regulators and stakeholders are raising concerns about the actual implementation of AI in clinical practice, as AI failures can cause harm to patients. The World Health Organization (WHO) is calling for safe and ethical AI in healthcare [15], and the European Commission is promoting the ALTAI list as a tool to assess whether the AI system that is being developed, deployed, procured, or used, complies with the requirements of Trustworthy AI. Two major requirements are reliability of predictions and transparency. Yet, while several approaches have been developed to tackle these two issues independently, to the best of our knowledge their interplay has not been thoroughly studied.

Here, we assess how predictive unreliability issues, caused by dataset drift, impact transparency. Tested on a shifted dataset of MDS, we used the reliability method implemented in [7] to identify reliable and unreliable predictions, which respectively showed better and worse performance (Fig. 1), proving that the classifier performs poorly on OOD samples. We then applied AraucanaXAI to get the explanations of both the reliable and unreliable predictions and we computed their *fidelity*, which showed to be significantly higher when the surrogate explainer tree is used to explain reliable predictions, while it considerably drops when dealing with unreliable ones; this is true for any combination of neighborhood size and oversampling size we tested (Fig. 3). Keeping a very low neighborhood size seems to be the best choice for achieving a high *fidelity*: in truth, training the surrogate tree on an overly low number of samples would result in overfitting, which guarantees the concordance between the XAI model prediction and the original classifier's one, but does not provide a meaningful explanation. From Fig. 3 it is also evident that performing oversampling leads to an improvement of *fidelity* on the unreliable subset with respect to the non-oversampling case: AraucanaXAI algorithm allows to introduce an oversampling step to enrich the local neighborhood, useful for training the explainer tree with more data points and making the decision boundary more explicit; this strategy is then particularly advantageous when explaining predictions that are unreliable due to the distance of the samples from the training population (i.e. they lay in a region not populated by training samples) or due to the low local accuracy of the classifier.

5 Conclusion

The predictive reliability of ML models strongly impacts the explanations provided by XAI; within reliable regions the explanations are more accurate (i.e. their *fidelity* is higher) and more robust to changes in the hyperparameters, while they get worse and more unstable where the classifier is unreliable. ML models are intended to be used on samples distributed as the training set, thus their reliability is subject to decrease in cases of dataset shift [6]. If accurately explaining reliable predictions is necessary to guarantee the transparency of a model, the application of a classifier out of the training distribution, leading to performance degradation [16], is a situation where it would be worthwhile to have an accurate local explanation for model inspection and debugging; in this context, reliability can be used to guide the hyperparameters tuning of AraucanaXAI to provide better explanations also in regions of low performances of the classifier.

Our study focused on a specific tool for explainability, namely AraucanaXAI. Future works will include further evaluation on other popular XAI approaches, such as SHAP [17], and benchmarking of different XAI under different reliability scenarios. Additionally, other performance measures along with *fidelity* will be investigated.

Acknowledgments. This work was supported by "Fit4MedRob-Fit for Medical Robotics" Grant B53C22006950001 and PNRR Ricerca DM 118/2023 CUP F13C23000490007.

Disclosure of Interests. The authors declare that they have no known competing financial interests or personal relationships that could have appeared to influence the work reported in this article.

References

1. Parimbelli, E., Buonocore, T.M., Nicora, G., Michalowski, W., Wilk, S., Bellazzi, R.: Why did AI get this one wrong? — Tree-based explanations of machine learning model predictions. Artif. Intell. Med. **135**, 102471 (2023). https://doi.org/10.1016/j.artmed.2022.102471
2. El Shawi, R., Sherif, Y., Al-Mallah, M., Sakr, S.: Interpretability in healthcare a comparative study of local machine learning interpretability techniques. In: Proceedings of the 2019 IEEE 32nd International Symposium on Computer-Based Medical Systems (CBMS), p. 275–80 (2019). https://doi.org/10.1109/CBMS.2019.00065
3. Brankovic, A., Cook, D., Rahman, J., Huang, W., Khanna, S.: Evaluation of popular XAI applied to clinical prediction models: can they be trusted? (2023). https://doi.org/10.48550/arXiv.2306.11985
4. Krishna, S., Han, T., Gu, A., Pombra, J., Jabbari, S., Wu, S., et al.: The disagreement problem in explainable machine learning: a practitioner's perspective (2022). https://doi.org/10.48550/arXiv.2202.01602
5. Wong, A., Otles, E., Donnelly, J.P., Krumm, A., McCullough, J., DeTroyer-Cooley, O., et al.: External validation of a widely implemented proprietary sepsis prediction model in hospitalized patients. JAMA Intern. Med. **181**, 1065–1070 (2021). https://doi.org/10.1001/jamainternmed.2021.2626
6. Moreno-Torres, J.G., Raeder, T., Alaiz-Rodríguez, R., Chawla, N.V., Herrera, F.: A unifying view on dataset shift in classification. Pattern Recogn. **45**, 521–530 (2012). https://doi.org/10.1016/j.patcog.2011.06.019
7. Peracchio, L., Nicora, G., Parimbelli, E., Buonocore, T.M., Bergamaschi, R., Tavazzi, E., et al.: Evaluation of predictive reliability to foster trust in artificial intelligence. A case study in multiple sclerosis (2024). https://doi.org/10.48550/arXiv.2402.17554
8. Nicora, G., Rios, M., Abu-Hanna, A., Bellazzi, R.: Evaluating pointwise reliability of machine learning prediction. J. Biomed. Inform. **127**, 103996 (2022). https://doi.org/10.1016/j.jbi.2022.103996
9. Nicora, G., Bellazzi, R.: A reliable machine learning approach applied to single-cell classification in acute myeloid leukemia. AMIA Annu. Symp. Proc. **2020**, 925–932 (2021)
10. Greenberg, P.L., Tuechler, H., Schanz, J., Sanz, G., Garcia-Manero, G., Solé, F., et al.: Revised international prognostic scoring system for myelodysplastic syndromes. Blood **120**, 2454–2465 (2012). https://doi.org/10.1182/blood-2012-03-420489
11. Greenberg, P.L., Young, N.S., Gattermann, N.: Myelodysplastic syndromes. Hematology **2002**, 136–161 (2002). https://doi.org/10.1182/asheducation-2002.1.136
12. Lopes, P., Silva, E., Braga, C., Oliveira, T., Rosado, L.: XAI systems evaluation: a review of human and computer-centred methods. Appl. Sci. **12**, 9423 (2022). https://doi.org/10.3390/app12199423
13. Belle, V., Papantonis, I.: Principles and practice of explainable machine learning. Front Big Data **4**, 688969 (2021). https://doi.org/10.3389/fdata.2021.688969
14. Joshi, G., Jain, A., Araveeti, S.R., Adhikari, S., Garg, H., Bhandari, M.: FDA-approved artificial intelligence and machine learning (AI/ML)-enabled medical devices: an updated landscape. Electronics **13**, 498 (2024). https://doi.org/10.3390/electronics13030498
15. World Health Organization: Ethics and governance of artificial intelligence for health: WHO guidance. World Health Organization (2021). https://iris.who.int/handle/10665/341996
16. Kelly, C.J., Karthikesalingam, A., Suleyman, M., Corrado, G., King, D.: Key challenges for delivering clinical impact with artificial intelligence. BMC Med. **17**, 195 (2019). https://doi.org/10.1186/s12916-019-1426-2
17. Lundberg, S.M., Lee, S.-I.: A unified approach to interpreting model predictions. Adv. Neural Inf. Process. Syst. **30** (2017)

Explainable AI for Fair Sepsis Mortality Predictive Model

Chia-Hsuan Chang, Xiaoyang Wang, and Christopher C. Yang(✉)

College of Computing and Informatics, Drexel University, Philadelphia, PA 19104, USA
{cc3859,xw388,chris.yang}@drexel.edu

Abstract. Artificial intelligence supports healthcare professionals with predictive modeling, greatly transforming clinical decision-making. This study addresses the crucial need for fairness and explainability in AI applications within healthcare to ensure equitable outcomes across diverse patient demographics. By focusing on the predictive modeling of sepsis-related mortality, we propose a method that learns a performance-optimized predictive model and then employs the transfer learning process to produce a model with better fairness. Our method also introduces a novel permutation-based feature importance algorithm aiming at elucidating the contribution of each feature in enhancing fairness on predictions. Unlike existing explainability methods concentrating on explaining feature contribution to predictive performance, our proposed method uniquely bridges the gap in understanding how each feature contributes to fairness. This advancement is pivotal, given sepsis's significant mortality rate and its role in one-third of hospital deaths. Our method not only aids in identifying and mitigating biases within the predictive model but also fosters trust among healthcare stakeholders by improving the transparency and fairness of model predictions, thereby contributing to more equitable and trustworthy healthcare delivery.

Keywords: Fairness · Explainable artificial intelligence · MIMIC-IV · Sepsis mortality prediction

1 Introduction

Artificial intelligence (AI) has the potential to revolutionize clinical decision-making by providing support to clinicians. Predictive modeling, a common AI technique, automates and accelerates various healthcare tasks, such as readmission, mortality, and disease recognition. However, it is crucial to prioritize fairness in these algorithms to ensure that the benefits of AI decision-making are distributed equitably across diverse patient populations. This helps to minimize the risk of perpetuating biases and inequalities in healthcare outcomes. Additionally, explainable AI (XAI) is one of the promising research topics in

C.-H. Chang and X. Wang—Both authors contributed equally.

J. Finkelstein et al. (Eds.): AIME 2024, LNAI 14845, pp. 267–276, 2024.
https://doi.org/10.1007/978-3-031-66535-6_29

healthcare informatics [5]. XAI explains the model's prediction, assisting clinicians and practitioners in gaining insights from data rather than mindlessly following the prediction outcomes [13]. Enhancing fairness and explainability gains more trust from the users [1,14] and is a key element of trustworthy AI designed for the healthcare system [5].

While various XAI techniques, such as SHapley Additive exPlanations (SHAP) [8], are available to explain the contribution of features to predictive performance, there remains a gap in explaining feature contributions to fairness enhancement. Understanding how specific features impact fairness is crucial, as this can help identify areas where biases may exist within the model and enable targeted interventions to mitigate these disparities. Therefore, exploring methods that provide explainability and facilitate the evaluation on fairness enhancement within clinical predictive models is essential.

Sepsis is one of the foremost life-threatening medical emergencies encountered by critically ill patients in the United States, arising as a complex syndrome that epitomizes the intricate interplay among the infecting microorganism, and the host's immune and coagulation responses. Annually, sepsis afflicts over 1.7 million adults in the United States, leading to a stark mortality rate wherein more than 350,000 of these individuals succumb during their hospital stay or are transitioned to hospice care [4]. Remarkably, sepsis is implicated in one out of every three hospital fatalities, underscoring its critical impact on healthcare outcomes. Given the pressing need to enhance clinical decision-making in the management of sepsis-a condition with profound implications for patient mortality-this study proposes a method, which integrates the principles of fairness and explainability into machine learning models tailored for sepsis mortality predictions.

The contributions of our proposed method are twofold: (1) we learn a fair sepsis mortality predictive model for different races from a performance-optimized predictive model by applying transfer learning, and (2) we introduce a novel permutation-based feature importance algorithm. Our method not only enables a nuanced understanding of how each individual feature contributes to enhancing fairness across different races but also provides healthcare practitioners with new insights and assistance. As far as we know, this is the first attempt to explain how features contribute to improving fairness. In our research, we extend the conventional focus on bias mitigation within predictive models to encompass a detailed examination of the relationship between features and fairness. By doing so, we bridge the realms of model explainability and fairness. Our findings contribute to the improvement of clinical outcomes, the reinforcement of ethical standards in healthcare AI, and the promotion of trust and transparency among medical professionals and patients alike.

2 Related Work

The process of explaining the rationale behind predictive modeling is crucial in ensuring fairness in human judgment [14]. AI-generated explanations can help identify factors that contribute to unfair outcomes. Both fairness and explanation are crucial aspects of creating trustworthy and responsible AI, and research shows that they are positively associated with user trust in scenarios where AI is used to make decisions [1]. However, the current approaches, LIME [9] and SHAP [8], only explain the contribution of features to predictive performance. As far as we know, no study yet proposes a method for explaining the contribution of features to fairness.

Currently, the researchers have devoted considerable efforts toward developing predictive models for sepsis mortality. Bao et al. [2] and Taylor et al. [10] explored a diverse array of models and features with the aim of identifying the model that exhibits optimal performance in predicting sepsis-related mortality and implemented feature importance analysis that elucidates the influence of individual features on the overall prediction performance. Wang et al. [12] have elucidated that disparities rooted in social determinants among groups of sepsis patients by various currently available diagnostic criteria and the efficacy of risk prediction models for sepsis is notably diminished when a universally trained model is indiscriminately applied across different subpopulations. However, these endeavors have predominantly aimed at enhancing prediction accuracy and have not conducted a thorough and precise analysis of explainability and fairness, leading to a situation where the predictive results, though accurate, may lack transparency and equitable consideration across diverse patient demographics.

3 Materials and Methods

3.1 Dataset Preparation

Our study focused on the critical task of predicting mortality among sepsis patients. The dataset we used is extracted from the MIMIC-IV database, which encompasses comprehensive medical records and demographic information of individuals diagnosed with sepsis. The Medical Information Mart for Intensive Care IV(MIMIC-IV) dataset [7] is a comprehensive collection of critical care records, detailing patient admissions to the Intensive Care Units (ICU) at Beth Israel Deaconess Medical Center. We preprocessed the dataset to make the experimental result reasonable and robust. In this study we measured the fairness by considering the predictive performance difference between races, so we removed the samples without a definitive race record and transformed the categorical race variable into a binary format: White vs Non-White. To avoid potential timely treatment bias, for multiple-time medical measurement records, we used the average of the first 24-hour records after the first ICU admission. To eliminate the extreme cases, the patient record exhibiting a value for at least one feature that fell outside the interpercentile range of [2%, 98%] for the total population was removed. Therefore, the remaining features are Age, Gender, Race, Temperature, Weight, Heart Rate, Glucose, Systolic Blood Pressure(SBP), Diastolic Blood Pressure(DBP), SpO2, Respiratory rate, Renal Replacement Therapy (RRT), Sequential Organ Failure Assessment (SOFA), Charlson Comorbidity Index (CCI), and APACHE-III score. The target label is the death of the patient, positive means the patient died, otherwise the patient survived. Table 1 presents the descriptive statistics of the dataset employed in our study, offering the detailed distribution of target variables.

Table 1. The MIMIC-IV Dataset Distribution

Characteristic	Overall (N = 9,349)	Neg. Class (N = 7,806)	Pos. Class (N = 1,543)
White	7,797 (83.4%)	6,546 (83.9%)	1,251 (81.1%)
Non-White	1,552 (16.6%)	1,260 (16.1%)	292 (18.9%)

3.2 Predictive Model and Fair Model

Figure 1 illustrates the process to obtain performance-optimized predictive model and fair model, detailed as follows.

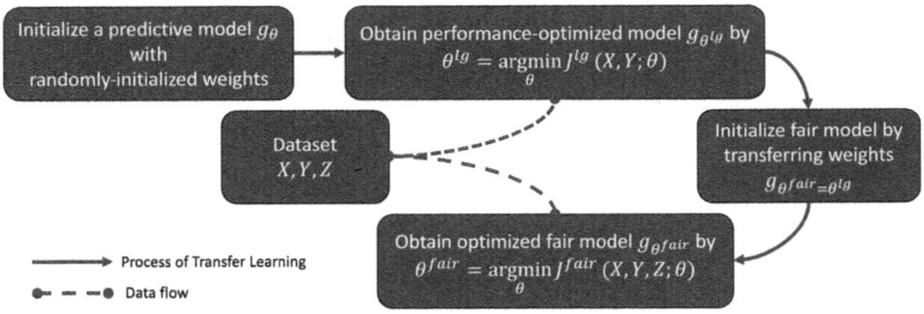

Fig. 1. The proposed transfer learning process for performance-optimized and fair model

Given the dataset $X \in \mathbb{R}^{n \times m}$, we train a logistic regression model as the mortality predictive model $g_{\theta^{lg}}$, where n is the number of septic patients and m is the number of features. We optimize feature weights $\theta^{lg} \in \mathbb{R}^m$ to minimize the binary cross-entropy between ground truths and predictions:

$$J^{lg}(X, Y; \theta) = -\sum_{i=1}^{n} [Y_i log(g_\theta(X_i)) + (1 - Y_i) log(1 - g_\theta(X_i))], \quad (1)$$

where ground truth $Y \in \mathbb{R}^n$ is al a list of dichotomous values, and Y_i indicates whether the patient i is dead (1) or not (0), and $g_\theta(X_i)$ is a outcome prediction to patient i. By minimizing Eq. 1, the resultant model, $g_{\theta^{lg}}$, is only performance-optimized for predictive performance. To make it generate the fair prediction for different races, we implement a transfer learning process that initializes a fair model by transferring weights θ^{lg} as fair model's initial weights $g_{\theta^{fair}=\theta^{lg}}$. The idea behind the weight transferring is that fair model inherits the pre-learned capabilities, ensuring its predictive ability is close to $g_{\theta^{lg}}$. Furthermore, we design a different optimization goal for fair model to make its predictions become fair across races. Specifically, we optimize the fair model for minimizing the Equalized Odds Disparity (EOD) between races. EOD is a comprehensive fairness measurement [6], denoted $\mathcal{E}$, that sums the square difference in true positive rate (TPR) and false positive rate (FPR) between races:

$$\begin{aligned} \mathcal{E}(g_\theta(X), Y, Z) &= TPR_{Diff} + FPR_{Diff} \\ TPR_{Diff} &= (Pr(g_\theta(X) = 1|Y = 1, Z = 1) - Pr(g_\theta(X) = 1|Y = 1, Z = 0))^2 \\ FPR_{Diff} &= (Pr(g_\theta(X) = 1|Y = 0, Z = 1) - Pr(g_\theta(X) = 1|Y = 0, Z = 0))^2 \end{aligned} \quad (2)$$

where $Z = 1$ for White and $Z = 0$ for Non-White, respectively. To prevent losing the predictive performance, we add a regularization term Ω, allowing overall TPR performance reached by performance-optimized predictive model $g_{\theta^{lg}}$ be changed in a small range: $\Omega(X, Y; \theta) = TPR - \epsilon < TPR < TPR + \epsilon$, where $TPR = Pr(g_\theta(X) = 1|Y = 1)$. As a result, the loss function turns out to be:

$$J^{fair}(X, Y, Z; \theta) = \mathcal{E}(g_\theta(X), Y, Z) + \Omega(X, Y; \theta), \quad (3)$$

we treat the optimized weights θ^{fair} as the final weights for the fair model, $g_{\theta^{fair}}$.

3.3 Permutation Fairness Importance Algorithm

Permutation-based feature importance method [3] was developed to quantify how each feature contributes to the predictive performance, but it does not measure how the features contribute to fairness when a fair model is compared with a base model.

To quantify the influence of each feature on the fairness improvement, we first define the fairness improvement, which calculates the difference of EOD performance ($\mathcal{E}$, Eq. 2) between $g_{\theta^{lg}}$ and $g_{\theta^{fair}}$:

$$\mathcal{F}(X, Y, Z, g_{\theta^{lg}}, g_{\theta^{fair}}) = \mathcal{E}(g_{\theta^{fair}}(X), Y, Z) - \mathcal{E}(g_{\theta^{lg}}(X), Y, Z) \quad (4)$$

Note that the smaller EOD, the smaller the difference in predictive performance between different races. In our experimental results, we observe that fair model $g_{\theta^{fair}}$ always has a smaller EOD than performance-optimized model $g_{\theta^{lg}}$. Therefore, the more negative $\mathcal{F}(\cdot)$ suggests more fairness improvements achieved by the fair model $g_{\theta^{fair}}$.

To quantify the contribution of each feature i on the fairness improvement, we extend the permutation-based feature importance algorithm and propose the Algorithm 1. Specifically, we define a *permute* function which randomly shuffles the values for the feature i (i.e., column i in X) and keeps the values of the other features unchanged. Since the *permute* function involves randomness, we reiterate it on the feature i for J times, resulting in J different permuted $\tilde{X}^j, j \in [1, ..., J]$ (line 3–4). We then calculate the influence of the feature i

on fairness improvement, denoted δ_i, by averaging $\mathcal{F}(\tilde{X}^j, Y, Z, g_{\theta^{lg}}, g_{\theta^{fair}})$ over J permuted inputs (line 5). The larger magnitude of δ_i suggests that feature i contributes more to the fairness improvement, also demonstrating its importance on fairness.

Algorithm 1: Permutation Fairness Importance Algorithm

Require: Dataset X, ground truths Y, race indicators Z, performance-optimized predictive model $g_{\theta^{lg}}$, and fair predictive model $g_{\theta^{fair}}$

1 initialize empty list of importance score δ
2 **for** *each feature i (column of X)* **do**
3   **for** *each repetition* $j \in 1, ..., J$ **do**
4         $\tilde{X}^j = permute(X[:, i])$
5     $\delta_i = \frac{1}{J}\sum_{j=1}^{J} \mathcal{F}(\tilde{X}^j, Y, Z, g_{\theta^{lg}}, g_{\theta^{fair}})$
6 **return** δ

3.4 Evaluation Setup

We employed the 5-fold cross-validation approach, and in each fold we conducted feature permutation 100 ($J = 100$) times to ensure the robustness of our findings. For the evaluation of model performance and fairness, we utilized a comprehensive suite of metrics including the Area Under the Receiver Operating Characteristic curve (AUC), Accuracy (ACC), group differences in TPR (TPR diff) and FPR (FPR diff), and the EOD. To choose the classification threshold for $g_{\theta^{lg}}$ and $g_{\theta^{fair}}$, we adopted the "The Closest to (0,1) Criteria (ER)", which selects the threshold that minimizes the Euclidean distance between ROC curve and the ideal point of (0,1) on the ROC space [11]. Based on these settings, our result analysis comprises several pivotal elements: Firstly, the feature importance is assessed using SHAP and we analyze the coefficient changes between θ^{lg} and θ^{fair}, which provides the detailed changes of each feature from the performance-optimized model to fair model. Furthermore, our analysis extends to examining the differential fluctuations in EOD across different races with feature permutation, identifying the extent to which specific features contribute to the model fairness.

4 Results and Discussions

Fair model has better fairness performance than performance-optimized predictive model while having a decreased predictive performance. Table 2 reports predictive and fairness performance comparison between performance-optimized predictive model $g_{\theta^{lg}}$ and fair model $g_{\theta^{fair}}$. The fair model has smaller TPR and FPR differences across demographic groups, leading to a smaller EOD compared with the performance-optimized predictive model. We observe that for fair model both predictive performances, including AUC and Accuracy, drop for exchange the fairness improvement.

Table 2. Predictive and fairness performance comparison

	AUC	ACC	TPR_{Diff}	FPR_{Diff}	EOD ($\mathcal{E}$)
$g_{\theta^{lg}}(X)$	0.746	0.675	0.071	0.064	0.068
$g_{\theta^{fair}}(X)$	0.723	0.637	0.002	0.015	0.008

What are the important predictive variables for performance-optimized predictive model and fair model? Figure 2 reports SHAP values and feature importance rankings for performance-optimized predictive model (upper two subplots) and fair model (lower two subplots). We notice that two most important variables, including *aspiii* and *charlson comorbidity index*, remain the same between two models while the importance of few variables change significantly. For example: (1) *age* and *spo2* contribute more in fair model than in performance-optimized predictive model, and (2) fair model depends more on *sbp* than *dbp*; on the contrary, performance-optimized predictive model relies more on *dbp* than *sbp* in prediction. These observations suggest that when optimizing performance-optimized predictive model for better fairness performance, the model will alter its reliance on the features, resulting in changes of feature importance.

What are the important features that influence the fairness? SHAP values only tell us how each feature of the model contributes to predictive performance. Based on our proposed algorithm (Algorithm 1), Fig. 3a shows how each feature contributes to the EOD difference. The feature of the more negative EOD difference suggests it has a higher influence on reducing the prediction disparity between races (Whites and Non-Whites). We observe that there are four sets of features grouped by the value gaps. For features having EOD Difference < -0.06, the first feature set has the two most influential variables on fairness performance. They are *temperature* (rank 10) and *sofa* (rank 7) with an average rank of 8.50, where rank is referred their SHAP importance rankings in the lower-right subplot of Fig. 2. For those having EOD Difference between $[-0.05, -0.06]$, the second set has *gender* (rank 8), *glucose* (rank 11), *rrt* (rank 15), and *sbp* (rank 9) with average rank 10.75. The third feature set, EOD Difference falls between $[-0.04, -0.05]$, consists of *spo2* (rank 6), *race* (rank 13), and *heart_rate* (rank 11), and *resp_rate* (rank 4) with average rank 8.50. The last set contains the least influential variables for fairness performance because their EOD differences are > -0.03. They are *weight_admit* (rank 5), *dbp* (rank 13), *age* (rank 3), *charlson comorbidity index* (rank 1), and *aspiii* (rank 2) with average rank 4.80. Our rank analysis suggests that the most influential variables for the fairness performance (i.e., the first feature set) are not the variables that

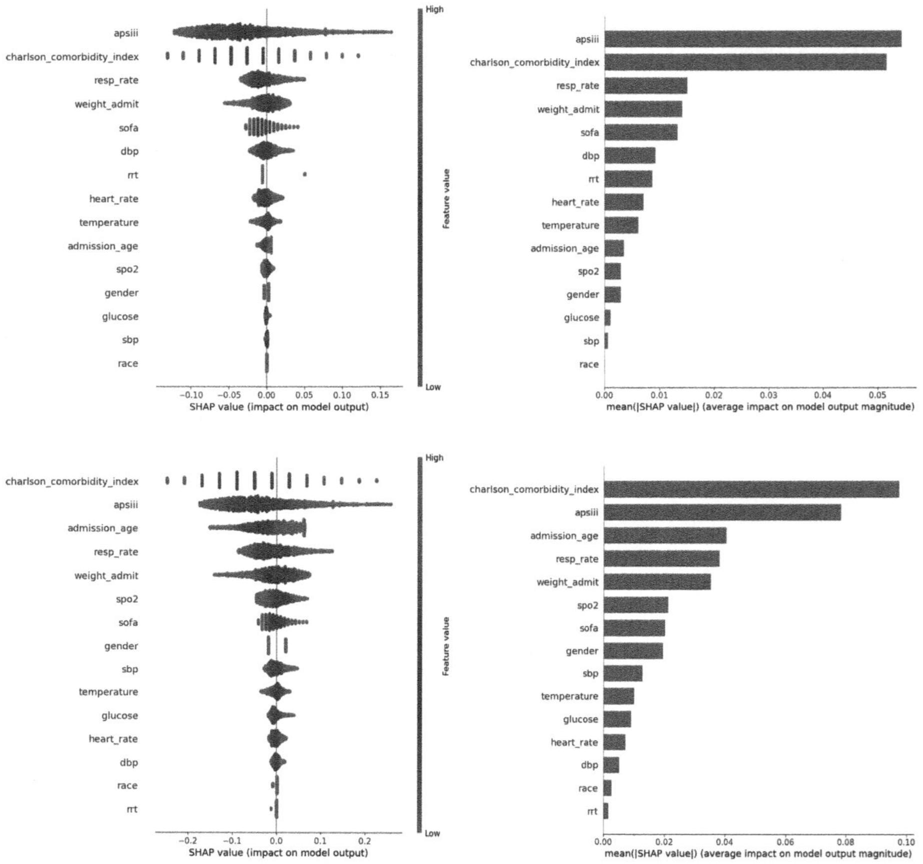

Fig. 2. The SHAP values and importance for performance-optimized predictive model (two upper subplots) and fair model (two lower subplots).

highly contribute to the predictive performance; on the contrary, those least influential variables for fairness performance (i.e., the last feature set) are the variables that mostly contribute to the predictive performance.

To further explore this correlation, Fig. 3b shows the changes of feature coefficient between performance-optimized predictive model (red; θ^{lg}) and fair model (green; θ^{fair}). We notice that both temperature and sofa have the most impact on EOD difference but have very little coefficient changes. Race is the sensitive attribute; however, the coefficient for race is only changed to a small extent and is very close to 0. Moreover, fair model stresses more weights on those least influential variables (e.g., aspiii, charlson comorbidity index, and age), which also reflects why those least influential variables are the most contributed predictive variables identified using SHAP (two lower subplots in Fig. 2). The reason for this may be because we add a performance regularization when optimizing the performance-optimized predictive model into fair model; under this given condition, altering the coefficient on important predictive variables will decrease the performance, so in order to reduce the EOD the other variables are the choices during the optimization. This observation aligns with our aforementioned observations, where features with less influence on prediction performance contribute more to improving fairness.

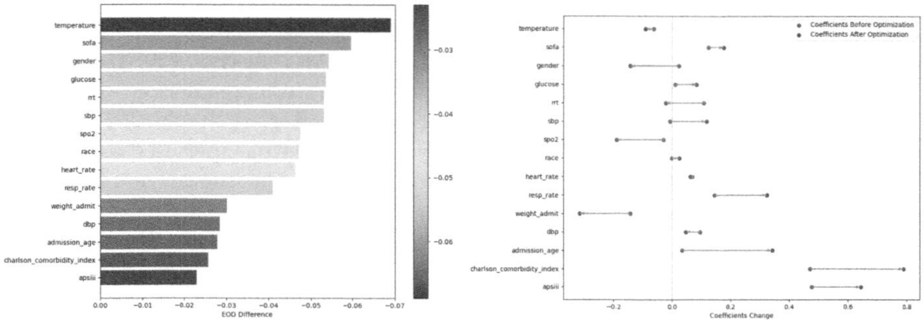

(a) Features vs EOD Differences (b) Coefficients Changes

Fig. 3. Comparison of fairness contribution (left subplot) and coefficients change (right subplot) of each feature.

5 Conclusion and Future Work

In this study, we employ the transfer learning technique to optimize a performance-optimized model for the prediction of sepsis mortality, aiming to achieve fairness in the model's predictions. Crucially, we introduce a novel permutation-based algorithm designed to elucidate the role of each individual feature in contributing to the fairness of the model. Through a detailed analysis of feature importance and the differential impact of feature permutation on the model's fairness, this approach enables us to effectively bridge the observed gap between model interpretability and fairness. Our findings reveal an under-investigated yet significant relationship between feature importance and the enhancement of model fairness: the more a feature contributes to the predictive performance, the less the feature contributes to improving fairness.

Acknowledgement. This work was supported in part by the National Science Foundation under the Grants IIS-1741306 and IIS-2235548, and by the Department of Defense under the Grant DoD W91XWH-05-1-023. This material is based upon work supported by (while serving at) the National Science Foundation. Any opinions, findings, and conclusions or recommendations expressed in this material are those of the author(s) and do not necessarily reflect the views of the National Science Foundation.

References

1. Angerschmid, A., Theuermann, K., Holzinger, A., Chen, F., Zhou, J.: Effects of fairness and explanation on trust in ethical AI. In: Holzinger, A., Kieseberg, P., Tjoa, A.M., Weippl, E. (eds.) Machine Learning and Knowledge Extraction, LNCS, vol. 13480, pp. 51–67. Springer, Cham (2022). https://doi.org/10.1007/978-3-031-14463-9_4
2. Bao, C., Deng, F., Zhao, S.: Machine-learning models for prediction of sepsis patients mortality. Medicina Intensiva (English Edition) **47**(6), 315–325 (2023)
3. Breiman, L.: Random forests. Mach. Learn. **45**(1), 5–32 (2001)
4. CDC: Sepsis Is The Body's Extreme Response To An Infection. (2023)
5. Combi, C., et al.: The IHI Rochester report 2022 on healthcare informatics research: resuming after the CoViD-19. J. Healthcare Inform. Res. **7**(2), 169–202 (2023)
6. Hardt, M., Price, E., Srebro, N.: Equality of opportunity in supervised learning. In: Proceedings of the 30th International Conference on Neural Information Processing Systems, pp. 3323–3331 (2016)
7. Johnson, A., Bulgarelli, L., Pollard, T., Horng, S., Celi, L.A., Mark, R.: MIMIC-IV. 10.13026/6MM1-EK67
8. Lundberg, S.M., Lee, S.I.: A unified approach to interpreting model predictions. In: Proceedings of the 31st International Conference on Neural Information Processing Systems, pp. 4768–4777 (2017)
9. Ribeiro, M., Singh, S., Guestrin, C.: "Why Should I Trust You?": explaining the predictions of any classifier. In: Proceedings of the 2016 Conference of the North American Chapter of the Association for Computational Linguistics: Demonstrations, pp. 97–101 (2016)
10. Taylor, R.A., et al.: Prediction of in-hospital mortality in emergency department patients with sepsis: a local big data-driven. Mach. Learn. Approach. Acad. Emerg. Med. **23**(3), 269–278 (2016)
11. Unal, I.: Defining an optimal cut-point value in ROC analysis: an alternative approach. Comput. Math. Methods Med. **2017**, 3762651 (2017)
12. Wang, H., Li, Y., Naidech, A., Luo, Y.: Comparison between machine learning methods for mortality prediction for sepsis patients with different social determinants. BMC Med. Inform. Decis. Mak. **22**(S2), 156 (2022)
13. Yang, C.C.: Explainable artificial intelligence for predictive modeling in healthcare. J. Healthcare Inform. Res. **6**(2), 228 (2022)
14. Zhou, J., Chen, F., Holzinger, A.: Towards explainability for AI fairness. In: Holzinger, A., Goebel, R., Fong, R., Moon, T., Müller, KR., Samek, W. (eds.) xxAI - Beyond Explainable AI: International Workshop, Held in Conjunction with ICML. LNCS, vol. 13200, pp. 375–386. Springer, Cham (2022). https://doi.org/10.1007/978-3-031-04083-2_18

Explanations of Augmentation Methods for Deep Learning ECG Classification

Nikil Sharan Prabahar Balasubramanian and Sagnik Dakshit(✉)

The University of Texas at Tyler, Tyler, TX 75799, USA
{nbalasubramanian,sdakshit}@uttyler.edu

Abstract. Recent progress in deep learning has sparked widespread interest in their development and adoption of diverse applications. The effectiveness of deep neural networks, particularly in extracting meaningful patterns from multimedia data, has resulted in their extensive use across various domains such as healthcare. Augmentation using synthetic data and transformation of real world data provides possible solutions to address prevalent data challenge and to consequently improve model robustness, improve generalization and reduce overfitting in development of high-performance DL models. While it has been documented that the choice of appropriate traditional augmentation method is task and data dependent, there is a dearth of explanations on how these augmentations affect the model learning and decision outcomes. This problem is exacerbated in healthcare domain where black-box methods are met with resistance in acceptance and for 1D time-series signals where some augmentation methods can be detrimental to performance. In this paper, we investigate and explain the effect of various augmentation methods for the modality of 1D time-series healthcare Electrocardiogram (ECG) signals. Our results demonstrate that various label in-variant augmentation methods can not only lead to improvement or degradation in model performance but also skew the activation features learned by the model.

Keywords: Deep Learning · CNN · Augmentation · Explainable AI · ECG · Healthcare Informatics

1 Introduction

Deep Learning (DL), a subset of machine learning, has emerged as a transformative paradigm in artificial intelligence research, demonstrating unparalleled capabilities in handling complex tasks and extracting intricate patterns from various healthcare tasks such as disease diagnosis, drug discovery, and personalized treatment planning. The hierarchical representation learning inherent in deep neural networks enables them to automatically discover intricate features and dependencies from multimedia data, alleviating the need for manual feature engineering. The success of deep learning models in achieving state-of-the-art performance has led to a significant rise in development and deployment of DL

J. Finkelstein et al. (Eds.): AIME 2024, LNAI 14845, pp. 277–287, 2024.
https://doi.org/10.1007/978-3-031-66535-6_30

models as decision support systems in healthcare for different modalities such as 2D images such as X-rays, 3D scans such as Optical Coherence Tomography and 1D time-series signals such as Electrocardiograms. Electrocardiogram (ECG) signals are graphical representations of the electrical activity occurring in the heart over time. These structured repetitive signals traces the intricate patterns of the heart's electrical impulses, enabling clinicians to detect abnormalities and assess cardiac health. The standard ECG waveform consists of several distinct clinical features namely P wave, the QRS complex, T, U waves. The intervals and amplitudes of these components serve as features conveying clinical information about the heart's rhythm, rate, and potential abnormalities such as arrhythmia, myocardial infarction (heart attacks), and other cardiovascular disorders.

Despite the advancements, the most prevalent challenges which hinder the development of state-of-the-art models and their widespread deployment for tasks such as 1D time-series ECG classification can be attributed to data availability. Generative AI and traditional Augmentation techniques present promising avenues to address these data challenges in healthcare and enhance the capabilities of deep learning models. Generative models, such as Generative Adversarial Networks (GANs) [5] and Variational Autoencoders (VAE) [2] can be utilized to generate synthetic healthcare data. This addresses the challenge of limited access to quality data by creating realistic data samples that mimic real-world conditions. In contrast, traditional data augmentation [18], involve creating new training samples by applying transformations to the existing data. General augmentation methods can be classified into label-invariant methods and Mixup methods [6]. Though prevalently used for mitigating issues related to limited data size, improving model generalization and robustness, the reason for selection of one augmentation method over others is not documented and is mostly selected at random or based on established success in literature. The transformations for augmentation can alter the characteristics or features of the data which influence DL model learning and performance making it a crucial problem for time-series healthcare data such as Electrocardiograms, where characteristics such as amplitude and frequency with clinical relevance can have adverse effect on healthcare outcomes. The critical nature of tasks where augmentation is prevalently used requires a functional understanding of the effect of various augmentation methods. To address this research gap, in this paper, we investigate the effect of various label in-variant augmentation methods on 1D time-series Electrocardiogram (ECG) signals through explanations on model learned features. To the best of the author's knowledge, this is the first work on explanations of the effect of 1D time-series augmentation methods on ECG. Our contribution in this paper can be summarized as:

- We investigate the effect of label in-variant augmentation methods on Convolutional Neural Network (CNN) model learning and performance for multi-class and multi-label classification of 1D time-series ECG signals.
- We provide visual explanations for functional understanding of the effect of different label in-variant augmentation methods on class distinguishing features of CNNs when used in developing ECG classifiers.

2 Related Works

In this section, we discuss closely related works to understand the effect of transformational augmentation on model performance, particularly in the context of ECG-based healthcare diagnosis. Transformational augmentations apply transformations to existing samples and not create new synthetic samples. ECG signals have been used extensively for various automated healthcare tasks as comprehensively reviewed in surveys [1,8] as different modalities namely raw time-series signals [3,9], Signals plotted as 2D images [12], Wavelet [11,17]. We consider only transformational augmentation methods and not synthetic data methods [2,7] which has been discussed in [10].

A large body of works have compared different augmentation strategies and presented empirical evidence for model performance improvement [13,14,16]. Various ECG specific augmentation methods have also been proposed in literature [16,17]. Despite the widespread use of data augmentation, there is a research gap in explaining their effect on model learning and classifying features particularly in context of time-series ECG signals [14]. Authors Pan et al. [14] empirically demonstrated that performance difference making it difficult to select augmentation strategy leading to random choice of augmentation. To address this gap, we provide explanations for functional understanding the difference in performance affects the learned features. In [4], the authors explain through activation maps their novel augmentation technique but is limited to only single beats and does not compare other prevalent augmentation methods. The effect of augmentation on multibeat ECG rhythms has been studied [7] with a focus on oversampling using synthetic data unlike our investigation with transformational augmentation. [20] presents similar performance comparison for different augmentation methods on 1D time-series signals with dicussions on challenging and faithfulness levels but lacks explanations or visual interpretation for functional understanding of learned clinical features. In this work we not only benchmark CNN model training with various label in-variant augmentation strategies for 1D time-series ECG signals, but also to the best of the author's knowledge is the first work on visual explanations to understand the impact of seven most prevalently used augmentation strategies on CNN model internals.

3 Methodology for Explaining Time-Series ECG Signals Augmentation

In literature, ECG signals have been used in multiple modalities of 1D time-series signal, 2D image plots, Heatmaps as well as feature engineered modalities. Though augmentation has been successfully applied to all modalities, we discuss the popular augmentation methods for 1D time-series signal modality of ECG namely Jitter, Scaling, Flipping, Magnitude Warping, Time Warping, Window Slicing and Window Warping methods.

3.1 Experimental Data

Dataset A: We use the MIT-BIH Arrhythmia database [15], of which top four classes in terms of number of data samples, namely the heart conditions of Normal Sinus Rhythm, Atrial Fibrillation, Peri-Ventricular Contraction, and Left Bundle Bunch Block with 198, 95, 94, and 73 training samples respectively following a 70:30 train test split.

Dataset B: The PTB-XL [19] dataset is used to explain multilabel classification with conduction disturbance, hypertrophy, myocardial infarction, normal, and ST-T change classes with 4402, 2387, 4919, 8551, 4714 training samples after a 70:30 train-test split which are passed through a Savgol filter, standardized, and labels are binarized. To illustrate the effect of data balancing, we augment samples from the respective classes to match the class with the highest number of samples.

3.2 Deep Neural Network Architectures

Architecture A: For multiclass classification experiment, our architecture is based of [10] on the same MIT BIH dataset. The 1D CNN architecture has with multiple 1-dimensional convolution layers with *Tanh* activation functions, intertwined with 1 dimensional maxpooling layers. The progression of the layers includes blocks of two convolution layers of 32, 64, 128, and 256, filters followed by maxpooling and two blocks of two convolution layers with 32, 64 filters. Lastly two layers of 128, and 256 filters are followed by a maxpooling and softmax classification layer.

Architecture B: For multilabel classification, our Architecture B uses 32, 64, 128, 256, 512, 1024, and 2048 filters, each with a kernel size of 3, and ReLU activation. Each convolution layer is followed by a batch normalization, max pooling layers with a pool size of 2, and dropout layers. The final convolution output is passed through global average pooling and sigmoid classification layer. Both our model are trained with Adam optimizer and learning rate was initially set at 0.001 with a decay rate of 0.01 for 100 epochs as shown in Fig. 1.

3.3 Experimental Setup

In this paper, we explain the effect of augmentation methods on model feature learning for ECG time-series signals using two popular imbalanced ECG datasets to perform class balancing for two tasks of Multiclass and Multilabel classification. For both experiments, we train two CNN architectures as shown in Fig. 1 on respective datasets using the same model architecture and hyperparameters. Deep learning models are trained on seven popular augmentation methods as observed in the literature and a baseline model without any data augmentation. We discuss our observations and present visual explanations for functional understanding of the effect of these augmentation methods as follows:

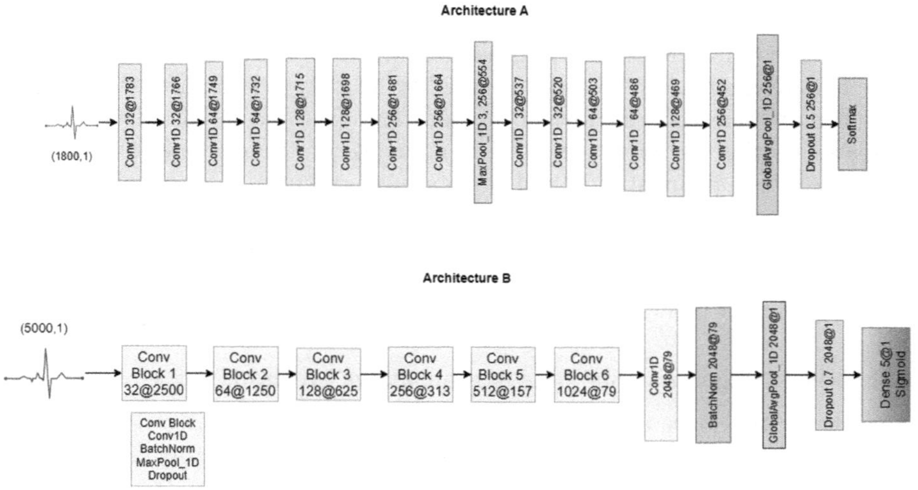

Fig. 1. Deep Neural Network Architectures.

Experiment 1: Explain features learned for multiclass classification with model architecture A by augmentation of Dataset A for three crucial cases of Performance Improvement (Case 1A), Performance Degradation (Case 1B), and Performance No-Change (Case 1C).

Experiment 2: Explain features learned for multilabel classification with model architecture B by augmentation of Dataset B for two cases of only correct predictions (Case 2A) and correct and incorrect predictions (Case 2B).

4 Experimental Results for Multiclass Classification

In this section, we present results for our trained baseline model without augmentation and our seven augmented models for dataset A on CNN architecture A as illustrated in Sect. 3.2. Moreover, through visual explanations, we present the model internals for functional understanding of the features activated for model predictions as an effect of the considered augmentation methods on 1D time-series ECG signals for the task of multiclass classification of Arrhythmia.

4.1 Empirical Model Performance on Dataset A

We record the empirical deep learning model performance aimed to assess the classification performance of the deep learning models subjected to different augmentation methods and report standard deep learning metrics namely Accuracy, Precision, Recall, and AUC score in Table 1. We observe that all augmentation methods except jitter lead to an improvement in test accuracy, precision and recall with scaling demonstrating the highest improvement by 3.13%, 2.65%, and 2.6% respectively. The results demonstrate empirically the benefit of balancing data is subjected to the augmentation method used. We can also observe that some augmentation methods can reduce model performance which makes it prudent to select augmentation method appropriately. While jitter leads to decreased testing performance, the requirement for trade-off between robustness from jitter augmentation and drop in performance is subjected to expert intervention and domain expert which is beyond the scope of this paper. The empirical results also highlight the need for an explainable functional understanding of the effect on the model parameters that leads to the improvement or degradation of model performance for selection of the appropriate method.

4.2 Visual Explanations of Augmentation Methods for Dataset A

We leverage the concept of activation maps to generate 1D activation maps for time-series ECG signals by explaining the focus on discriminating class features for the three cases to understand the difference in features that the model activates to change classification outcomes when trained on balanced data through augmentation. We do not provide clinical interpretation of the activation maps and clinical features which is beyond the scope of this paper. In Fig. 2, we identify test sample where the classification outcome is changed from incorrectly classified by baseline model without augmentation to correctly classified by augmentation and present activation feature maps for Scaling augmentation which achieves highest improvement in performance. For Experiment 1B, we present the activation feature map for a signal which is classified correctly by the baseline model as class 1 but misclassified by the scaling augmentation as class 0 (Fig. 3). Lastly for Experiment 1C, in Fig. 4, we present the activation feature map of a sample which is classified incorrectly by both the baseline no-augmentation model and the best performing scaling augmentation. The visual explanations not only highlight that augmentation methods affect the model internals and learned features which influence the features activated during prediction but also motivates expert intervention for validating domain relevance in selection of augmentation method.

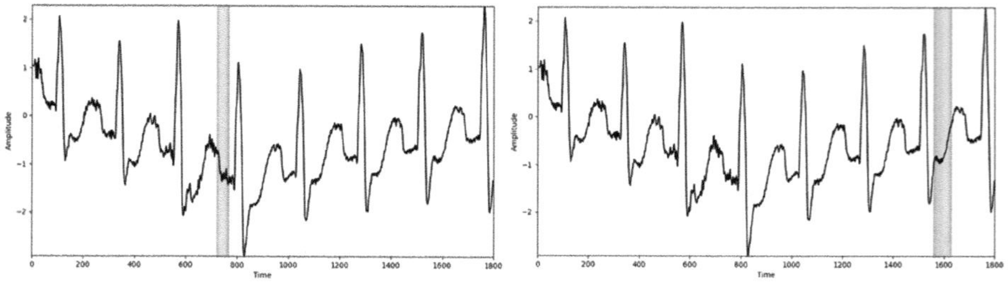

Fig. 2. Visualizations for Experiment 1 Case A. LEFT: Baseline Model Correctly Classified Output as Class 0, RIGHT: Scale Augmented Model Incorrectly Classified Output as Class 3.

5 Experimental Results for MultiLabel Classification

5.1 Empirical Model Performance on Dataset B

We present our empirical results in terms of traditional deep learning metrics for the task of multilabel Arrhythmia classification with Dataset B on Deep

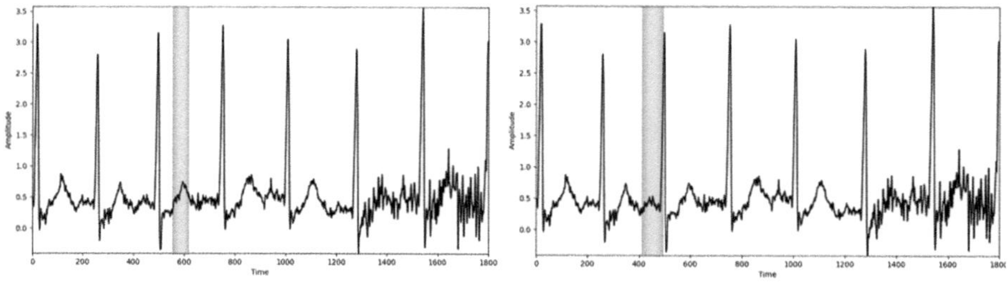

Fig. 3. Visualizations for Experiment 1 Case B. LEFT: Baseline Model Incorrectly Classified Output as Class 1, RIGHT: Scale Augmented Model Correctly Classified Output as Class 0.

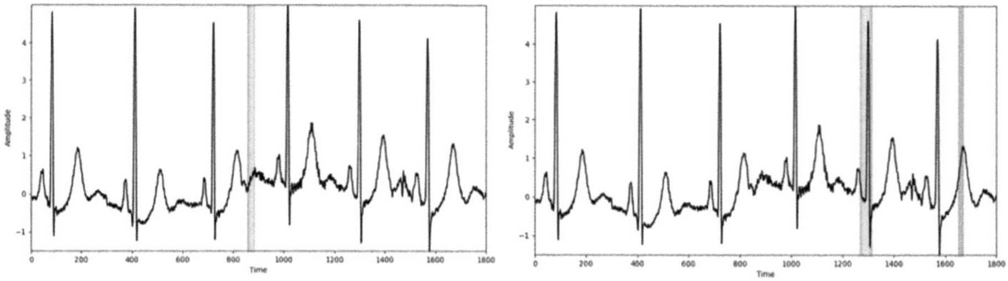

Fig. 4. Visualization for Experiment 1 Case C. LEFT: Baseline Model Incorrectly Classified Output as Class 2, RIGHT: Scaling Incorrectly Classified Output as Class 2.

Table 1. Performance for Multiclass (Experiment 1: Dataset A on Architectures A) and Multilabel (Experiment 2: Dataset B on Architectures B)

	Experiment 1				Experiment 2			
Augmentation	Accuracy	Precision	Recall	AUC	Accuracy	Precision	Recall	AUC
No-Augmentation	90.10%	0.9058	0.901	0.9864	83.38%	0.6676	0.6891	0.8527
Flip	91.15%	0.9115	0.9115	0.9780	82.41%	0.6464	0.6791	0.8395
Jitter	89.06%	0.8942	0.8802	0.9798	82.97%	0.6644	0.6658	0.8486
Magnitude Warp	92.71%	0.9319	0.9271	0.9839	83.32%	0.6643	0.6945	0.8504
Scaling	93.23%	0.9323	0.9271	0.9784	82.02%	0.6332	0.6945	0.8405
Time Warp	92.71%	0.9271	0.9271	0.9694	82.24%	0.6345	0.7095	0.8433
Window Slice	91.15%	0.9115	0.9115	0.9676	80.38%	0.5968	0.7024	0.8298
Window Warp	92.17%	0.9217	0.9217	0.9868	83.13%	0.6599	0.6934	0.8504

Learning Architecture B similar to Sect. 4 to demonstrate generalizability of our observations. As reported in Table 1, Window Slice augmentation methods was observed to show the most significant drop in performance and Magnitude Warp as the least derogatory method. We observe a sparing improvement in recall values for Magnitude Warp, Scale, Time Warp, and Window Slice augmentation methods with the highest improvement of 0.0204 for Time Warp. Our observation bolsters the need for explanations and expert intervention in selection of augmentation strategy but also establishes data dependence on the augmentation strategy in-line with observations from Experiment 1.

5.2 Visual Explanations of Augmentation Methods for Dataset B

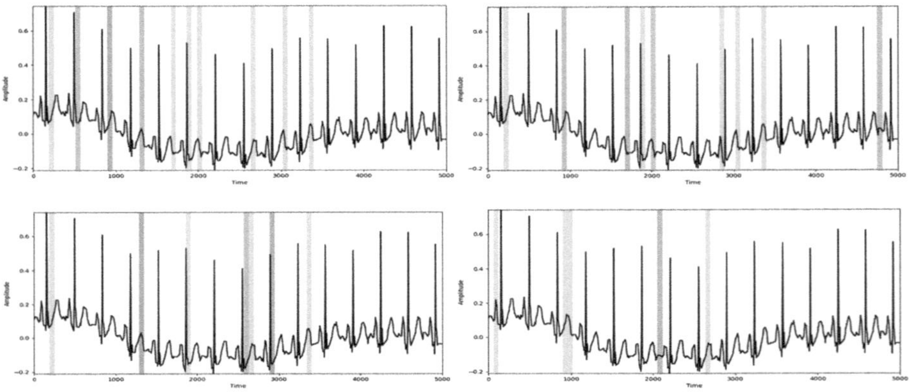

Fig. 5. Explanations for Multilabel Classification for Experiment 2A. Top Left: No-Augmentation Correctly Classifies Output as Label 3, Top Right: Time Warp Model Correctly Classifies Output as Label 3, Left Bottom: Magnitude Warp Model Correctly Classifies Output as Label 3, and Right Bottom: Window Slice Model Correctly Classifies Output as Label 3.

While empirical results demonstrate the dependency of data and the correct augmentation strategy, in this section, we present visual explanations for task of multilabel classification of 1D time-series signals. To this effect, we present feature activation maps of sample signal for which the baseline model and all the augmentation models predict the correct set of labels for Experiment 2A in Fig. 5 and sample signal where baseline model predicts correct set of labels but all other augmentation models predict an incorrect set of labels for Experiment 2B as shown in Fig. 6. Each model has distinct activation features for each predicted label which we overlay together as a single activation map for visualization. We present activation maps for Baseline (control experiment), Time Warp (highest improvement in Recall), Magnitude Warp (lowest drop in test accuracy), and Window Slice (highest drop in accuracy). We observe that some of the activation

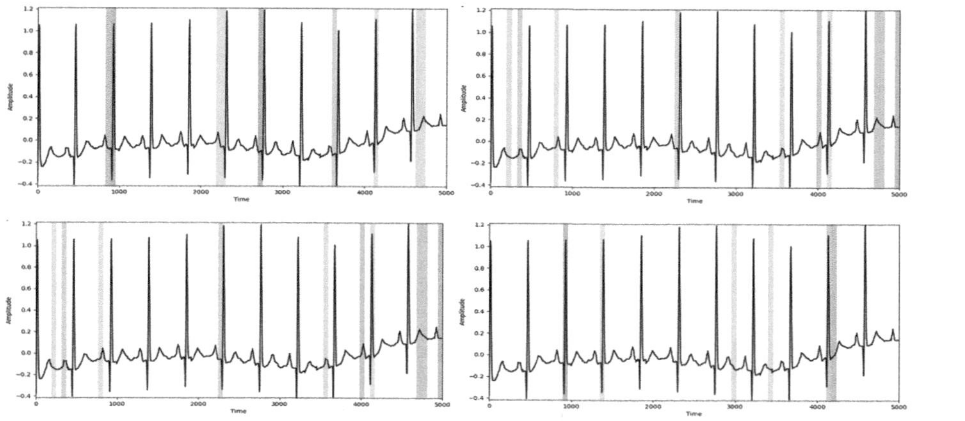

Fig. 6. Explanations for Multilabel classification for Experiment 2B. Top Left: No-Augmentation Correctly Classifies Output as Labels 1,2, Top Right: Magnitude Warp Model Incorrectly Classifies Output as Labels 0,1,2, 4, Left Bottom: Time Warp Model Incorrectly Classifies Output as Labels 0,1,2,3, and Right Bottom: Window Slice Model Incorrectly Classifies Output as Label 3.

features are similar for all the models for Experiment 2A explaining the correct set of predicted labels while a significant difference in the activation features are observed for Experiment 2B explaining the difference in label set predictions. Please note incorrect predicted set of label means the set can have both the correct and incorrect labels as defined by the multilabel nature of the task.

6 Conclusion

Deep Learning has been successful in various domains owing to its ability to process raw multimedia including the critical domain of healthcare. Augmentation is one of the strategies that has aided in the addressing some of the prevalent data challenges leading to state-of-the-art deep learning models. Literature review establishes performance improvement through traditional augmentation methods, there is a lack of understanding how traditional augmentation methods effect the model internals and consequently the learned features used for classification. To the best of the author's knowledge, this is the first investigation to provide a functional understanding of the effect of prevalent label invariant augmentation methods on the task of time-series multiclass and multilabel classification. Our empirical results demonstrate that each of the considered augmentation methods lead to different model performances demonstrating the relation between the method and the data, and the importance of selection of appropriate augmentation method. We present visual explanations through class activation maps for functional understanding of the classifying features that ensure class separability. Our investigation not only demonstrates that selection of the appropriate augmentation method is data specific but also demonstrates the importance of expert intervention in selection. As future work, we would like to explore automatic selection of the appropriate augmentation method using domain knowledge.

References

1. Chen, S.W., Wang, S.L., Qi, X.Z., Samuri, S.M., Yang, C.: Review of ECG detection and classification based on deep learning: coherent taxonomy, motivation, open challenges and recommendations. Biomed. Signal Process. Control **74**, 103493 (2022)
2. Dakshit, S., Prabhakaran, B.: CVAE-based generator for variable length synthetic ECG. In: 2023 IEEE 11th International Conference on Healthcare Informatics (ICHI), pp. 235–244. IEEE (2023)
3. Dakshit, S., Prabhakaran, B.: Twelve lead double stacked generalization for ECG classification. In: 2023 IEEE 11th International Conference on Healthcare Informatics (ICHI), pp. 245–251. IEEE (2023)
4. Do, E., Boynton, J., Lee, B.S., Lustgarten, D.: Data augmentation for 12-lead ECG beat classification. SN Comput. Sci. **3**, 1–17 (2022)
5. Goodfellow, I., et al.: Generative adversarial networks. Commun. ACM **63**(11), 139–144 (2020)
6. Guo, P., Yang, H., Sano, A.: Empirical study of mix-based data augmentation methods in physiological time series data. In: 2023 IEEE 11th International Conference on Healthcare Informatics (ICHI), pp. 206–213 (2023). https://doi.org/10.1109/ICHI57859.2023.00037
7. Hatamian, F.N., Ravikumar, N., Vesal, S., Kemeth, F.P., Struck, M., Maier, A.: The effect of data augmentation on classification of atrial fibrillation in short single-lead ECG signals using deep neural networks. In: ICASSP 2020-2020 IEEE International Conference on Acoustics, Speech and Signal Processing (ICASSP), pp. 1264–1268. IEEE (2020)
8. Hoffmann, J., et al.: A survey on machine learning approaches to ECG processing. In: 2020 Signal Processing: Algorithms, Architectures, Arrangements, and Applications (SPA), pp. 36–41. IEEE (2020)
9. Iglesias, G., Talavera, E., González-Prieto, Á., Mozo, A., Gómez-Canaval, S.: Data augmentation techniques in time series domain: a survey and taxonomy. Neural Comput. Appl. **35**(14), 10123–10145 (2023)
10. Maweu, B.M., Dakshit, S., Shamsuddin, R., Prabhakaran, B.: CEFES: a CNN explainable framework for ECG signals. Artif. Intell. Med. **115**, 102059 (2021)
11. Mewada, H.: 2d-wavelet encoded deep CNN for image-based ECG classification. Multimedia Tools Appl. 1–17 (2023)
12. Nonaka, N., Seita, J.: Data augmentation for electrocardiogram classification with deep neural network. arXiv preprint arXiv:2009.04398 (2020)
13. Nonaka, N., Seita, J.: RandECG: data augmentation for deep neural network based ECG classification. In: Takama, Y., et al. (eds.) JSAI 2021. vol. 1423, pp. 178–189. Springer, Cham (2021). https://doi.org/10.1007/978-3-030-96451-1_16
14. Pan, Q., Li, X., Fang, L.: Data augmentation for deep learning-based ECG analysis. In: Feature Engineering and Computational Intelligence in ECG Monitoring, pp. 91–111 (2020)
15. Plawiak, P.: Ecg signals (1000 fragments). Mendeley Data, v3 (2017)

16. Raghu, A., Shanmugam, D., Pomerantsev, E., Guttag, J., Stultz, C.M.: Data augmentation for electrocardiograms. In: Conference on Health, Inference, and Learning, pp. 282–310. PMLR (2022)
17. Safdar, M.F., Pałka, P., Nowak, R.M., Al Faresi, A.: A novel data augmentation approach for enhancement of ECG signal classification. Biomed. Signal Process. Control **86**, 105114 (2023)
18. Van Dyk, D.A., Meng, X.L.: The art of data augmentation. J. Comput. Graph. Stat. **10**(1), 1–50 (2001)
19. Wagner, P., et al.: Ptb-xl, a large publicly available electrocardiography dataset. Scientific data **7**(1), 154 (2020)
20. Yang, H., Yu, H., Sano, A.: Empirical evaluation of data augmentations for biobehavioral time series data with deep learning. arXiv preprint arXiv:2210.06701 (2022)

Exploring the Possibility of Arrhythmia Interpretation of Time Domain ECG Using XAI: A Preliminary Study

Sunghan Lee[1], Jeonghwan Koh[1,2], Guangyao Zheng[3], Vladimir Braverman[3], and In cheol Jeong[1,2,4,5](✉)

[1] Cerebrovascular Disease Research Center, Hallym University, Chuncheon 24252, Republic of Korea
{sh.lee,jeonghwan.koh}@hallym.ac.kr, incheol1231@gmail.com
[2] Department of Artificial Intelligence Conversions, Hallym University, Chuncheon 24252, Republic of Korea
[3] Department of Computer Science, Rice University, Houston, TX 77005, USA
{tz30,vb21}@rice.edu
[4] Department of Neurology, College of Medicine, Hallym University, Chuncheon 24252, Republic of Korea
[5] Department of Population Health Science and Policy, Icahn School of Medicine at Mount Sinai, New York, NY 10029, USA

Abstract. Recent advancements in deep learning have spurred active research in arrhythmia detection and automatic electrocardiogram (ECG) labeling. However, the labels provided by deep learning often lack transparent justification, making real-world applications challenging. Here, we utilized signals from a portable single-lead ECG device to classify eight arrhythmia classes with convolutional neural networks (CNN), achieving 79.91% accuracy with a single heartbeat. Additionally, employing the layer-wise relevance propagation (LRP), explainable artificial intelligence (XAI) algorithm, we analyzed the distinctive features of each arrhythmia, particularly focusing on significant differences in the heartbeat. Our XAI approach revealed strong activation in the QRS complex and T-wave for VPC, consistent with medical interpretations that wide QRS complex and an opposite direction of large T-wave in VPC. This research contributes to interpretable deep learning for arrhythmia diagnosis, potentially enhancing decision support systems and biomarker discovery in arrhythmia management. Further exploration of arrhythmia biomarkers is anticipated.

Keywords: Arrhythmia · CNN · LRP · VPC · XAI

1 Introduction

Cardiovascular disease (CVD) is one of the widespread causes of death worldwide, with a continuously increasing prevalence. One of the major examples of

J. Finkelstein et al. (Eds.): AIME 2024, LNAI 14845, pp. 288–295, 2024.
https://doi.org/10.1007/978-3-031-66535-6_31

CVD is arrhythmia characterized by irregularities in cardiac electrical activity. Arrhythmia could be primarily diagnosed through electrocardiography (ECG), with echocardiography and stress test also being used [5,8,10]. The ECG provides information about the patient's heart rhythm and surrounding conditions, such as electrophysiological characteristics, anatomical structure, and tissue characteristics of the heart [7]. For example, an enlargement of the QRS complex indicates abnormal ventricular depolarization, which can help predict atrial fibrillation due to changes and heart valve disease due to hypertension [4]. Recently, methods of collecting ECGs simply with smartwatches or patch-type devices have been introduced.

Many of these arrhythmia ECG data remain unlabeled due to a shortage of medical personnel. Manual analysis and classification of ECG signals for arrhythmia detection traditionally demand significant human resources and time, prone to errors. The advent of deep learning has facilitated automated analysis, aiding clinicians in identifying arrhythmias. However, prevailing deep learning approaches often prioritize accuracy while neglecting interpretability due to their black-box nature, limiting their deployment in medical settings.

In this study, we developed a convonutional neural networks (CNN)-based single beat classification model to detect eight arrhythmias and validated it using explainable artificial intelligence (XAI). We aimed to investigate whether XAI techniques in the time domain could effectively visualize differences among arrhythmia types. Furthermore, we explored the potential of these results to align with existing clinical studies or to discover new biomarkers based on deep learning approaches.

2 Methods

2.1 Dataset

The data utilized in this study were obtained from wireless portable single-lead ECG monitoring devices, Hicardi (Mezoo Co., Ltd., Republic of Korea) with a sampling rate of 250 Hz. The study protocol was approved by the ethics committee of Kosin University Gospel Hospital (IRB No. 2022-08-029). A total of 192 patients were participated, comprising 94 females and 98 males, with an average age of 53.95 years and a standard deviation of 17.36. The data were fully anonymized before being provided. The ECG signals were recorded for a duration ranging from a minimum of 13 h to a maximum of 3 days during daily activities. Subsequently, two cardiologists manually annotated the recordings with arrhythmia labels. Each cardiologist annotated for different sets. These labels encompassed 27 categories, including artifacts and lead-offs in the ECG. Detailed labeling information is presented in Table 1.

2.2 Preprocessing

The acquired ECG signals initially were eliminated artifacts and detached signals. Subsequently, a 0.5 Hz Butterworth high-pass filter was applied to remove

Table 1. Description of the 27 types of arrythmia labels

Label ID	Label Name	Label ID	Label Name
0	Normal	13	Atrial Flutter
1	Pause	14	Atrial Premature Contraction
2	Ventricular Fib./Tach	15	Run VPCs
3	Ventricular Premature Contraction	16	Sinus Tachycardia
4	Ventricular Tachycardia	20	Block
5	Ventricular Bigeminy	21	Ventricular Ectopic Beat
6	Ventricular Trigeminy	22	Supraventricular Ectopic Beat
7	VPC Couplet	23	Run APCs
8	Bradycardia	95	Main-Rhythm
9	Atrial Tachycardia	96	Data-loss
10	Supraventricular Tachycardia	97	Artifact
11	Paced Rhythm	98	Lead-Off
12	Atrial Fibrillation/Flutter	99	In-progress
		100	Unknown

baseline wander. Following this, segmentation was performed based on R peaks, dividing the signals into single beats. The length of each beat varies among participants and even within the same participant. Therefore, all beat signals were resampled to a length of 125 samples. Following that, normal heartbeats were removed. Additionally, except for the eight most frequent classes, arrhythmia beats were removed. The distribution of each arrhythmia after preprocessing is shown in Fig. 1

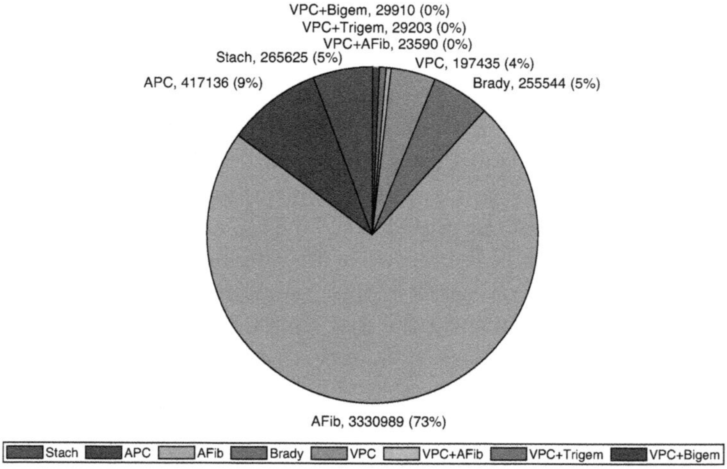

Fig. 1. Pie chart showing the number of each arrhythmia beat and its proportion.

2.3 CNN

After the preprocessing, there are eight classes: sinus tachycardia (Sinus Tach), atrial premature contraction (APC), atrial fibrillation (AFib), bradycardia (Brady), premature ventricular contraction (VPC), VPC+AFib, VPC+ Trigeminy, and VPC+Bigeminy. To classify arrhythmias, a 1D CNN model was employed, comprising four CNN blocks consisted of 1D Convolutional layers, max pooling, batch normalization, and dropout layers. CNN layers used rectified linear unit (ReLU) as an activation function, pooling size was 4 and dropout ratio was 0.2. The final layer utilized a Dense layer. Since our ECG signal was multi-label, sigmoid was used for activation in the last Dense layer. Adam optimizer and binary focal loss [9] with learning rate of 0.0005 was utilized for training.

2.4 Focal Loss

Focal loss is a loss function developed for object detection and classification tasks, particularly to address the challenge of imbalanced class distributions [9]. Its formula is given by (1)

$$FL(p_t) = -(1 - p_t)^{\gamma} \log(p_t), \tag{1}$$

where t represents labels, p_t represents the model's predicted probability and γ is a tuning parameter. Unlike conventional cross-entropy loss, focal loss allows for more precise adjustment of weights for misclassified samples. It assigns higher weights to challenging samples, encouraging the model to learn them better.

2.5 XAI

We compared the contribution in the time domain to classifying arrhythmia classes in 1D CNN models using a modified version of layer-wise relevance propagation (LRP) [3], known as LRP-z. For this analysis, the iNNvestigate [2] library was used. To visualize the principal contributions for each class, activation heatmaps were aggregated and averaged across classes. Notably, the positive and negative 5% intervals exhibiting the highest contributions at each timestamp were considered significant.

3 Results

Table 2 presented the classification accuracy of the proposed 1D CNN model during training and testing. The proposed CNN model achieved a classification accuracy of 90.50% on the training dataset and 79.91% on the testing dataset for the eight arrhythmia classes.

Figure 2 depicts the averaged ECG signals for each arrhythmia class along with the time-domain ECG component contributing significantly to the classification for each class as determined by XAI. Interestingly, the major contributions identified by LRP-z in the time-domain exhibited remarkable similarities with the positions of the P-wave, QRS complex, and T-wave.

Table 2. Train and test accuracy of the proposed 8-class CNN classifier.

	Train Accuracy (%)	Test Accuracy (%)
Proposed CNN	90.50	79.91

In the averaged ECG signals, Sinus Tach and Brady classes preserved the morphological features of the ECG waveform relatively well, whereas other classes showed deteriorated waveform shapes. Regarding the contributions from XAI, in the Sinus Tach class, high contributions were observed in the latter part of the T-wave, while in APC, it was observed around the RS-wave; in Brady, it was around the QRS complex. For VPC+Bigem and VPC+Trigem classes, high contributions were observed in the early part of the T-wave. VPC exhibited higher contributions in the P-wave segment compared to other classes, although the contributions were scattered. AFib and VPC+AFib classes also displayed considerable dispersion in contributions.

4 Discussions

The accuracy in the training set was 90.5% while testing is 79.9%, which may suggests overfitting and require additional network tuning in the future. We also used binary focal loss [9], which assigns small weights to well-classified examples while giving large weights to some examples that are difficult to classify. These concentrate learning on complex examples and thus improve performance in the class imbalance problem. The focal loss was adopted since our data is also unbalanced, as shown in Fig. 1.

Utilizing XAI techniques, particularly focusing on time-domain contributions, provided insightful results regarding classifying arrhythmia classes. Interestingly, our findings revealed that the major contributions identified by the LRP-z algorithm in the time-domain often coincided with the positions of well-known ECG components such as the P-wave, QRS complex, and T-wave. However, there were instances where our observations deviated from traditional clinical interpretations. For example, since atrial contraction is generally known to be related to the P-wave [6], we expected that APC would also have neuronal activation in the P-wave. However, our XAI analysis highlighted significant activations around the QRS complex. VPCs are known to exhibit characteristics such as a wide QRS complex and a large T-wave in the opposite direction [1]. Similarly, our LRP analysis revealed contributions across the P-wave, QRS complex, and T-wave, aligning to some extent with existing literature findings. In addition, the position of the positive and negative activation from LRP is inconsistent. This discrepancy may be a distortion caused by resampling different heartbeat signals to the same length, or a biomarker found through XAI analysis, suggesting the need for further investigation.

In light of these observations, future research directions could explore alternative XAI methodologies such as Grad-CAM [12] and DeepTalyor [11] or analyze

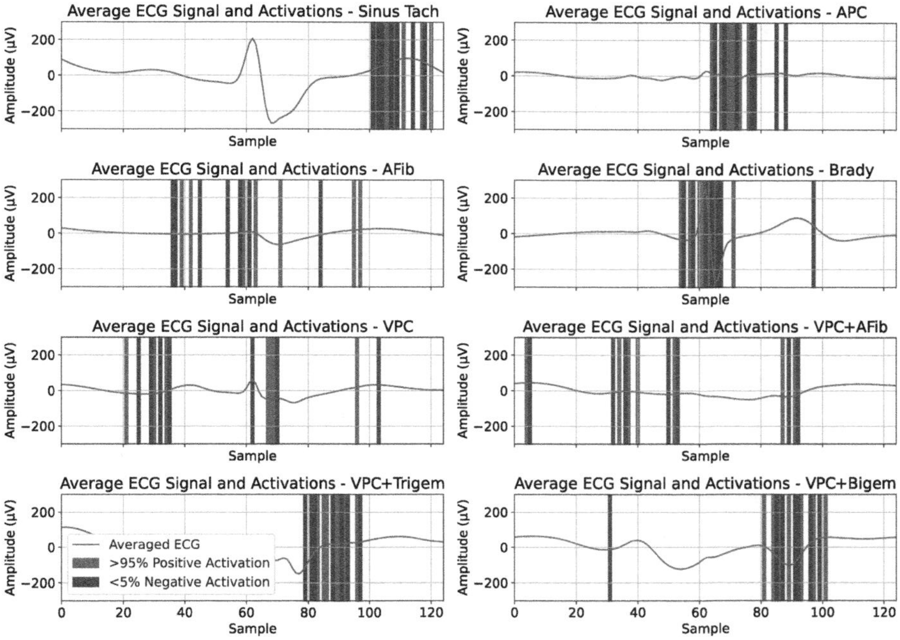

Fig. 2. Average signal of sigle-beat ECG signal and main activation time from LRP for each class.

ECG signals time-frequency domains to complement our current findings. In addition to ECG, physiological signals such as body impedance, body temperature, and respiration must also be applied in a multimodal manner to improve performance and find other biomarkers. Incorporating such analyses may offer a more comprehensive understanding of the underlying physiological processes associated with arrhythmia and enhance the interpretability of deep learning models in medical applications.

5 Conclusion

We developed and evaluated a 1D CNN model for classifying eight arrhythmia classes using single-beat ECG signals. The proposed model achieved a promising classification accuracy of 79.91%. By employing XAI techniques, mainly focusing on time-domain contributions, we gained valuable insights into the distinctive features of each arrhythmia class. Our analysis revealed strong activations in the QRS complex and T-wave for VPC, aligning with previous studies. However, some observations deviated from conventional clinical knowledge, suggesting the need for further exploration and validation. Importantly, our study has the potential to significantly advance interpretable deep learning for arrhythmia diagnosis, thereby enhancing decision support systems and facilitating biomarker discovery in arrhythmia management.

Acknowledgments. This work was supported by the Ministry of Trade, Industry and Energy(MOTIE) and the Korea Institute for Advancement of Technology(KIAT) (Project Number: P0019781) and also supported by a National Research Foundation of Korea (NRF) grant funded by the Korean Government (MSIT) (No. 2022R1A5A8019303).

Disclosure of Interests. The authors have no competing interests to declare that are relevant to the content of this article.

References

1. Alamatsaz, N., Tabatabaei, L., Yazdchi, M., Payan, H., Alamatsaz, N., Nasimi, F.: A lightweight hybrid CNN-LSTM explainable model for ECG-based arrhythmia detection. Biomed. Signal Process. Control **90**, 105884 (2024). https://doi.org/10.1016/j.bspc.2023.105884
2. Alber, M., et al.: iNNvestigate neural networks! J. Mach. Learn. Res. **20**(93), 1–8 (2019)
3. Bach, S., Binder, A., Montavon, G., Klauschen, F., Müller, K.R., Samek, W.: On pixel-wise explanations for non-linear classifier decisions by layer-wise relevance propagation. PLoS ONE **10**(7), e0130140 (2015). https://doi.org/10.1371/journal.pone.0130140, publisher: Public Library of Science
4. Gosse, P., Jan, E., Coulon, P., Cremer, A., Papaioannou, G., Yeim, S.: ECG detection of left ventricular hypertrophy: the simpler, the better? **30**(5). https://journals.lww.com/jhypertension/fulltext/2012/05000/ecg_detection_of_left_ventricular_hypertrophy__the.25.aspx
5. Guess, M., Zavanelli, N., Yeo, W.H.: Recent advances in materials and flexible sensors for arrhythmia detection. Materials **15**(3) (2022). https://doi.org/10.3390/ma15030724, https://www.mdpi.com/1996-1944/15/3/724
6. Kabutoya, T., Imai, Y., Ishikawa, S., Kario, K.: Association between P wave polarity in atrial premature complexes and cardiovascular events in a community-dwelling population. BMJ Open **10**(11), e033553 (2020). https://doi.org/10.1136/bmjopen-2019-033553, publisher: British Medical Journal Publishing Group Section: Cardiovascular medicine
7. Kashou, A., May, A., DeSimone, C., Noseworthy, P.: The essential skill of ECG interpretation: How do we define and improve competency? Postgraduate Med. J. **96**(1133), 125–127 (12 2019). https://doi.org/10.1136/postgradmedj-2019-137191, https://doi.org/10.1136/postgradmedj-2019-137191
8. Kawakami, H., Ramkumar, S., Pathan, F., Wright, L., Marwick, T.H.: Use of echocardiography to stratify the risk of atrial fibrillation: comparison of left atrial and ventricular strain. Eur. Heart J. - Cardiovascular Imaging **21**(4), 399–407 (2019). https://doi.org/10.1093/ehjci/jez240, https://doi.org/10.1093/ehjci/jez240
9. Lin, T.Y., Goyal, P., Girshick, R., He, K., Dollar, P.: Focal loss for dense object detection. In: Proceedings of the IEEE International Conference on Computer Vision (ICCV) (2017)
10. Lüderitz, B., de Luna, A.B.: The history of electrocardiography. J. Electrocardiol. **50**(5), 539 (2017)

11. Montavon, G., Lapuschkin, S., Binder, A., Samek, W., Müller, K.R.: Explaining nonlinear classification decisions with deep Taylor decomposition. Pattern Recogn. **65**, 211–222 (2017). https://doi.org/10.1016/j.patcog.2016.11.008
12. Selvaraju, R.R., Cogswell, M., Das, A., Vedantam, R., Parikh, D., Batra, D.: Grad-CAM: visual explanations from deep networks via gradient-based localization. In: 2017 IEEE International Conference on Computer Vision (ICCV), pp. 618–626 (2017).https://doi.org/10.1109/ICCV.2017.74, iSSN: 2380-7504

Improving XAI Explanations for Clinical Decision-Making – Physicians' Perspective on Local Explanations in Healthcare

Ulf Lesley[1(✉)] and Alejandro Kuratomi Hernández[2]

[1] Department of Learning, Informatics, Management and Ethics, Karolinska Institutet, Stockholm, Sweden
ulf.lesley@gmail.com

[2] Department of Computer and System Science, Stockholm University, Stockholm, Sweden

Abstract. Healthcare faces significant global challenges due to an aging population and the surge of chronic health conditions. Artificial Intelligence (AI) has emerged as a promising tool to address these issues, but the lack of transparency hampers clinicians' trust. Explainable AI (XAI) is a method to explain the AI's logic behind its predictions, thereby improving transparency. However, very little research has been done on XAI from the clinicians' perspective. Our research aims to understand how clinicians want the XAI explanations of medical AI predictions to be presented. A total of 30 physicians from 9 medical specialties evaluated two of the most used XAI methods on AI predictions on patients: Local Interpretable Model-agnostic Explanations (LIME) and Diverse Counterfactual Explanations (DiCE). A mixed method approach was used with web-based questionnaires, and the results were analyzed with statistical and thematic methods. We found a significant disparity: the explanations generated by XAI methods often fail to align with physicians' expectations and requirements for reducing uncertainty and providing enough depth of explanation in AI predictions. Moreover, there is a paradox where the areas rated most important by the physicians performed the poorest, whereas XAI excels in areas not rated equally essential. The physicians also highlighted additional areas that need to be addressed, including general upskilling of clinical staff in AI and XAI, ensuring the AI and XAI tools are integrated into the normal healthcare processes, and the ability to personalize the XAI explainability presentations.

Keywords: Artificial Intelligence · Explainable AI · Physicians · Trust · Physician-Patient Relations

1 Introduction

Healthcare globally faces formidable challenges: an expanding elderly population requiring comprehensive care, the escalation of chronic health conditions, and the advent of new, costly treatments [1]. This has led to an unprecedented demand for resources, while government spending, including healthcare, is under increasing pressure to be curtailed

J. Finkelstein et al. (Eds.): AIME 2024, LNAI 14845, pp. 296–312, 2024.
https://doi.org/10.1007/978-3-031-66535-6_32

[2]. In addition, healthcare staff is in short demand. In some countries, like Sweden, the situation is so acute that providing adequate care for older adults would necessitate an unattainable number of new healthcare professionals [3].

In light of these challenges, Artificial Intelligence (AI) has emerged as an encouraging supporting tool, particularly in image-based diagnostics [4], clinical data interpretation [5], and personalized treatment plans [6]. Concurrently, AI's role in Clinical Decision Support Systems (CDSS) has become increasingly significant [7], aiding clinicians in patient-specific decision-making.

However, a critical challenge remains: the black-box nature of AI algorithms [8]. These models' opacity conflicts with healthcare's transparency requirements, breeding skepticism and hindering AI's adoption [9]. Additionally, regulations such as the EU's GDPR [10] demand meaningful explanations of AI's logic and implications.

Explainable AI (XAI) is a suite of algorithms developed to clarify AI models' decision-making processes for users and designers. XAI has been warmly received in healthcare [11], particularly in CDSS applications, due to its potential to bridge this trust gap. Despite its importance, most XAI research has been from a purely technical perspective, e.g., creating entirely new or improved XAI models [12]. This technical orientation starkly contrasts a user-centric approach, which places the needs and perspectives of the user at the forefront, prioritizing ease of use, trust in results, and integration into existing healthcare workflows. Furthermore, to the best of our knowledge, no dedicated metric scale exists, specifically for healthcare, that would aid XAI developers in focusing their efforts on improving XAI within this sector.

The main contributions of this research are:

- Identifying the gap between the outputs of current XAI methods and the practical needs of healthcare professionals in terms of explainability for AI models.
- Demonstrating the necessity for XAI solutions that align with clinicians' expectations and requirements in their daily practice.
- Proposing the Clinical Explanation Satisfaction Scale (CESS), an improved metric scale for evaluating XAI explanations in clinical settings.

By focusing on these aspects, this research not only addresses the immediate challenge of AI's black-box in healthcare, but we believe it will also set a foundation for more practical, clinician-friendly AI applications, ultimately leading to improved patient results and more efficient healthcare delivery.

2 Related Work

A literature review, including articles from peer-reviewed magazines between 2019 and 2022, revealed that indications or anecdotes can mainly be found to answer how physicians would like the AI results to be presented to them [13–16]. Additionally, most of the research papers focus on the algorithmic aspect of these tools, i.e., they first develop an XAI method and then show it to physicians instead of applying a user-centered design philosophy. There is a research gap in studies that address the human aspects of the explanation requirements, the assessment of these qualities, and how this should be attended to increase the understandability of healthcare clinicians and XAI designers.

Building upon the identified gaps, our study seeks to understand comprehensively how physicians perceive AI explainability. To the best of our knowledge, very few studies, if any, have systematically examined how different physicians perceive and interpret AI-generated explanations within their clinical practice. This exploration is vital, as understanding these perceptions is critical to tailoring explainable AI systems that meet the real-world needs of healthcare professionals. By adopting a methodology that prioritizes the insights and feedback of physicians themselves, this research seeks to illuminate the actual reception and impact of AI explanations in the medical field.

3 Methodology

3.1 Methodology Overview

In response to the gaps identified, this research aims to evaluate the effectiveness of XAI and identify opportunities to provide physicians with understandable medical predictions. The key components of our methodology include i) selecting appropriate AI techniques and datasets, ii) training the AI model with the datasets, iii) choosing XAI models, iv) developing a comprehensive questionnaire, v) engaging a diverse group of physicians, and vi) statistically and thematically analyzing their responses.

3.2 Detailed Methodology

AI Technique and Dataset Selection. Recent studies using gradient boosting AI techniques have shown promising results in healthcare applications [17]. XGBoost is one of the most popular algorithms using gradient boosting techniques and was chosen as the AI algorithm used in the study. Publicly available datasets were used to train the AI model, chosen to align with the specialties of the participating physicians. These included datasets on heart disease [18], dementia [19], prostate cancer [20], diabetes [21], and emergency department triage [22].

XAI Model Selection. To foster the generalization of the study result, we selected two of the main and most used local explanation methods in XAI [23]: Local Interpretable Model-agnostic Explanations (LIME) and Diverse Counterfactual Explanations (DiCE). LIME is an interpretability framework designed to elucidate the decision-making process by approximating the original model's behavior around a specific instance with a locally interpretable model, such as a linear model or decision tree [24]. Conversely, DiCE adopts a counterfactual reasoning approach to interpretability by generating a set of nearest possible worlds' scenarios, where minimal alterations to the input features result in a change in the model's output prediction. This "what-if" approach highlights the sensitivity of the predictive model to changes in input variables, and identifies the smallest changes needed to alter the AI's prediction [25].

LIME outputs were presented to physicians as a probability figure for the AI's prediction, along with a list of the influential variables and their respective contribution percentages (e.g., 'Patient's gender is male – 14%'). DiCE, on the other hand, provided a prediction accompanied by a sentence detailing the influential variables and specifying the conditions that would alter the prediction (e.g., 'The patient is predicted as having a diagnosis of heart disease because the patient has serum cholesterol above 190 mg/dL...').

Questionnaire Development. The Explanation Satisfaction Scale (ESS) by Hoffman et al. [26], forming the questionnaire's basis, evaluates eight aspects of the XAI explainability of an AI prediction. To specifically tailor the ESS for healthcare, we incorporated 'patients' as a ninth aspect, reflecting their central role in healthcare settings. Each of these aspects is assigned a label name, which will be used to describe the results graphically. The label names in this adapted ESS, with brief descriptions, are: (i) **understand:** the understandability of the AI's prediction, (ii) **satisfaction**: satisfaction with the XAI explanation, (iii) **details:** sufficiency of explanation details, (iv) **complete:** completeness of the explanation, (v) **usage:** guidance on using the prediction, (vi) **goals:** usefulness of the explanation for user goals, (vii) **patient:** helpful when explaining to a patent, (viii) **accuracy:** perceived accuracy of the prediction, and (ix) **trust:** enabling user judgment on trusting the AI prediction. The ESS was further adapted to focus more on XAI improvements, adding queries about improving the explanations and rating the importance of each aspect. This adapted version was vetted for readability by 5 physicians across various specialties. See Appendix 1 for details of the adapted ESS.

Participant Recruitment and Data Collection. 30 physicians across 9 different specialties participated in the study. Each received an email containing an information letter, a consent form, a survey guide, and a link to the web-based questionnaires. Participants reviewed case studies of two patients subjected to AI predictions: one predicted to have the disease, and another predicted not to have it. Appendix 2 details the case of a patient with a predicted disease using LIME for explanation, and Appendix 3 presents the same case using DiCE. Data collected from the survey were analyzed using graphical representation, descriptive statistics, and thematic analysis.

Participant Information. The ages of the surveyed medical specialists ranged from 33 to 73, with a mean value of 51, and the average working experience as a physician was 24 years, varying between 4 and 50 years.

Saturation Principle for Interviews. The principle of saturation was employed to determine the number of interviews, typically achieved after 16–24 interviews in diverse groups [27]. Our research involved 30 physicians, ensuring a comprehensive understanding of their perspectives.

3.3 Ethical Considerations

This study ensured participant privacy by anonymizing sensitive questionnaire data, such as age and years of professional experience, using random numbers within the same range. Further, AI and XAI models were trained on publicly available datasets that had already been anonymized, thereby minimizing personal data privacy concerns. The risk of biased datasets used in AI and XAI was considered [28]. However, the focus of this study is on utilizing AI predictions to generate XAI explanations rather than for direct clinical decision-making, which mitigates the potential ethical impact of such biases.

4 Results and Discussion

4.1 Evaluation of LIME and DiCE for Clinical Use

The results from the questionnaires evaluating LIME and DiCE revealed moderate levels of satisfaction among physicians. On a Likert scale of 1 to 5 (1 = Disagree strongly, 2 = Disagree somewhat, 3 = I am neutral about it, 4 = I agree somewhat, 5 = I strongly agree) evaluating the explanations from LIME and DiCE, the average scores were 3.5 (range 3.2 – 3.6) for LIME and 3.4 (range 3.0 – 4.0) for DiCE. DiCE was preferred for its enhanced understandability and overall satisfaction ("DiCE" bars in Fig. 1), indicating that its sentence-based explanation approach resonated better with the clinicians. In contrast, LIME was rated higher for accuracy and guiding when to trust the AI's predictions ("LIME" bars in Fig. 1), possibly due to its presentation of a probability value associated with the AI predictions, a feature absent in DiCE.

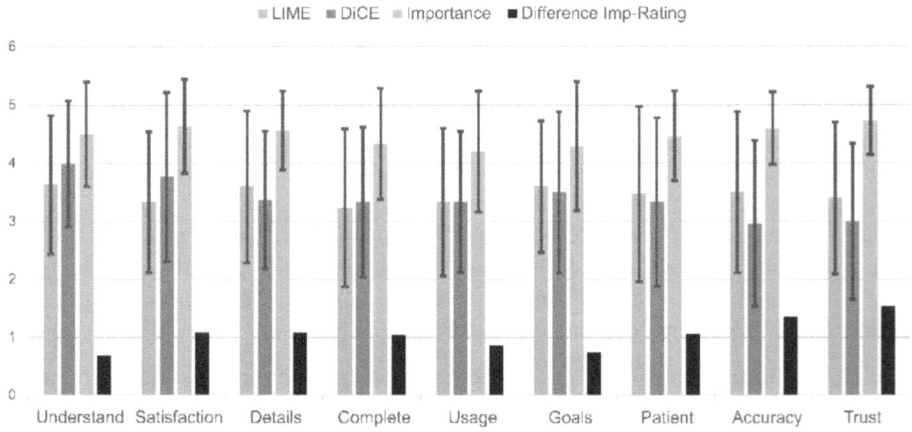

Fig. 1. Physicians' rating of LIME and DiCE according to the XAI explanation aspects in the adapted ESS. Mean and Standard Deviation per question (Rating scale is 1 to 5).

4.2 Important Aspects of XAI

The survey revealed the crucial importance of AI explainability in clinical settings, with physicians consistently rating the importance of all aspects above 4.0 ("Importance" bars in Fig. 1). Notably, 'Accuracy' and 'Trust' were central, as they address managing uncertainty, a significant challenge in clinical decision-making. These aspects not only scored highly for importance but also showed low standard deviation values (0.58, 0.62), indicating a strong consensus among physicians.

However, a noticeable discrepancy was observed in the performance of XAI models. In aspects deemed most crucial by clinicians, such as 'Accuracy' and 'Trust', the models were less effective. This discrepancy is illustrated in Fig. 1 by the "Difference Imp-Rating" bars, which represent the gap between the importance rating assigned by the physicians ("Importance" bars in Fig. 1) and the average scores they gave to LIME and DiCE ("LIME" and "DiCE" bars in Fig. 1), calculated as the Importance scores minus the average of the LIME and DiCE scores. In contrast, LIME and DiCE performed better in areas less prioritized by physicians. This finding highlights the need for future XAI development to focus on minimizing uncertainty to enhance their utility and clinicians' trust.

The aspect of 'Usage' received relatively lower importance scores, suggesting that physicians either find the XAI explanations intuitive or have other resources for guidance. Furthermore, as seen in Fig. 2, the analysis revealed minimal differences between younger (age < 50) and older (age ≥ 50) physicians in terms of prioritizing different aspects. However, slight variations in 'Satisfaction' and 'Usage' suggest that older physicians may require more guidance in using AI, likely due to a generational gap in AI familiarity.

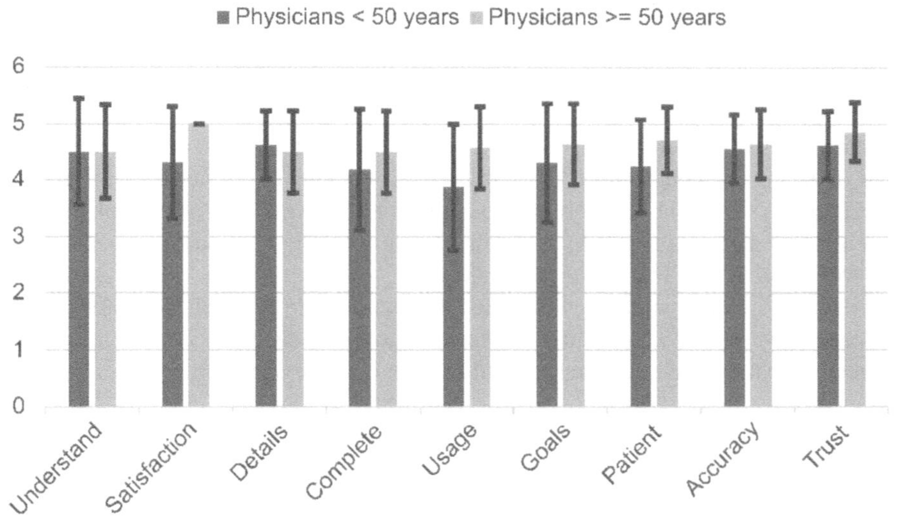

Fig. 2. Physicians in age group <50 and >=50 rating the importance of the explanation aspects. Mean and Standard Deviation per question (Rating scale is 1 to 5).

4.3 Improvements in Explanations

The third group of questions focused on physicians' suggestions for enhancing the LIME and DiCE explanations, including general enhancements for XAI explanations. The findings were categorized into seven themes, each highlighting specific areas for improvement:

AI and XAI Understanding. 67% of the responses indicated a need for better education in AI and XAI among physicians. Understanding AI's functionality and limitations will reduce apprehension about integrating AI into clinical practice [29].

Reference Interval. 43% of the physicians throughout all specialties requested information about normal value ranges for variables in XAI explanations and how close the presented values are to these ranges. This would aid in contextualizing the data.

Judging Uncertainty. A significant challenge in clinical work is assessing uncertainty. 67% of the physicians requested additional information on the reliability of the predictions and more thorough explanations from XAI, and suggested including scientific references to minimize uncertainty and increase trust.

Personalized Presentations. 63% of the clinicians expressed a need for more personalized user interfaces in XAI systems, including customizable visualization options and detail levels.

Integration. 23% of the responses highlighted the issue of non-integrated healthcare systems. Physicians suggested that better integration of XAI tools into existing systems would facilitate diagnosis and treatment, particularly for complex cases.

How to Use. Most physicians understood how to use XAI explanations in their practice, with only 10% indicating difficulties in the area. However, there is a demand for broader education on AI and XAI use rather than solely on the XAI model.

Explanations to Patients. There was mixed sentiment about presenting XAI explanations directly to patients, with 20% of physicians opposing it. Conversely, 13% foresaw it becoming a reality, but the format needs to be more accessible and empathetic. Nevertheless, the idea of using XAI as a tool for physicians to communicate with patients was well-received by 87% of the respondents.

Also, 7% of the physicians highlighted that the demand for XAI explanations might decrease if AI shows clear superiority over traditional methods in clinical settings, suggesting that proven effectiveness could render detailed explanations less critical.

5 Limitations

This research was subject to certain limitations that may affect the interpretation and applicability of its findings:

Use of Structured Data Only. To ensure comparability across various physician specialties, the study exclusively utilized structured data. This approach is unlikely to impact the evaluations significantly, as even complex medical information, like CT scan results for dementia diagnosis, can be represented as structured data [30].

Reliance on Public Datasets. The study used publicly available datasets due to the lengthy approval process required for accessing proprietary datasets. However, these public datasets may only encompass some of the variables typically used by the healthcare specialists in the survey, potentially leading to biases in the evaluation.

Lack of Integration into Clinical Workflows. Information was presented to physicians in a PDF format, not integrated into their regular medical work processes. This presentation method may not have fully captured the complexities of real-world clinical scenarios or accurately reflected the context in which XAI systems are used. Consequently, the evaluations might not completely represent the physicians' experiences in a practical clinical setting.

6 Conclusion and Future Work

This study deployed an adapted ESS in a survey with 30 physicians to evaluate their perceptions of the interpretability of XAI algorithms across various datasets related to different medical fields. The findings underscore the crucial role of explainability in building clinicians' trust and confirming the practical utility of AI in clinical settings. Despite this, existing XAI models failed to fully meet physicians' needs, particularly in diminishing uncertainty surrounding AI predictions. While DiCE marginally outperformed LIME in understandability and completeness, LIME was preferred for its accuracy and insights into the reliability of AI predictions. A notable imbalance was observed where XAI models excelled in less critical areas but underperformed in aspects deemed crucial by physicians, such as minimizing uncertainty in AI predictions. To bridge this gap, it is imperative for XAI designers to refine approaches, prioritizing alignment with vital clinical needs. This involves a shift away from broader aspects like 'Usage' and 'Goals' towards specifically enhancing trust and the practical applicability of XAI in real-world healthcare environments.

Given these insights, future research should directly address the identified gaps, particularly by developing more effective XAI explanations to minimize the uncertainty that physicians face. Developing frameworks for personalized, tailor-made XAI explanation presentations and exploring patient perspectives on XAI explanations are two other vital areas for investigation. Further studies could involve live XAI systems integrated into clinicians' workflows, thereby covering broader factors such as organizational dynamics and time constraints in everyday work environments. Additionally, the "moving target" of XAI, i.e., the potential diminishing need for XAI as established AI proves more effective than traditional methods while new areas for AI require explanations, also warrant investigation.

The study introduces the Clinical Explanation Satisfaction Scale (CESS), a healthcare-focused version of the ESS, refined with the findings from this research to specifically evaluate XAI explanations in clinical settings (see Appendix 4). Incorporating elements crucial to clinician-patient communication while excluding directions on XAI usage, the CESS is anticipated to substantially advance the evaluative frameworks for XAI within healthcare. While yet to be validated, CESS is part of the future work planned for the project.

As the integration of AI in healthcare progresses, the importance of XAI is underscored not only for its role in trust-building but also due to legal mandates like the GDPR that require transparent AI explanations. Beyond XAI, fostering a comprehensive understanding of AI and facilitating positive clinical experiences through its repeated use are also crucial for its successful adoption.

Acknowledgment. For using the OASIS data for dementia, these grant numbers are recognized: P50 AG05681, P01 AG03991, R01 AG021910, P20 MH071616, and U24 RR0213.

Appendix 1 - Questionnaire

Questionnaire for evaluating satisfaction with XAI explanations.
Adapted from Hoffman et al. [26].

*= mandatory to answer

Q1a. From the explanation, I **understand** how the AI prediction was made.*

5	4	3	2	1
I strongly agree	I agree some-what	I'm neutral about it	I disagree somewhat	I disagree strongly

Q1b. How can **understanding** be improved?

Q1c. **Understanding** the explanations is very important to me.*

5	4	3	2	1
I strongly agree	I agree some-what	I'm neutral about it	I disagree somewhat	I disagree strongly

Q2a. This explanation of how the AI prediction was made is **satisfying**.*

5	4	3	2	1
I strongly agree	I agree some-what	I'm neutral about it	I disagree somewhat	I disagree strongly

Q2b. What changes in the explanation would lead to a higher level of **satisfaction**?

Q2c. Being given a **satisfying** explanation is very important to me.*

5	4	3	2	1
I strongly agree	I agree some-what	I'm neutral about it	I disagree somewhat	I disagree strongly

Q3a. This explanation of how the AI prediction was made has **sufficient detail.***

5	4	3	2	1
I strongly agree	I agree some-what	I'm neutral about it	I disagree somewhat	I disagree strongly

Q3b. If not sufficient, how can the given **details** be improved?

Q3c. Being given **sufficient detail** about the explanation is very important to me.*

5	4	3	2	1
I strongly agree	I agree some-what	I'm neutral about it	I disagree somewhat	I disagree strongly

Q4a. This explanation of how the AI prediction was made seems **complete**.*

5	4	3	2	1
I strongly agree	I agree some-what	I'm neutral about it	I disagree somewhat	I disagree strongly

Q4b. If not **complete**, what is missing?

Q4c. Being given a **complete** explanation is very important to me.*

5	4	3	2	1
I strongly agree	I agree some-what	I'm neutral about it	I disagree somewhat	I disagree strongly

Q5a. This explanation of how the AI prediction was made **tells me how to use** it.*

5	4	3	2	1
I strongly agree	I agree some-what	I'm neutral about it	I disagree somewhat	I disagree strongly

Q5b. If not, what could be added to the AI prediction to **tell you how to use** it?

Q5c. Being **told how to use** the AI prediction is very important to me.*

5	4	3	2	1
I strongly agree	I agree some-what	I'm neutral about it	I disagree somewhat	I disagree strongly

Q6a. This explanation of how the AI prediction was made is **useful to my goals.***

5	4	3	2	1
I strongly agree	I agree some-what	I'm neutral about it	I disagree somewhat	I disagree strongly

Q6b. If not, how can it be made **more useful to your goals**?

Q6c. Being given an explanation that is **useful to my goals** is very important to me.*

5	4	3	2	1
I strongly agree	I agree some-what	I'm neutral about it	I disagree somewhat	I disagree strongly

Q7a. This explanation of how the AI prediction was made is useful when **explaining to a patient**.*

5	4	3	2	1
I strongly agree	I agree some-what	I'm neutral about it	I disagree somewhat	I disagree strongly

Q7b. If not, how can it be more useful when **explaining to a patient**?

Q7c. Being given an explanation that is useful when **explaining to a patient** is very important to me.*

5	4	3	2	1
I strongly agree	I agree somewhat	I'm neutral about it	I disagree somewhat	I disagree strongly

Q8a. This explanation of the AI prediction shows me how **accurate** the AI prediction is.*

5	4	3	2	1
I strongly agree	I agree somewhat	I'm neutral about it	I disagree somewhat	I disagree strongly

Q8b. If not, how can it better show you how **accurate** the AI prediction is?

Q8c. Being shown the **accuracy** of the AI prediction is very important to me.*

5	4	3	2	1
I strongly agree	I agree somewhat	I'm neutral about it	I disagree somewhat	I disagree strongly

Q9a. This explanation lets me judge when I should **trust and not trust** the AI prediction.*

5	4	3	2	1
I strongly agree	I agree somewhat	I'm neutral about it	I disagree somewhat	I disagree strongly

Q9b. If not, how can it better let you judge when to **trust and not trust** the AI prediction?

Q9c. Judging when to **trust and not trust** the AI prediction is very important to me.*

5	4	3	2	1
I strongly agree	I agree somewhat	I'm neutral about it	I disagree somewhat	I disagree strongly

Q10. What else should be included in the explanation of the AI prediction for it to be useful and trusted by you?

Appendix 2 – Example of Survey Information of a Patient to Participants - LIME

PATIENT A	
PATIENT INFORMATION	
Age (years)	64
Gender	Male
Chest Pain Type	Asymptomatic
Resting blood pressure (mmHg)	120
Serum cholesterol (mg/dL)	246
Fasting blood sugar > 120 mg/dL	No
Resting electrocardiographic result	Probable or definite left ventricular hypertrophy by Estes' criteria
Maximum heart rate achieved (bpm)	96
Exercise-induced angina	Yes
ST depression induced by exercise relative to rest	2.2
Slope of the peak exercise ST segment	Downsloping
Number of major vessels colored by fluoroscopy	1
Thalassemia (defect type)	Normal
AI PREDICTION	
Diagnosis of heart disease	**Yes**

XAI EXPLANATION - Probability	
Probability of AI prediction	91%

XAI EXPLANATION - Detailed (The explanations can support different outcomes of the AI prediction, like "on the one hand – on the other hand")	The explanation's contribution to the prediction (%)	Contribution towards:
One or more major vessels colored by fluoroscopy	25%	Disease
Chest pain type is asymptomatic	18%	Disease
Patient does not have a reversible defect (thalassemia)	16%	No disease
Patient's gender is male	14%	Disease
ST depression induced by exercise relative to rest is > 1.60	14%	Disease

Appendix 3 – Example of Survey Information of a Patient to Participants - DiCE

PATIENT A	
PATIENT INFORMATION	
Age (years)	64
Gender	Male
Chest Pain Type	Asymptomatic
Resting blood pressure (mmHg)	120
Serum cholesterol (mg/dL)	246
Fasting blood sugar > 120 mg/dL	No
Resting electrocardiographic result	Probable or definite left ventricular hypertrophy by Estes' criteria
Maximum heart rate achieved (bpm)	96
Exercise-induced angina	Yes
ST depression induced by exercise relative to rest	2.2
Slope of the peak exercise ST segment	Downsloping
Number of major vessels colored by fluoroscopy	1
Thalassemia (defect type)	Normal
AI PREDICTION	
Diagnosis of heart disease	**Yes**

XAI EXPLANATION		
The patient is predicted as HAVING a diagnosis of heart disease, because the patient is above/below the limits where this particular patient would have been diagnosed with no heart disease These limits specific to this patient are:		
Serum cholesterol (mg/dL)	190	
ST depression induced by exercise relative to rest	2.0	
Number of major vessels colored by fluoroscopy	0	
To summarize the explanation: The patient is predicted as having a diagnosis of heart disease because the patient has serum cholesterol above 190 mg/dL, has ST depression induced by exercise relative to rest above 2.0, and has one or several major vessels colored by fluoroscopy		

Appendix 4 – Clinical Explanation Satisfaction Scale (CESS)

Q1. From the Explanation, I **Understand** How the AI Prediction Was Made.

5	**4**	**3**	**2**	**1**
I strongly agree	I agree somewhat	I'm neutral about it	I disagree somewhat	I disagree strongly

Q2. This Explanation of How the AI Prediction Was Made is **Satisfying**.

5	**4**	**3**	**2**	**1**
I strongly agree	I agree somewhat	I'm neutral about it	I disagree somewhat	I disagree strongly

Q3. This Explanation of How the AI Prediction Was Made Has **Sufficient Detail.**

5	**4**	**3**	**2**	**1**
I strongly agree	I agree somewhat	I'm neutral about it	I disagree somewhat	I disagree strongly

Q4. This Explanation of How the AI Prediction Was Made Seems **Complete**.

5	**4**	**3**	**2**	**1**
I strongly agree	I agree somewhat	I'm neutral about it	I disagree somewhat	I disagree strongly

Q5. This Explanation of How the AI Prediction Was Made is **Useful to My Goals.**

5	**4**	**3**	**2**	**1**
I strongly agree	I agree somewhat	I'm neutral about it	I disagree somewhat	I disagree strongly

Q6. This Explanation of How the AI Prediction Was Made is Useful When I **Explain It to a Patient**.

5	**4**	**3**	**2**	**1**
I strongly agree	I agree somewhat	I'm neutral about it	I disagree somewhat	I disagree strongly

Q7. This Explanation of the AI Prediction Shows Me How **Accurate** the AI Prediction is.

5	4	3	2	1
I strongly agree	I agree somewhat	I'm neutral about it	I disagree somewhat	I disagree strongly

Q8. This Explanation Lets Me Judge When I Should **Trust and not Trust** the AI Prediction.

5	4	3	2	1
I strongly agree	I agree somewhat	I'm neutral about it	I disagree somewhat	I disagree strongly

References

1. Smith, J., Walshe, K.: Introduction: the current and future challenges of healthcare management. In: Walshe, K., Smith, J. (eds.) Healthcare Management, pp. 1–10. McGraw-Hill Education, New York (2016)
2. World Health Organization: Global spending on health: rising to the pandemic's challenges. https://www.who.int/publications-detail-redirect/9789240064911. Accessed 27 Feb 2023
3. Dagens Medicin: Vården slåss med försvaret om personal. https://www.dagensmedicin.se/arbetsliv/bemanning/varden-slass-med-forsvaret-om-personal/. Accessed 27 Feb 2023
4. Forghani, R., Savadjiev, P., Chatterjee, A., Muthukrishnan, N., Reinhold, C., Forghani, B.: Radiomics and artificial intelligence for biomarker and prediction model development in oncology. Comput. Struct. Biotechnol. J. **17**, 995–1008 (2019). https://doi.org/10.1016/j.csbj.2019.07.001
5. Davenport, T., Kalakota, R.: The potential for artificial intelligence in healthcare. Future Healthc. J. **6**, 94–98 (2019). https://doi.org/10.7861/futurehosp.6-2-94
6. Kirk, M.D., Pires, S.M., Black, R.E., Caipo, M., Crump, J.A., Devleesschauwer, B., et al.: Correction: world health organization estimates of the global and regional disease burden of 22 foodborne bacterial, protozoal, and viral diseases, 2010: a data synthesis. PLOS Med. **12**, e1001940 (2015). https://doi.org/10.1371/journal.pmed.1001940
7. Spooner, S.A.: Mathematical foundations of decision support systems. In: Berner, E.S. (ed.) Clinical Decision Support Systems. HI, pp. 19–43. Springer, Cham (2016). https://doi.org/10.1007/978-3-319-31913-1_2
8. Ahmad, M.A., Eckert, C., Teredesai, A., McKelvey, G.: Interpretable machine learning in healthcare. In: Proceedings of the 2018 ACM International Conference on Bioinformatics, Computational Biology, and Health Informatics, pp. 559–560 (2018). https://doi.org/10.1145/3233547.3233667
9. Naiseh, M., Al-Thani, D., Jiang, N., Ali, R.: Explainable recommendation: when design meets trust calibration. World Wide Web-Internet Web Inf. Syst. **24**, 1857–1884 (2021). https://doi.org/10.1007/s11280-021-00916-0
10. Wachter, S., Mittelstadt, B., Floridi, L.: Why a right to explanation of automated decision-making does not exist in the general data protection regulation. Int. Data Priv. Law. **7**, 76–99 (2017). https://doi.org/10.1093/idpl/ipx005
11. Islam, M.R., Ahmed, M.U., Barua, S., Begum, S.: A systematic review of explainable artificial intelligence in terms of different application domains and tasks. Appl. Sci. **12**, 1353 (2022). https://doi.org/10.3390/app12031353

12. Rong, Y., et al.: Towards human-centered explainable AI: a survey of user studies for model explanations. IEEE Trans. Pattern Anal. Mach. Intell. **46**, 2104–2122 (2024). https://doi.org/10.1109/TPAMI.2023.3331846
13. Meyer, J., Khademi, A., Têtu, B., Han, W., Nippak, P., Remisch, D.: Impact of artificial intelligence on pathologists' decisions: an experiment. J. Am. Med. Inform. Assoc. JAMIA **29**, 1688–1695 (2022). https://doi.org/10.1093/jamia/ocac103
14. El-Sappagh, S., Alonso, J.M., Islam, S.M.R., Sultan, A.M., Kwak, K.S.: A multilayer multimodal detection and prediction model based on explainable artificial intelligence for Alzheimer's disease. Sci. Rep. **11**, 2660 (2021). https://doi.org/10.1038/s41598-021-82098-3
15. Jaber, D., Hajj, H., Maalouf, F., El-Hajj, W.: Medically-oriented design for explainable AI for stress prediction from physiological measurements. BMC Med. Inform. Decis. Mak. **22**, 38 (2022). https://doi.org/10.1186/s12911-022-01772-2
16. Deperlioglu, O., Kose, U., Gupta, D., Khanna, A., Giampaolo, F., Fortino, G.: Explainable framework for Glaucoma diagnosis by image processing and convolutional neural network synergy: analysis with doctor evaluation. Future Gener. Comput. Syst. **129**, 152–169 (2022). https://doi.org/10.1016/j.future.2021.11.018
17. Ryu, S.-E., Shin, D.-H., Chung, K.: Prediction model of dementia risk based on XGBoost using derived variable extraction and hyper parameter optimization. IEEE Access **8**, 177708–177720 (2020). https://doi.org/10.1109/ACCESS.2020.3025553
18. UCI Machine Learning Repository: Heart failure clinical records Data Set. https://dx.plos.org/https://doi.org/10.1371/journal.pone.0181001. Accessed 29 Jan 2023
19. Battineni, G.: Data for: Machine Learning in Medicine: Classification and Prediction of Dementia by Support Vector Machines (SVM) (2019). https://doi.org/10.17632/TSY6RBC5D4.1
20. Kaggle: Prostate Cancer. https://www.kaggle.com/datasets/sajidsaifi/prostate-cancer. Accessed 02 Apr 2023
21. National Institute of Diabetes and Digestive and Kidney Diseases: Diabetes Dataset. https://www.kaggle.com/datasets/mathchi/diabetes-data-set. Accessed 05 Apr 2023
22. Hong, W.S., Haimovich, A.D., Taylor, R.A.: Emergency service - triage application. https://www.kaggle.com/datasets/ilkeryildiz/emergency-service-triage-application. Accessed 05 Apr 2023
23. Bhattacharya, A.: Applied Machine Learning Explainability Techniques: Best Practices for Making ML Algorithms Interpretable in the Real-World Applications Using LIME, SHAP and Others. Packt Publishing, Birmingham (2022)
24. Ribeiro, M.T., Singh, S., Guestrin, C.: "Why Should I Trust You?": Explaining the predictions of any classifier. In: Proceedings of the 22nd ACM SIGKDD International Conference on Knowledge Discovery and Data Mining, pp. 1135–1144. Association for Computing Machinery, New York (2016). https://doi.org/10.1145/2939672.2939778
25. Mothilal, R., Sharma, A., Tan, C.: Explaining machine learning classifiers through diverse counterfactual explanations. In: Proceedings of the 2020 Conference on Fairness, Accountability, and Transparency (2020). https://doi.org/10.1145/3351095.3372850
26. Hoffman, R.R., Mueller, S.T., Klein, G., Litman, J.: Metrics for explainable AI: challenges and prospects. http://arxiv.org/abs/1812.04608 (2019). https://doi.org/10.48550/arXiv.1812.04608
27. Hennink, M.M., Kaiser, B.N., Marconi, V.C.: Code saturation versus meaning saturation: how many interviews are enough? Qual. Health Res. **27**, 591–608 (2017). https://doi.org/10.1177/1049732316665344
28. Ntoutsi, E., et al.: Bias in data-driven artificial intelligence systems—An introductory survey. WIREs Data Min. Knowl. Discov. **10**, e1356 (2020). https://doi.org/10.1002/widm.1356

29. Castagno, S., Khalifa, M.: Perceptions of artificial intelligence among healthcare staff: a qualitative survey study. Front. Artif. Intell. **3** (2020). https://doi.org/10.3389/frai.2020.578983
30. Wahlund, L.-O., Westman, E., van Westen, D., Wallin, A., Shams, S., Cavallin, L., et al.: Imaging biomarkers of dementia: recommended visual rating scales with teaching cases. Insights Imaging **8**, 79–90 (2016). https://doi.org/10.1007/s13244-016-0521-6

Manually-Curated Versus LLM-Generated Explanations for Complex Patient Cases: An Exploratory Study with Physicians

Martin Michalowski[1(✉)], Szymon Wilk[2], Jenny M. Bauer[1], Marc Carrier[3], Aurelien Delluc[3], Grégoire Le Gal[3], Tzu-Fei Wang[3], Deborah Siegal[3], and Wojtek Michalowski[4]

[1] University of Minnesota, Minneapolis, MN, USA
martinm@umn.edu
[2] Poznan University of Technology, Poznan, Poland
[3] The Ottawa Hospital, Ottawa, ON, Canada
[4] University of Ottawa, Ottawa, ON, Canada

Abstract. Multimorbdity guideline-based clinical decision support systems (MGCDSSes) have emerged to optimize outcomes for multimorbid patients by generating personalized treatment plans that consider many clinical data sources. The success of these systems relies on their ability to explain treatment rationale, fostering trust in their outcomes among physicians. While traditionally developing treatment explanations required significant manual effort from physicians, the emergence of large language models (LLMs) offers potential to automate and simplify this process. LLMs like Meditron70B have shown promise in generating treatment explanations, saving time and resources for physicians. However, questions remain regarding the accuracy and depth of LLM-generated explanations. In this work, we evaluate the performance of Meditron70B in generating treatment explanations within our MitPlan MGCDSS using a physician-focused survey. We highlight both the promise and potential limitations of using LLMs for this purpose.

Keywords: Clinical Decision Support · Explainable AI · LLM

1 Introduction

Multimorbidity guideline-based clinical decision support systems (MGCDSSes) [17] have emerged as tools for optimizing multimorbid patient management. They are able to formulate personalized treatment plans that take into account information contained across multiple clinical data sources including computer-interpretable guidelines (CIGs), adverse drug interactions repositories, electronic health records, and others. An important driver of the success of these systems is their ability to explain the rationale behind recommended treatments [3]. These

J. Finkelstein et al. (Eds.): AIME 2024, LNAI 14845, pp. 313–323, 2024.
https://doi.org/10.1007/978-3-031-66535-6_33

explanations foster trust between physicians and the MGCDSS, improving the chances a MGCDSS is used in clinical practice [5].

Traditionally, providing treatment explanations relied on physicians. While this manual approach ensures a high degree of accuracy and relevance, it is not without its challenges. One notable hurdle is the considerable time and effort required from physicians to craft comprehensive explanations, diverting resources from direct patient care [1]. Furthermore, the manual creation of explanations may inadvertently overlook pertinent evidence or fail to adequately contextualize clinical actions with respect to the patient encounter, potentially compromising the quality and effectiveness of the provided information.

The emergence of large language models (LLMs) presents a promising avenue for automating the generation of treatment explanations within MGCDSSes and saving valuable time and resources for physicians. However, questions linger regarding the sufficiency of detail and context provided by LLM-generated explanations, as well as their accuracy in reflecting evidence-based clinical practices. In this paper, we aim to address these critical considerations by evaluating the performance of an LLM, specifically Meditron70B [4], in generating treatment explanations within the context of a MGCDSS. To this end we have conducted an exploratory, survey-based study with physicians comparing manually-curated explanations and automatically generated ones using Meditron70B. Through empirical analysis of results we demonstrate that using an LLM shows significant potential but the generation process is susceptible to hallucinations and clinical inaccuracies.

2 Related Work

2.1 Large Language Models in Medical Applications

LLMs have shown potential in improving interpretability for end-users, extracting medical knowledge from unstructured sources, supporting clinical decisions and diagnoses, and summarizing interpretable clinical information. Recent research has explored LLM performance to generate answers to real-life medical queries [13,14], generate diagnoses from LLM-patient dialogue [16,19] and support clinical care by extracting patient care information from narrative text contained in electronic health records [6]. LLMs have also been combined with Markov models to improve the reasoning logic for medical diagnoses while still leveraging the ability of an LLM to retain interpretability, and acquire and formalize knowledge [19]. However, the use of LLMs in complex patient cases such as multimorbdity, polypharmacy, or longitudinal care appeared limited in the current works.

2.2 Evaluation of Explanations by Physicians

Given the high risks inherent in the healthcare domain, physician evaluation of LLMs has extended the theoretical evaluation to further assess safety and appropriateness from a holistic clinical context. Physician evaluation of LLMs are inconclusive. On one hand they have shown that LLMs can perform comparable or outperform physicians in measures related to utility, alignment with domain knowledge, reading comprehension, omission of important information, and risk of harm [14]. On the other hand, several studies reported that physicians outperformed LLMs in reducing risk of harm and in having less inaccurate or irrelevant information contained [2,13,14]. Current studies conflicted in evaluation of bias, where one reported comparable risks of bias between LLMs and physicians [13], and another reported bias in LLMs related to race and ethnicity demographics [6] pointing to the importance of selecting appropriate training data. The current work suggests performance was comparable between LLMs and physicians overall, although these findings were not generalizable, and few of these differences were statistically significant.

3 Generating Explanations for MGCDSS

We compare the generation of explanations for clinical actions in a treatment plan in the context of an MGCDSS we developed called MitPlan [10,11]. Below we briefly describe MitPlan and provide two alternatives for generating explanations: manually (Sect. 3.2) and using an LLM (Sect. 3.3). The two resulting explanations are compared in a physician-led survey described in Sect. 4.

3.1 MitPlan

MitPlan is designed to generate personalized treatment plans tailored to individual patient encounters using automated planning [11]. It operates under the assumption that CIGs are represented following the task-network model formalism. Additionally, MitPlan formalizes secondary clinical knowledge capturing adverse interactions and their mitigation strategies, as revision operators (ROs), which are applied through graph rewriting rules [10]. Leveraging patient data, MitPlan identifies applicable ROs for a patient case, utilizes them to revise formalized CIGs via graph rewriting, and constructs a unified internal model amalgamating primary (guidelines) and secondary (revisions) clinical knowledge into a coherent graphical structure. Applying a planner to the unified internal model, it finds an optimized plan based on an objective function defined by the physician. This plan is post-processed for readability and represents a patient treatment plan with explanations for each clinical action within it.

3.2 Manual Generation

MitPlan's explainability component operates through a structured approach encompassing three distinct aspects: (1) *clinical* – providing the reasons behind specific clinical actions within the treatment plan, (2) *revision* – explaining the application of specific ROs aimed at mitigating adverse interactions, and (3) *customization* – clarifying how the selection of clinical actions is influenced by the objective function used to optimize the plan. These aspects serve to address critical questions concerning the necessity of each clinical action, the rationale behind the application of ROs, and the customization of the treatment plan based on physician and patient preferences. Importantly, MitPlan's approach to explanation differs from traditional eXplainable AI (XAI) methods, as it does not focus on explaining how input features influence predicted outcomes but associates clinically relevant justifications with each action in the treatment plan.

MitPlan relies on explanation templates to organize textual explanations related to clinical actions. These templates are prepared in advance by a domain expert and are filled in at run-time using text retrieved for each potential clinical action from a repository of textual summaries created by a physician. As such, MitPlan creates each explanation from manually curated text (not from LLM outputs) and these explanations have demonstrated high fidelity and interpretability when evaluated by physicians [12].

3.3 LLM Generation

To automate the process of generating explanations of clinical actions, we replaced the repository of manually curated summaries described in Sect. 3.2 with an LLM. Specifically, we use the Meditron 70 billion parameter model [4]. Meditron70B has been adapted to the medical domain from the open source Llama-2-70B LLM through continued pre-training on a comprehensively curated medical corpus. We use zero-shot learning by creating a prompt input that provides the patient context using the full description of the patient encounter, a detailed question asking the model to explain why a clinical action was taken (*clinical*), why a clinical action was replaced by another in the case of an adverse interaction (*revision*), or why a certain clinical action was selected given a set of physician and patient preferences (*customization*). Using this prompt engineering approach allows us to tailor the LLM output to address the three aspects of explanations described in Sect. 3.2.

The LLM output is post-processed for clarity and to avoid repetition (potentially leading to easy detection of which explanation in the survey was generated by an LLM) by applying three simple rules: (1) *if* the last sentence in the output starts with "In summary," *or* "Overall" *then* remove a sentence, (2) *if* a sentence starts with "In this case study," *then* remove the phrase *and* capitalize the first word in the sentence, (3) *if* a sentence contains "in this case study" *or* "for this case study," *then* remove the phrase from the sentence. Prior to generating the explanations used in our survey, we went through an iterative process with our physician collaborator testing different prompt engineering strategies in order to refine the prompts.

4 Evaluation

A physician-focused survey was used to determine physicians' preference of explanations generated manually or automatically using an LLM. The survey included two realistic complex use cases describing patients managed according to multiple CPGs. These use cases were developed in close collaboration with respective clinical specialists and were additionally vetted by our physician collaborator. Due to space limitations, we only describe the first case in order to illustrate the case's language, depth, and complexity. The case is as follows:

Case 1: A 70-year-old male patient with a past medical history of coronary artery disease (recent stents placed during a percutaneous coronary intervention) and unprovoked proximal deep venous thrombosis of the lower extremity. He is known to have low risk of bleeding. He is also known for alcohol-related induced liver cirrhosis. The patient has been on single anti-platelet therapy (SAPT: clopidogrel, 75mg QD) and a direct oral anticoagulant (DOAC: rivaroxaban, 10mg QD) for secondary prevention of recurrent arterial and venous thrombosis, respectively.

The patient has been newly diagnosed with resectable and castration-sensitive metastatic prostate cancer (PC). Upon diagnosis, his renal and hepatic functions were assessed with the following results: creatinine clearance (CC) of 35 cc/min and a Child-Pugh score of A. Patient was prescribed androgen deprivation therapy (ADT: abiraterone, 1000mg QD) combined with adjuvant steroid therapy (steroid: prednisone, 5mg BID). The patient recently retired, and lost health insurance provided by his employer – he became concerned about the monthly cost of rivaroxaban and asked for a cheaper alternative. Additionally, based on past experience with rivaroxaban, the physician was concerned with the patient's adherence. As a result of shared decision-making, the patient's treatment was optimized for cost, adherence to therapy, and patient's convenience (where cost was the most important, and convenience was the least important factor) and rivaroxaban was replaced by warfarin.

Treatment of PC with abiraterone, combined with prednisone, clopidogrel, and warfarin is associated with adverse interactions between abiraterone and warfarin and between abiraterone and clopidogrel. To mitigate the former, the patient needs to be scheduled for reassessment and subsequent continuous monitoring. The latter is mitigated by replacing clopidogrel with ASA (low dose aspirin, 81mg QD). The final treatment for this 70-year-old male is the following: abiraterone (1000mg QD), prednisone (5mg BID), warfarin (10mg QD), reassessment and continuous monitoring, and low dose aspirin (81mg QD).

Case 2 was a middle-aged man diagnosed with stable coronary artery disease (CAD) and treated for an unprovoked proximal deep vein thrombosis (DVT), while also developing gastrointestinal bleeding requiring hospitalization. Our evaluation study focused on several dimensions of quality as described in Sect. 4.1. Section 4.2 presents results and Sect. 5 discusses their implications.

4.1 Design

In both use cases, devising a safe treatment plan required mitigating adverse interactions between clinical actions recommended by different CIGs. Moreover, the plans need to be customized according to shared preferences (e.g., minimization of the total financial cost and maximization of medication adherence in case 1). The treatment plans included 11 actions – five for case 1, and six for case 2. Since one of the actions (low-dose aspirin) appeared in both treatment plans, it was considered only once, resulting in 10 unique actions. For each unique action MitPlan provided two explanations – one was created from manually provided text based on published evidence using the process described in Sect. 3.2 and the other one generated using an LLM as described in Sect. 3.3. Each clinical action included an explanation covering the clinical aspect, and some actions had additional explanations covering the aspects related to revisions and customization (see Sect. 3.2 and [12] for more details).

In the survey, the respondents were presented both explanations (marked as A and B) for each action from the treatment plan and were asked six questions pertaining to specific evaluation dimensions. The dimensions and questions were inspired by the System Causability Scale [8] and by the evaluation of the Med-PaLM 2 model [14], and they are presented in Table 1. Given 10 unique actions, there were 60 questions in total. They were answered by selecting one of the following options: explanation A, explanation B, both, or none. The order of explanations was shuffled – for some actions the manual one was presented as A, and the LLM one as B, and for others the order was reversed.

Table 1. Questions used in the survey to evaluate each dimension.

Dimension	Evaluation Question
Evidence	Better reflects current scientific evidence?
Understanding	Is easier to understand?
Detailed	Is presented at an appropriate level of details?
Irrelevant	Contains information irrelevant to the case?
Inaccurate	Contains inaccurate information?
Self-contained	Is more self-contained or requires consulting fewer sources?

An example of manually and LLM-generated explanations for the clinical aspect of warfarin associated with case 1 is given below:

Manually-Curated: *Warfarin sodium remains an effective option for the secondary prevention of venous thromboembolism (VTE) despite a narrow therapeutic index, wide inter-patient dosing variability, predisposition to drug and food interactions, and need for close monitoring of the intensity of anticoagulation effect using the international normalized ratio (INR).*

LLM-Generated: *Warfarin was prescribed as a direct oral anticoagulant (DOAC) for the secondary prevention of recurrent venous thrombosis. Warfarin inhibits the activity of vitamin K-dependent clotting factors, which are involved in the coagulation cascade and the formation of blood clots. By inhibiting these clotting factors, warfarin helps prevent the formation of new blood clots and reduces the risk of recurrent thrombosis. Warfarin is a commonly prescribed anticoagulant therapy for patients with a history of venous thrombosis, particularly in cases where there is a high risk of recurrent thrombosis. The use of warfarin is* *based on the patient's clinical history and the need for ongoing anticoagulation therapy to manage the risk of recurrent venous thrombosis.*

4.2 Results

The survey was administered to five physicians specializing in the clinical presentations described in the use cases. Four responded to all 60 questions, and one was unable to answer the question related to the *inaccurate* dimension for four actions. The distribution of responses to specific questions is presented in Fig. 1. In this figure and subsequent analysis we decoded A and B responses and mapped them to actual sources of associated explanations (manual, LLM-generated – later referred to as manual and LLM, respectively). The figure is split into two – in Fig. 1a we show *discriminating* responses that point at a specific explanation source, while in Fig. 1b the *non-discriminating* ones. For the *inaccurate* and *irrelevant* dimensions most responses are non-discriminating, indicating no differences in evaluation of explanations. For the remaining four dimensions, the pattern is different and the discriminating responses are more frequent, thus for most of the dimensions the respondents were able to point at a preferred explanation.

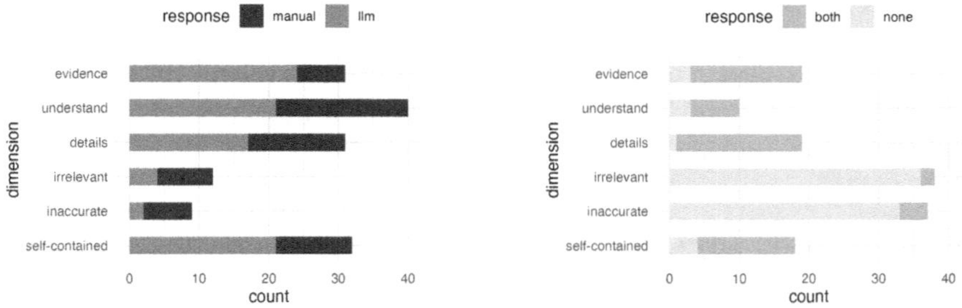

(a) Discriminating (manual or LLM) (b) Non-discriminating (both or none)

Fig. 1. Distribution of responses

The inter-rater reliability was measured using Fleiss' kappa and has a value of 0.254. The associated p-value is 4e–28 indicating that the agreement is significantly better than observed by chance. The respondents were generally consistent in evaluating explanations, regardless of their source, as relevant and accurate as illustrated in Fig. 2. Specifically, the most frequent response on the *irrelevant* and *inaccurate* dimensions was none which indicates that neither a manual nor an LLM explanation was found to be irrelevant or inaccurate, though some inaccuracies in LLM explanations due to hallucinations were identified (see Sect. 5).

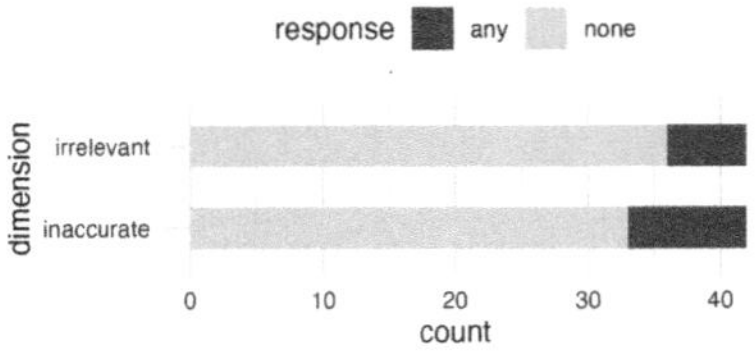

Fig. 2. Distribution of aggregated responses on the *irrelevant* and *inaccurate* dimensions (any = manual or LLM or both)

In order to determine if explanations from a particular source were preferred by the respondents, we used a paired sign test that tests "whether one condition is preferable to another in the context of paired responses" (in our study the condition refers to the source of an explanation). The sign test ignores non-discriminating responses (Fig. 1b) and focuses solely on the discriminating ones (Fig. 1a). The sign test was applied to individual dimensions and we used a two-tailed test to check if there were any statistically significant differences in evaluations of manual and LLM explanations. Obtained p-values are reported in the first row of Table 2 and they reveal there was a significant difference for the *evidence* dimension. Then we performed a one-tailed test to see if there was a significant difference in evaluations in favor of LLM explanations – our goal was to check if LLM explanations were at least as good as the manual ones, as ultimately we plan to generate them automatically using an LLM. These results are given in the second row of Table 2. As previously, there is a statistically significant difference on the *evidence* dimension that indicates that respondents believed LLM explanations better reflected current evidence.

Table 2. Results (p-values) of the sign test for all explanations. Cells marked with gray background indicate statistically significant differences (p-value < 0.05).

	Dimension					
	Evidence	Understanding	Detailed	Irrelevant	Inaccurate	Self-contained
two-tailed	0.0033	0.8746	0.7201	0.3877	0.1797	0.1102
one-tailed	0.0017	0.4373	0.3601	0.1938	0.0898	0.0551

We also applied a one-tailed test to various combinations of aspects. The resulting p-values are given in Table 3. In case of explanations covering only the clinical aspect (*cli*), there were significant differences in evaluations on the *evidence* and *self-contained* dimensions – LLM explanations better reflected evidence and were better self-contained than manual ones. In case of explanations covering clinical and revision aspects ($cli + rev$), there were significant differences on the *evidence* and *inaccurate* dimensions, i.e., LLM explanations better reflected evidence and contained less inaccurate information according to the participants. Finally, for the remaining two combinations no significant differences in evaluations were observed, which may be partially explained by the relatively small number responses within these combinations.

Table 3. Results (p-values) of the one-tailed sign test for subsets of explanations covering particular aspects (*cli* = clinical, *rev* = revision, *cust* = customization). Cells with gray background indicate statistically significant differences (p-value < 0.05).

Covered factors	Dimension					
	Evidence	Understanding	Detailed	Irrelevant	Inaccurate	Self-contained
cli	0.0318	0.1796	0.2905	0.9688	0.5000	0.0384
cli + rev	0.0020	0.3953	0.2744	0.0313	0.1250	0.2744
cli + cust	0.5000	0.9375	0.8125	0.2500	0.5000	0.8750
cli + rev + cust	1.0000	1.0000	1.0000	1.0000	1.0000	0.7500

5 Discussion and Future Work

We presented our work on physicians' attitudes towards manual and LLM-generated explanations of actions in treatment plans for multimorbid patients as generated by a MGCDSS (MitPlan). We evaluated using LLMs versus manually-generated explanations and explored clinical significance as a key determining factor in using LLM-generated explanations. To the best of our knowledge, our study is the first that evaluates an LLM for a complex patient case (multimorbidity). Our physician-focused evaluation indicates that explanations generated by an LLM fine-tuned on a medical corpus are as good or preferred to those manually-curated by a physician. This initial finding is promising but several observations from our evaluation show that LLMs need to be used with caution.

While the results indicated that most of the LLM-generated explanations were considered accurate, relevant to the patient's case, and grounded in evidence, this was not always the case. One of the LLM-generated clinical explanations for warfarin (provided in Sect. 4.1) incorrectly categorized it as a direct oral anticoagulant (DOAC), when warfarin belongs to a class of anticoagulants known as vitamin K antagonists (VKAs). Most, but not all, of the physicians caught this inaccuracy. A more subtle falsehood was the LLM hallucinated results of a study included in its explanation. In the explanation for why apixaban was prescribed to mitigate the patient's gastrointestinal bleeding, the text stated that "*... The lower risk of major bleeding with apixaban compared to warfarin has been demonstrated in several clinical trials. In the AMPLIFY trial, which compared apixaban and warfarin in patients with VTE, apixaban was associated with a significantly lower rate of major bleeding compared to warfarin (1.4% vs. 2.8%, respectively)...*" While the reference to this trial is correct, the results in the text are wrong (correct numbers are 1.7% versus 2.3% [18]. Therefore, it is important end-users are informed of the risk of LLM hallucinations and use these explanations *within* a broader decision support framework that includes clinical validation and vetting steps.

The research presented in this paper points to several key clinical practice implications. The first is that using LLMs can increase the risk of over reliance (also known as automation bias) [9, 15] where clinicians become over-reliant and bypass basic validation steps and clinical judgement. This back slide can result in risky behaviors, errors, and patient harm. Secondly, LLM use could help address healthcare literacy concerns when an explanation is provided at a literacy level that is challenging for physicians to comprehend given their experience or specialization. LLMs can produce explanations at a literacy level that improves accessibility when their outputs are tailored to individual physicians making them easier to understand [7].

Our future work aims to address limitations in this study. Firstly, we used an off-the-shelf LLM (Meditron70B) that while very effective in the medical domain was not fine-tuned to our specific task. We will explore the utility of fine-tuning this model further with a corpus of explanations to improve its output across the measured dimensions and we will consider more advanced prompting techniques (e.g., retrieval augmented generation). Secondly, the study was small in size. Based on respondents' feedback we are currently redesigning the survey instrument and recruiting physicians across Canada for a large survey sample. Finally, we used Meditron70B's output directly as explanations while the effective use of LLMs in the medical domain requires clinical verification and vetting of their outputs. We are addressing this limitation by integrating a medical knowledge base (ontology) and logical reasoner for medical concepts that we developed for a different research project.

References

1. Cabitza, F., Rasoini, R., Gensini, G.F.: Unintended consequences of machine learning in medicine. JAMA **318**(6), 517–518 (2017)
2. Cai, Z.R., Chen, M.L., Kim, J., et al.: Assessment of correctness, content omission, and risk of harm in large language model responses to dermatology continuing medical education questions. J. Invest. Dermatol. S0022–202X(24)00088–5 (2024)
3. Chen, Z., Liang, N., Zhang, H., et al.: Harnessing the power of clinical decision support systems: challenges and opportunities. Open Heart **10**(2), e002432 (2023)
4. Chen, Z., Hernández-Cano, A., Romanou, A., et al.: Meditron-70b: scaling medical pretraining for large language models (2023)
5. Diprose, W.K., Buist, N., Hua, N., et al.: Physician understanding, explainability, and trust in a hypothetical machine learning risk calculator. J. Am. Med. Inf. Assoc. JAMIA **27**(4), 592–600 (2020)
6. Guevara, M., Chen, S., Thomas, S., et al.: Large language models to identify social determinants of health in electronic health records. npj Dig. Med. **7**(6) (2024)
7. Hasani, A.M., Singh, S., Zahergivar, A., et al.: Evaluating the performance of generative pre-trained transformer-4 (gpt-4) in standardizing radiology reports. Eur. Radiol. (2023)
8. Holzinger, A., Carrington, A., Müller, H.: Measuring the quality of explanations: the system causability scale (scs). Künstl. Intell. **34**, 193–198 (2020)
9. Lyell, D., Coiera, E.: Automation bias and verification complexity: a systematic review. JAMIA **24**(2), 423–431 (2016)

10. Michalowski, M., Rao, M., Wilk, S., Michalowski, W., Carrier, M.: Using graph rewriting to operationalize medical knowledge for the revision of concurrently applied clinical practice guidelines. Artif. Intell. Med. **140**, 102550 (2023)
11. Michalowski, M., Wilk, S., Michalowski, W., Carrier, M.: A planning approach to mitigating concurrently applied clinical practice guidelines. Artif. Intell. Med. **112** (2021)
12. Michalowski, M., Wilk, S., Michalowski, W., Rao, M., , Carrier, M.: Explainability component for an automated planning-based approach to solving the multimorbidity problem (under review). J. Biomed. Inf. (2024)
13. Singhal, K., Azizi, S., Tu, T., et al.: Large language models encode clinical knowledge. Nature **620**, 172–180 (2023)
14. Singhal, K., Tu, T., Gottweis, J., et al.: Towards expert-level medical question answering with large language models (2023). https://arxiv.org/abs/2305.09617v1
15. Sujan, M., Furniss, D., Grundy, K., et al.: Human factors challenges for the safe use of artificial intelligence in patient care. BMJ Health Care Inf. **26**(1) (2019)
16. Tu, T., Palepu, A., Schaekermann, M., et al.: Towards conversational diagnostic AI (2024). https://doi.org/10.48550/arXiv.2401.05654
17. Van Woensel, W., Tu, S.W., Michalowski, W., et al.: A community-of-practice-based evaluation methodology for knowledge intensive computational methods and its application to multimorbidity decision support. J. Biomed. Inf. **142**, 104395 (2023)
18. Weycker, D., Li, X., Wygant, G.D., et al.: Effectiveness and safety of apixaban versus warfarin as outpatient treatment of venous thromboembolism in U.S. clinical practice. Thrombosis Haemostasis **118**(11), 1951–1961 (2018)
19. Zhang, H., Li, J., Wang, Y., Song, Y.: Integrating automated knowledge extraction with large language models for explainable medical decision-making. In: 2023 IEEE BIBM, pp. 1710–1717 (2023)

On Identifying Effective Investigations with Feature Finding Using Explainable AI: An Ophthalmology Case Study

Rathika Suresh Kumar[1], Kelvin Zhenghao Li[3], Si Yin Charlene Chia[1], Li Rong Wang[1], Vivien Cherng Hui Yip[3], Wei Kiong Ngo[3], Yih Yng Ng[2,3], and Xiuyi Fan[1,2(✉)]

[1] School of Computer Science and Engineering, Nanyang Technological University, Singapore, Singapore
xyfan@ntu.edu.sg
[2] Lee Kong Chian School of Medicine, Nanyang Technological University, Singapore, Singapore
[3] Tan Tock Seng Hospital, Singapore, Singapore

Abstract. Over-investigation is a longstanding challenge in the contemporary healthcare system. Employing feature selection techniques from Explainable AI (XAI), we conceptualise medical investigation decisions as a "feature finding" problem in machine learning, utilising XAI feature attribution to select the most effective investigations for each patient. Focused on ophthalmology, our research applies this framework to effectively identify investigations for diagnosing eye conditions on an individual basis. Our results demonstrate the algorithm's proficiency in accurately identifying recommended investigations that align with clinical judgment. Our contributions include modelling the selection of optimal medical investigations as a feature finding problem and introducing an algorithm for computing optimal investigations.

Keywords: Medical Investigation · Feature Selection · Explainable AI

1 Introduction

Over-investigation in medicine has been a longstanding problem that pervades contemporary healthcare systems, impacting patient outcomes, resource allocation, and overall healthcare costs. A piece of correspondence published in The British Medical Journal 50 years ago [18] reported a case where a terminally ill patient had been given 24 blood and biochemical tests along with 20 X-rays, unnecessarily. Fast forward to 2024, a recent study [7] reported that out of 274 suspected transfusion reaction reports collected at The Canberra Hospital, Australia, 148 patients were unnecessarily investigated for suspected transfusion reactions.

This work is supported by the Ministry of Education, Singapore. (Grant RG17/22).

J. Finkelstein et al. (Eds.): AIME 2024, LNAI 14845, pp. 324–334, 2024.
https://doi.org/10.1007/978-3-031-66535-6_34

Although over-investigation has been widely acknowledged by clinicians [6], the excessive reliance on diagnostic tests and procedures, often driven by defensive medical practices, contributes to a cascade of consequences [6]. Over-investigation may lead to unnecessary interventions, exposing patients to potential risks and complications [16]. Moreover, the financial burden associated with superfluous tests strains healthcare budgets and contributes to escalating healthcare costs [7]. This trend not only diverts resources away from essential areas but also perpetuates health disparities by limiting access to critical services. Additionally, over-investigation can erode the patient-doctor relationship, fostering a culture of mistrust and anxiety.

In artificial intelligence, feature selection is a critical step in the machine learning pipeline aimed at enhancing model performance by identifying and prioritising the most relevant variables [9]. The process involves choosing a subset of features from the original set, eliminating irrelevant or redundant attributes [7]. Recently, feature attribution [21] developed in Explainable AI (XAI) [19] has been used in feature selection [20]. Feature attribution provides insights into the significance of individual features in machine learning models. Methods such as SHAP (SHapley Additive exPlanations) [10] or LIME (Local Interpretable Model-agnostic Explanations) [13] help quantify the contribution of each feature to model predictions. By assessing feature attribution on model outcomes, practitioners can make informed decisions about which variables are essential for predictions and which are redundant.

In this study, we conceptualise the selection of medical investigations as **feature finding**, a novel approach in clinical diagnosis within the realm of medicine. This concept is distinctive as clinicians must actively seek out specific diagnostic clues. Once a test is chosen, the next step is to identify precisely what information is needed from that test. Leveraging XAI techniques, we evaluate the relevance and efficacy of individual medical **investigations** by treating them as **features** in machine learning.

We introduce two key concepts: *Retrospective Features* and *Prospective Features*. Retrospective Features are essential for achieving a specific diagnosis and are utilised for auditing instances of over-investigation. Prospective Features, on the other hand, suffice for reducing diagnosis uncertainties and aid in clinical decision-making for individual patient cases. We aim to enhance clinical practices with better-informed decision-making processes, ultimately leading to improved patient outcomes.

Situated in an ophthalmology context, this work studies eye conditions diagnosis for Diabetic Retinopathy (DR), Diabetic Maculopathy (DM), and Age-related Macular Degeneration (AMD). We demonstrate that, solely from patient data, our proposed approaches identify features that are essential for correct diagnosis in up to 99.1% of all cases. In cases where identified features deviate from diagnosis guidelines, our user study reveals that clinicians consider them "reasonable" in 75% of instances.

The two main contributions of this work are as follows:

1. Modelling investigation selection as "instance-level" feature finding.
2. Introducing algorithms for auditing and decision-making, respectively.

2 Related Work

Over-investigation in healthcare has garnered significant research attention within the medical community, stemming from its pivotal role in achieving more streamlined and cost-effective healthcare. Previous research has proposed diverse methodologies for pinpointing instances of over-investigation, characterized by unnecessary diagnostic testing. The various definitions of over-investigation include deviations from established guidelines [3] and the ordering of tests that contribute minimally to patient care [11,17]. Existing studies have identified diagnostic practices that deviate from evidence-based clinical guidelines established by international governmental bodies and professional societies [1,5,12,15]. This has been achieved by scrutinizing clinician records in comparison with clinical guidelines [14]. Additionally, research has identified instances of inappropriate diagnostic procedures through intrusive methods such as operations or postmortem examinations [16] and through manual scrutiny of diagnostic results [8], resulting in a potentially time-consuming over-investigation identification process. Despite recent advancements in the application of efficient artificial intelligence models in healthcare [2,4] and the widespread acknowledgement of over-investigation as a pervasive concern among clinicians [6], scant literature exists on the utilisation of artificial intelligence models for the effective identification of over-investigation instances.

3 Methodology

In contrast to "feature selection" aimed at reducing the dimensionality of a data set globally as in E.g., [9], "feature finding" in our work pertains to identifying key features at the instance level. Intuitively, features that are crucial for diagnosing one condition in a given patient may not hold the same importance for another patient with different conditions. Similarly, features that are irrelevant to one patient's diagnosis can be critical for diagnosing another patient with different conditions. Therefore, it is essential to study feature importance in the context of individual patients.

We present two approaches for finding features that are key for diagnosis. We first describe the retrospective finding of key features: from complete instances, we identify a smaller subset of features for each instance that is sufficient for correct diagnosis. Then, we describe prospective feature finding: given an incomplete (or blank) instance, we incrementally add features until a diagnosis can be reliably made. Retrospective feature finding models the over-investigation auditing process, whereas prospective feature finding models the clinical diagnostic decision-making process.

Retrospective Feature Finding. We consider diagnosis as a classification problem. For a patient $\mathbf{x} = \langle x_1, \ldots, x_m \rangle$ characterised by his demographic information, medical history, vital signs, laboratory results, radiology reports, pathology reports, etc., there is a diagnostic function $f : \mathbb{R}^m \mapsto \mathbb{Z}$ that maps $\mathbf{x}$ to some diagnosis c, which can be denoted as an integer within some range $C = \{c_1, c_2, \ldots, c_n\} \subseteq \mathbb{Z}$. The function f is such that

$$f(\mathbf{x}) = \arg\max_{c_i \in C} \Pr(c_i|\mathbf{x}). \tag{1}$$

With this setting, the retrospective feature finding question is: **What is the minimum subset of x such that, with this subset, the diagnostic function still predicts the same diagnosis as when using the complete x?** Note that this formulation permits multiple diagnoses per patient by treating each diagnosis as an independent prediction.

Determining the minimum subset of features for diagnosis is crucial, as it not only streamlines the diagnostic process but also highlights the essential features contributing to accurate diagnosis. However, in medical decision-making, uncertainty is inevitable. E.g., vital signs may not be captured in real-time, and laboratory results may occasionally be erroneous. Consequently, it is rare to eliminate uncertainty in diagnosis, meaning arriving at a diagnosis $c_i \in C$ with certainty $\Pr(c_i|\mathbf{x}) = 1$ is challenging. Thus, for critical medical decision-making, it is imperative to quantify the level of uncertainty we encounter. Simply arriving at a diagnosis c that is marginally more preferable than another c', indicated by $\Pr(c|\mathbf{x}) = \Pr(c'|\mathbf{x}) + \epsilon$ for some infinitesimally small $\epsilon > 0$, is often unsatisfactory.

To quantify uncertainty for diagnosis, we express "uncertainty quality" of diagnosis $f(\mathbf{x})$ using the entropy of the probability distribution $\Pr(c_i|\mathbf{x})$:

$$H(C|\mathbf{x}) = -\sum_{c_i \in C} \Pr(c_i|\mathbf{x}) \cdot \log_2(\Pr(c_i|\mathbf{x})). \tag{2}$$

Intuitively, when entropy is used as a measure of uncertainty, if the entropy of a distribution is high, it indicates higher uncertainty as the distribution is spread out, as there is no dominant outcome and more features are likely needed. On the other hand, if the entropy is low, then the distribution is more concentrated, indicating there is less uncertainty about the possible outcomes and the diagnosis is likely robust. In this setting, we can model feature finding as solving the following optimisation problem:

$$\underset{\mathbf{z} \in \{0,1\}^m}{\text{minimize}} \quad \|\mathbf{z}\| \tag{3a}$$

$$\text{subject to}$$

$$f(\mathbf{x} \odot \mathbf{z} + (\mathbf{1} - \mathbf{z}) \odot \bar{\mathbf{x}}) = f(\mathbf{x}), \tag{3b}$$

$$H(C|\mathbf{x} \odot \mathbf{z} + (\mathbf{1} - \mathbf{z}) \odot \bar{\mathbf{x}}) \leq \sigma, \tag{3c}$$

Algorithm 1. Retrospective Feature Finding Algorithm

$\Phi \leftarrow$ SHAP($\mathbf{x}$) // Compute Feature Attribution Values $\Phi = \langle \phi_1, \ldots, \phi_m \rangle$ using SHAP
$\mathbf{I} \leftarrow \langle i_1, \ldots, i_m \rangle$, where i_j is the jth most important feature in Φ
$\mathbf{z} \leftarrow \langle 0, \ldots, 0 \rangle$ // $\mathbf{z}$ is initialised to an array with m zeros
$c \leftarrow f(\mathbf{x})$ // Compute the classification outcome with $\mathbf{x}$
for j from 1 to m **do**
  $\mathbf{z}[\mathbf{I}[j]] \leftarrow 1$ // Mark $\mathbf{z}$ based on feature importance computed in Φ
  if $f(\mathbf{x} \odot \mathbf{z} + (\mathbf{1} - \mathbf{z}) \odot \bar{\mathbf{x}}) = c$ and $H(C|\mathbf{x} \odot \mathbf{z} + (\mathbf{1} - \mathbf{z}) \odot \bar{\mathbf{x}}) \leq \sigma$ **then**
    return z // Return the mask $\mathbf{z}$ if Eq. 3b and 3c hold with respect to $\mathbf{z}$
return z

where $||\cdot||$ denotes the L2-norm; $\odot$ denotes the element-wise multiplication; $\sigma \in \mathbb{R}$ is the *uncertainty constraint*. Here, $\mathbf{z}$ is utilised as a *feature finding mask*, $\mathbf{x} \odot \mathbf{z} + (\mathbf{1} - \mathbf{z}) \odot \bar{\mathbf{x}}$ denotes the subset of $\mathbf{x}$ masked by $\mathbf{z}$. Equations 3b and 3c specify the constraints such that they enforce that the diagnosis is "correct" and "certain", respectively (σ quantifies the level of uncertainty).

To compute $\mathbf{z}$, one naive approach is to try all elements in $\{0, 1\}^m$, and select the smallest one that satisfies the constraints in Eqs. 3b and 3c. However, such an approach leads to computing Eqs. 3b and 3c 2^m times, thus it is not feasible for large m. To overcome this challenge, we solve the optimisation with a hill-climbing search, using feature attribution scores as the heuristics. As shown in Algorithm 1, we first let $\mathbf{z}$ be an empty mask. Iteratively, we add features based on their feature attribution scores found by SHAP by inserting "1"s into $\mathbf{z}$. The process stops once Eq. 3b and 3c are both satisfied or all features in $\mathbf{x}$ are used. "1"s in $\mathbf{z}$ indicate features identified as key to the diagnosis $f(\mathbf{x})$ under the uncertainty constraint σ.

Prospective Feature Finding. In addition to finding a subset of $\mathbf{x}$ representing key features from a complete instance, we can also "construct" the set of key features for an unknown instance by progressively uncovering its feature values. This mimics the investigation decision-making process, where one conducts a sequence of investigations until a certain diagnosis is reached. Thus, the prospective feature finding question we ask is: **For a subject with unknown feature values, how to find the minimum set of features, with which a diagnosis can be reached at some pre-defined "certainty level"** γ?

To answer this question, we again consider uncertainty as entropy. The process of "revealing feature values" is the process of reducing diagnosis entropy with the sequence $F^{(0)}, \ldots, F^{(n)}$ of the feature set. The starting point is a blank vector $F^{(0)} = \emptyset$ that contains no specific information about this subject. In each step, $F^{(j+1)}$ is created from $F^{(j)}$ by filling in the value x for some feature F^* from the set of all possible features $\mathcal{F}$, such that, in the context of $F^{(j)}$, knowing x reduces the entropy the most, i.e.,

Algorithm 2. Prospective Feature Finding Algorithm

```
F^(n) ← ∅                                          // F^(n) is initialised as an empty set
while H(C|F^(n)) > γ do
    {F*} ← argmax_{F∈𝓕\F^(n)} [H(C|F^(n)) − H(C|F^(n) ∪ {F})]      // Identify {F*}
    F^(n) ← F^(n) ∪ {F*}        // F^(n) is updated with the value x for some feature F*
    if F^(n) = 𝓕 then
        return F^(n)                // Return F^(n) if all features has been exhausted
return F^(n)
```

$$F^* = \underset{F \in \mathcal{F} \setminus F^{(i)}}{\arg\max} [H(C|F^{(i)}) - H(C|F^{(i)} \cup \{F\})]. \tag{4}$$

$$F^{(i+1)} = F^{(i)} \cup \{F^*\}. \tag{5}$$

The process terminates at $F^{(n)}$ when either $F^{(n)} = \mathcal{F}$, indicating that the entire set of features has been exhausted, or when $H(C|F^{(n)}) < \gamma$, signifying that there is sufficient information to reach a diagnosis with enough certainty. The process is outlined in Algorithm 2.

4 Ophthalmology Case Study with Experiments

Data Set. The dataset utilised was collected at Tan Tock Seng Hospital (TTSH), a tertiary hospital in Singapore, and approved by its Domain Specific Review Board. The dataset encompasses patient information from 848 patients, detailing symptoms presented and the diagnosed eye conditions. In our study, the eye conditions are categorised into three types:

1. Diabetic Retinopathy (DR): A diabetes-related condition that might lead to vision impairment. Diagnostic procedures include using image-based analysis to detect abnormalities in the retina.
2. Diabetic Maculopathy (DM): A complication of DR affecting the macula, resulting in vision distortion or loss. Diagnostic procedures include Optical Coherence Tomography to detect swelling or fluid accumulation.
3. Age-Related Macular Degeneration (AMD): A condition impacting the macula, leading to central vision loss. Diagnostic procedures include the Optical Coherence Tomography and home-based Amsler Grid Test.

To diagnose patients, clinicians rely on a set of diagnostic guidelines that outline symptoms such as the location of drusens for AMD, the presence of haemorrhages for DM, and the number of cotton-wool spots for DR. Additionally, clinicians classify the diagnosed conditions based on their severity. For example, a patient diagnosed with DM might have *Mild, Moderate*, or *Severe* DM. Diagnosing the condition's severity, not just the condition, is crucial, as it influences the treatment plan. In this study, the task of diagnosing each eye of the patient for each of these conditions. Feature finding was formulated as a multi-class classification problem, with a separate single-class classification model trained for each of the three conditions, and the severity of the conditions serving as the classes in each single-class classification model.

The initial dataset of textual diagnosis notes recorded by clinicians was cross-referenced with the diagnostic guidelines to extract 64 features, resulting in a tabular dataset with numerical values. For each condition, the guidelines outline a subset of features for diagnosis, referred to as the *diagnostic guideline features*. A simplified version of the diagnostic guidelines and diagnostic guideline features for Mild and Moderate DM is presented in Table 1. Referring to these guidelines, Patient 1, identified in green, with haemorrhages and a Visual Acuity (VA) better than 6/12, will be diagnosed with Mild DM. Meanwhile, Patient 2, identified in yellow, with haemorrhages in the outer zone of the fovea and a VA worse than 6/12, will be diagnosed with Moderate DM. For the respective severities, the features denoted by **'1'** are diagnostic guideline features and are considered essential for diagnosis.

Table 1. Two examples of diagnostic guidelines and encoded features.

Diagnostic guideline	Features	Example
Patient 1: Mild DM		
– Any haemorrhages **AND** VA better than 6/12	Haemorrhages	1
	Inner Zone	0
	Outer Zone	0
	VA better than 6/12	1
	VA worse than 6/12	0
Patient 2: Moderate DM		
– Haemorrhages within outer zone **AND** VA worse than 6/12	Haemorrhages	1
	Inner Zone	0
	Outer Zone	1
	VA better than 6/12	0
	VA worse than 6/12	1

Diagnostic Model. XGBoost (eXtreme Gradient Boosting), known for its performance in classifying tabular data served as the classification model in this study. To facilitate experiments, three distinct XGBoost models were trained on the training set with all 64 features present. The dataset was partitioned into train, validation, and test sets in a 6:2:2 ratio.

Finding Features. We chose σ for the retrospective algorithm as the lowest entropy (ranging from 0 to 0.05 with an increment of 0.0001) ensuring that at least 25% of training instances had at least one dropped feature[1].

[1] We also experimented with 50% and 75%, but the resulting concordance with diagnostic guidance remained consistent.

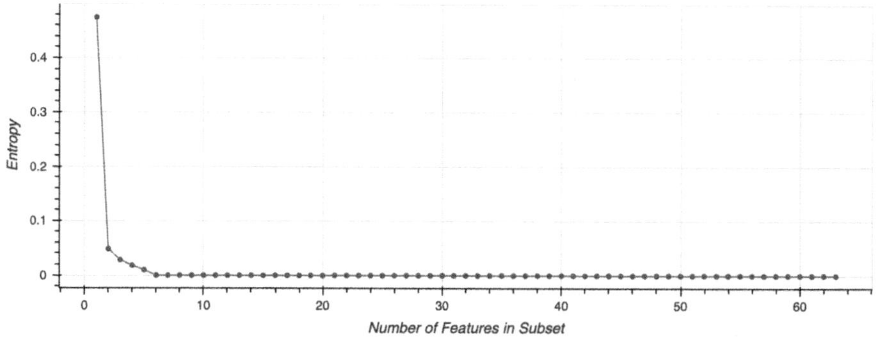

Fig. 1. Uncertainty with respect to the number of features.

To select γ for the prospective algorithm, we mask all features of the validation set and then incrementally unmask them. We select the next feature to unmask based on its ability to minimize the entropy of the resultant set $F^{(j+1)}$. This process concludes when $F^{(i)}$ ensures correct classification and we record the final entropy of $F^{(i)}$. This process was repeated on all data instances in the validation set, and γ is identified as the maximum of all recorded final entropies. Using this γ, we apply Algorithm 2 on the test set. Figure 1 shows that the uncertainty of classification decreases as the size of $F^{(i)}$ increases and plateaus beyond a certain point, suggesting that adding more features will not impact predictions.

Table 2. Example of features found by the diagnostic guidelines, the retrospective approach and the prospective approach for diagnosing *Mild DM*.

Features	Diagnostic Guidelines	Retrospective Features	Prospective Features
Dot haemorrhages	1	1	1
Blot haemorrhages	1	0	0
Haemorrhages within inner zone	1	0	0
Haemorrhages within outer zone	0	0	0
Visual Acuity better than 6/12	1	0	0
Visual Acuity 6/12 or worse	0	0	0
Hard exudates	0	0	0

An example of the features found by the two approaches and the diagnostic guidelines are presented in Table 2. While the features identified differ, all three feature sets enable correct classifications. Noting this difference, we conduct a focus group study in the next section to investigate if this feature is sufficient for diagnosis.

Table 3. Performance of Model on Test Sets ($n = 340$). "Avg. Subset Size" refers to the average number of selected features.

Test Set	Avg. Subset Size	Jaccard Similarity	Accuracy (%)
Performance for DR Grading Classification (98.2% for the complete data)			
Diagnostic Guideline Set	2.99		98.8
Retrospective Set ($\sigma = 0.875$)	0.67	0.97	95.6
Prospective Set ($\gamma = 0.137$)	2.84	0.70	97.4
Performance for DM Grading Classification (97.4% for the complete data)			
Diagnostic Guideline Set	2.99		98.2
Retrospective Set ($\sigma = 0.976$)	1.25	0.66	96.5
Prospective Set ($\gamma = 0.347$)	2.00	0.43	98.2
Performance for AMD Grading Classification (99.1% for the complete data)			
Diagnostic Guideline Set	2.99		99.1
Retrospective Set ($\sigma = 0.616$)	1.39	0.86	99.1
Prospective Set ($\gamma = 0.432$)	2.89	0.71	75.2

Model Results. Table 3 presents a comparative analysis of the results found by the two approaches against the diagnostic guideline features. We observe that (1) the XGBoost models are accurate in predicting grading for all three conditions on the complete feature set; (2) with diagnostic guideline features, the prediction performance improves slightly, this is achieved by masking features that are not relevant and deemed by the guideline in each instance so the model can focus on the selected key features; (3) retrospective finding gives comparable classification performance, demonstrating its ability to identify key features for all three conditions; (4) prospective feature finding performs well for DR and DM, but poorly on AMD. This discrepancy may be attributed to Algorithm 2's potential non-convergence when identifying the "certainty level" γ; (5) The prospective algorithm chose the least features, yielding an average subset size of less than one for DR classification as most class samples had insufficient information for diagnosis.

Table 3 also compares the consistency between the diagnostic guideline features and the features found by the two approaches. Despite achieving high classification accuracies, disparities exist between features that are deemed to be key for diagnosis between different methods as illustrated by the calculated weighted Jaccard Similarity. To gain deeper insights, a focus group study with three ophthalmologists at the consultant level from TTSH was conducted. Instances presenting different retrospective or prospective features compared to the diagnostic guidelines were identified, comprising of 12 unique sets of instances. Clinicians evaluated whether features found by the two approaches were deemed "sufficient" when all other features were unknown. In 75% of these cases, at least one clinician concurred with the approaches, highlighting their proficiency in accurately identifying features essential for precise diagnosis.

5 Conclusion

In this work, we explore an XAI-based technique for addressing the over-investigation challenge by modelling medical investigation selection as a "feature finding" problem that is solved by retrospective and prospective algorithms while considering uncertainties in medical decision-making. Focused on ophthalmology, our approaches successfully find features for diagnosing eye conditions. Our findings indicate that our algorithms not only align with diagnostic guidelines but also garner agreement with the evaluations made by practising clinicians. Through the application of feature finding in medical decision-making, we contribute to the development of a more effective decision support system. We acknowledge that this study is limited to a single medical domain. In future work, we aim to apply this technique to other medical domains, particularly those with a one-to-one correspondence between medical investigations and dataset features.

References

1. Dreisler, E., et al.: EMAS clinical guide: assessment of the endometrium in peri and postmenopausal women. Maturitas **75**, 181–190 (2013)
2. Du, Y., et al.: An explainable machine learning-based clinical decision support system for prediction of gestational diabetes mellitus. Sci. Rep. **12**, 1170 (2022)
3. Ferraro, S., Panteghini, M.: The role of laboratory in ensuring appropriate test requests. Clin. Biochem. **50**(10), 555–561 (2017)
4. Hardy, N., et al.: Intraprocedural artificial intelligence for colorectal cancer detection and characterisation in endoscopy and laparoscopy. Surg. Innov. **28**, 768–775 (2021)
5. Hill, M., et al.: ACOG practice bulletin. Diagnosis of abnormal uterine bleeding in reproductive-aged women. Obstet. Gynecol. **120**, 197–206 (2012)
6. Jankauskaite, L., et al.: Overuse of medical care in paediatrics: a survey from five countries in the European academy of pediatrics. Front. Pediatrics **10**, 945540 (2022)
7. Joel, C.T., Burgess, M., Crispin, P.: Costs and consequences of over-investigation of minor transfusion reactions. Internal Med. J. **54**, 301–306 (2023)
8. Khanduja, S., et al.: Retrospective review of magnetic resonance imaging of the lumbosacral spine: are we overinvestigating? Neurospine **15**, 383–387 (2018)
9. Li, J., et al.: Feature selection: a data perspective. ACM Comput. Surv. (CSUR) **50**(6), 1–45 (2017)
10. Lundberg, S.M., Lee, S.I.: A unified approach to interpreting model predictions. In: Advances in Neural Information Processing Systems, vol. 30 (2017)
11. Petrou, P.: Failed attempts to reduce inappropriate laboratory utilization in an emergency department setting in Cyprus: lessons learned. J. Emerg. Med. **50**(3), 510–517 (2016)
12. Renaud, M.C., et al.: Epidemiology and investigations for suspected endometrial cancer. J. Obstet. Gynaecol. Can. **35**(4), 380–381 (2013)
13. Ribeiro, M.T., Singh, S., Guestrin, C.: "Why should i trust you?" explaining the predictions of any classifier. In: Proceedings of SIGKDD, pp. 1135–1144 (2016)

14. Scrimin, F., et al.: Hysteroscopic chasing for endometrial cancer in a low-risk population: risks of overinvestigation. Arch. Gynecol. Obstet. **293**, 851–856 (2015)
15. Singh, S., et al.: Abnormal uterine bleeding in pre-menopausal women. J. Obstet. Gynaecol. Can. **35**(5), 473–475 (2013)
16. Stewart, J., et al.: Unknown primary adenocarcinoma: incidence of overinvestigation and natural history. BMJ **1**(6177), 1530 (1979)
17. Walraven, C.V., et al.: Do we know what inappropriate laboratory utilization is? A systematic review of laboratory clinical audits. JAMA **280**(6), 550–558 (1998)
18. Wright, A.D.: Overinvestigation. BMJ **1**(5641), 440 (1969)
19. Xu, F., Uszkoreit, H., Du, Y., Fan, W., Zhao, D., Zhu, J.: Explainable AI: a brief survey on history, research areas, approaches and challenges. In: Tang, J., Kan, M.-Y., Zhao, D., Li, S., Zan, H. (eds.) NLPCC 2019. LNCS (LNAI), vol. 11839, pp. 563–574. Springer, Cham (2019). https://doi.org/10.1007/978-3-030-32236-6_51
20. Zacharias, J., et al.: Designing a feature selection method based on explainable artificial intelligence. Electron. Mark. **32**(4), 2159–2184 (2022)
21. Zhou, Y., et al.: Do feature attribution methods correctly attribute features? In: Proceedings of the AAAI, vol. 36, pp. 9623–9633 (2022)

Towards Interactive and Interpretable Image Retrieval-Based Diagnosis: Enhancing Brain Tumor Classification with LLM Explanations and Latent Structure Preservation

Pranav Manjunath(✉), Brian Lerner, and Timothy Dunn

Duke University, Durham, NC 27005, USA
{pranav.manjunath,timothy.dunn}@duke.edu

Abstract. When evaluating patient scans, clinicians make use of their previous experience to make a diagnosis. However, for complex conditions such as brain tumors, the availability of relevant information beyond a clinician's personal context becomes more valuable. While the application of content-based image retrieval (CBIR) for medical images is not new, their introduction into practice has been relatively minimal. Where older CBIR systems relying on manually extracted image features suffered from poor performance, newer systems incorporating deep learning may have better performance but decreased interpretability. In this study, we present an interactive image retrieval system that enables accurate and interpretable brain tumor classification. We show that image encoders trained with supervised contrastive learning preserve latent structure within the retrieval space and exhibit classification performance on par with, or exceeding, that of conventional black box classifiers. We integrate off-the-shelf LLMs to enhance the system's accessibility through retrieval report summarization and user Q&A interactions. We recognize the importance of developing clinician-ML systems by providing a framework that clinicians can not only trust the performance of but can interact with. Our findings provide for the seed of a system that can augment the performance of human clinicians in a process that mirrors their natural thought patterns, while increasing the speed of their interactions with medical CBIR through LLMs.

Keywords: Medical CBIR · Brain Tumor Classification · Large Language Models · Deep Learning · Contrastive Learning

1 Introduction

While brain tumors rarely occur, when they do, they are one of the most deadly forms of cancer [1]. Magnetic resonance imaging (MRI) is an important modality employed to identify tumors, but disease complexity and overlapping phenotypes can slow or stymie diagnosis [2]. Machine learning (ML) has proven capable of solving a wide variety of important healthcare problems, but its clinical adoption has been slow [3], despite

J. Finkelstein et al. (Eds.): AIME 2024, LNAI 14845, pp. 335–349, 2024.
https://doi.org/10.1007/978-3-031-66535-6_35

evidence that it can often surpass the performance of human clinicians in the task of classifying diseases in medical imaging [4–6]. Hampered by a distrust of ML, attempts to translate research into the clinic [7] have been sparse. Conventional black box ML models usually fail to offer human-interpretable insight into their decision-making processes–a flaw poorly tolerated by patients, clinicians, and administrators alike. The implementation of interpretable frameworks remain essential to clinicians in cancer diagnosis [8]. When clinicians engage in diagnosis, they often relate past encounters to the present one–behavior that is reminiscent of content-based image retrieval (CBIR). Previous work has shown that retrieved images which are structurally like the query image instill more trust in the user [9]. In certain contexts, images are naturally accompanied by detailed descriptions, i.e. radiology reports.

In this study, we design and implement an interactive LLM-explained CBIR system that is interpretable and offers high performance in brain tumor classification (Fig. 1). The system follows a two-step process: i) it employs a deep-learning based image encoder to identify scans similar to the query scan from a vector database (CBIR) favoring latent structure preservation ii) it classifies the query image by running a neighborhood classifier on these retrieved images, thereby enhancing example-based interpretability. We improve upon neighborhood classifiers by introducing a hyperparameter-free version, nearest incorrect retrieval (NIR). To enable interactivity, we pioneer an approach that utilizes off-the-shelf LLMs to provide summarization and dynamic Q&A over text descriptions of retrieved scans.

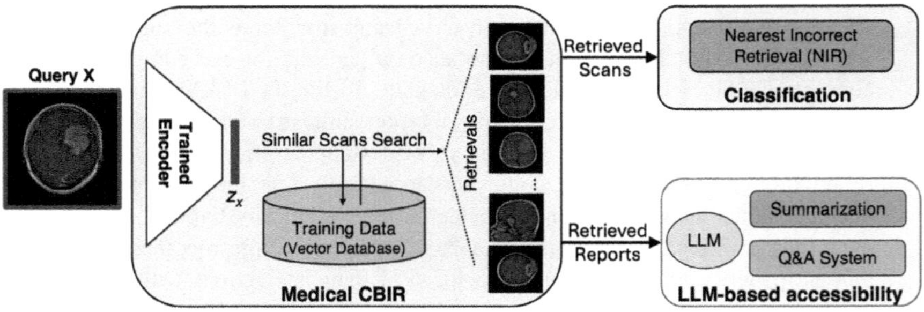

Fig. 1. The process of LLM-explained CBIR for interpretable medical image classification. A query image is encoded into z_x by a deep learning-based encoder. z_x is used to search the training database for similar images and retrieve them. The retrievals are used for downstream tasks like classification, while text descriptions (the report) of each retrieved image are passed to an LLM for summarization and question answering.

2 Related Work

CBIR has origins predating deep-learning algorithms, yet the advent of deep learning has significantly enhanced the effectiveness of CBIR in the medical domain [9, 10]. As we do here, previous work has framed CBIR in terms of example-based interpretability and posited that CBIR-based classification can instill trust in users [11]. [12, 13] trained a deep-learning based encoder to classify and retrieve various modalities of medical imaging, though in contrast to our approach they focus on training with cross-entropy (CE). Other work has explored using contrastive learning to train models for medical image retrieval [11, 14]. Here, we evaluate retrieval and classification using both CE and contrastive learning.

To provide explanations on why certain images are retrieved for a given query, [11] explore saliency-based explanations between the query and the retrieved images for chest X-rays. However, saliency maps have been shown to be inconsistent and untrustworthy for medical imaging [15]. Our approach focuses on creating inherent interpretability through CBIR, which is in turn made more accessible with LLMs.

3 Data

We utilize a brain tumor dataset [16] consisting of 3064 T1-weighted and contrast-enhanced 2D MRI slices. These slices span 233 patients, with a mixture of axial, coronal, and sagittal slices attributed to each. Each slice contains a segmented tumor that falls into one of three categories. Followed by the respective number of slices, these categories are meningioma (708), glioma (1426), and pituitary (930). Plane distribution within each tumor type is roughly uniform. We use train and test partitions from [16], comprising 186 and 47 unique patients respectively.

4 Preservation of Latent Structure in CBIR

We define latent structure as those aspects of an image which are relevant to visual understanding but are not explicitly used while training. As structural similarities in retrieved images correlates with user trust [9], we use the known values of anatomical plane and tumor size as a proxy for latent structural properties and hold that latent structure is preserved when these properties are highly similar for a query image and its top retrievals. For anatomical plane, we measure the fraction of retrievals that match the plane of the query and denote this as anatomical consistency. For tumor size, we measure the average size difference between query and retrievals. Latent structure is investigated for various training objectives.

5 Neighborhood Classification

As an alternative to K-Nearest Neighbors (KNN), we propose Nearest Incorrect Retrieval (NIR), which does not need hyperparameter tuning and is less likely to overfit on the validation set (Appendix 1). To get the NIR, we sort a query's retrievals in order of decreasing distance. Using c_n, the array of retrieval class labels, for each class c we obtain I, the index of the first neighbor that is not of class c. We store I for each class in I_c an array of $[1xC]$ where C is the total number of classes.

$$I_c = \left[argmin(c_n \neq c) | c \in C\right] \quad (1)$$

We convert I_c into probabilities using a softmax function σ and assign the class with the highest probability as the query class Q.

$$Q_c = argmax(\sigma(I_c)) \tag{2}$$

In contrast to KNN, the number of neighbors that NIR incorporates into its process is not bounded by a set number. Knowing I indicates how far within the neighborhood space the algorithm has to look to obtain the first incorrect neighbor. Thus, I can be considered a measure of model confidence. We also introduce Iterative-NIR (I-NIR), a variant of NIR that has a higher tolerance to error, offering a more lenient evaluation framework compared to the original NIR metric. We present results for both the standard NIR and I-NIR, with the latter examining the 10th nearest incorrect retrieval.

6 LLM-Explained CBIR

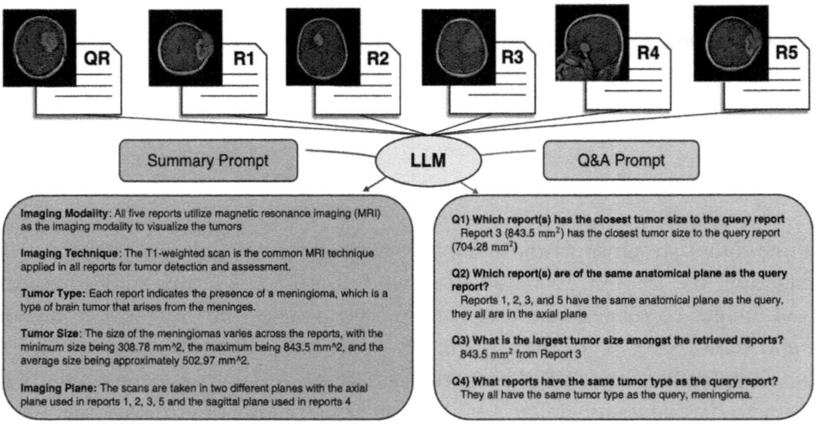

Fig. 2. Integration of LLM in CBIR for medical text summary and Q&A system. QR indicates the query report and R1–5 are the reports for each of the retrieved images. Summarization uses retrieved reports while Q&A uses both the query and retrieved report.

6.1 Image Description Generation

As there are no radiology reports present in the original dataset, we generate a pseudo-text description for each scan by extracting the following attributes: imaging modality, imaging technique, tumor size, tumor type, and anatomical plane. In this paper, we refer to these text descriptions as "reports". We generated three distinct report types of varying complexity to test the robustness of LLM interpretation which we use for the.

LLM-based summarization and Q&A tasks. Descriptions of examples are in Appendix 2.

1. Plain Report: Hardcoded sentence structure; reports only differ based on the extracted values of the respective scan.
2. Normal Report: Generated by GPT-4, this report gives a concise breakdown of the attributes roughly in the style of a radiologist. Natural variation in style occurs due to the nature of LLMs.
3. Alternative Report: This report takes the normal report and randomly performs the following swaps for roughly half of the reports: tumor size units ($mm^2 \rightarrow cm^2$), tumor type (meningioma → meningeal tumor, glioma → glial tumor, pituitary tumor → pituitary gland tumor), and anatomical plane (axial → transverse, sagittal → longitudinal, coronal → frontal).

6.2 Summarization

Summarization of query's retrievals is a common task in CBIR that is conventionally handled by the human user. Here we evaluate the capability of an LLM for generating a summary that is accurate, relevant and complete (RC), these metrics are explained in Appendix 4.2. After reports are generated for all scans, 50 sets containing the report of a query scan and those for the 5 nearest neighbors are fed into the LLM, which is prompted to generate a detailed and comprehensive summary (Fig. 2).

6.3 Interactive Question and Answer (Q&A) System

We also develop and probe an LLM-based Q&A system, designed to simulate interactivity with the retrieved data. We supply the LLM with both the query report and the top five retrieved reports. Subsequently, the LLM is prompted to answer questions that pertain directly to the content of these reports (Fig. 2). An interactive Q&A system allows users to investigate the specific aspects of reports that are most relevant to their needs. The questions asked in this experiment are:

1. Which report(s) has the closest tumor size to the query report?
2. Which report(s) are of the same anatomical plane as the query report?
3. What is the largest tumor size amongst the retrieved reports?
4. What reports have the same tumor type as the query report?

7 Experiment Design

7.1 Training Objectives

To determine the paradigm that maximizes both latent structure and classification performance, we train image encoders using a range of training objectives. We employ four different contrastive losses: triplet loss [17], Neighborhood Component Analysis (NCA) [18], InfoNCE [19], and Supervised Contrastive Loss (SupCon) [20]. All of these are supervised except for InfoNCE, which is essentially an unsupervised form of SupCon that we use to gauge the impact of supervision. We compare these contrastive losses with black-box models trained end-to-end with CE.

7.2 Image Encoders

We run the described training objectives on two CNN-based encoder architectures: EfficientNet-B2 (EN-B2) as [21] shows the B2 variant has better performance on this dataset, and ResNet50. Both use pretrained ImageNet weights and output a 128-length embedding used for downstream tasks. For CE trained models, we train an MLP on top of the embedding layer, after which we perform classification using i) the class probabilities output from the MLP, i.e. black box, and ii) the output of the embedding layer. Data augmentation and further training details are discussed in Appendix 5.

7.3 Evaluation

Query images from the test set are classified using NIR, I-NIR, and KNN. For KNN, we focus on K values of 5, 7, and 25, which fall within the upper bound that can be cognitively interpreted by humans [22]. For a given query image, similar images are retrieved based on the Euclidean distance between their embeddings and the query embedding. Classification performance is measured by AUROC and Kappa statistic (average and ensemble) and the retrieval space is assessed for anatomical consistency and average tumor size.

7.4 LLM

We query OpenAI's gpt-4–1106-preview (GPT-4) and gpt-3.5-turbo-1106 (GPT-3.5), using the same set of prompts (Appendix 3) for inference. We prioritize the use of off-the-shelf LLMs due to their ready-to-use nature and ease of accessibility, eliminating the need for fine-tuning. Report generation, summarization, and Q&A tasks are validated on 50 sets of retrievals and assessed for accuracy. Summarization is also assessed for RC. The data utilized for this project is open source, contains no PHI, and is thus safe for use with these LLMs.

8 Results

Given the superior classification performance of EN-B2 vs. ResNet50, only the former is used to generate the following results. Results for both encoders across training objectives on the classification task are shown in Appendix 6.

8.1 Retrieval

Figure 3 illustrates that SupCon preserves higher latent structure when compared to other supervised training objectives in both anatomical plane and average tumor size difference. However, we notice that InfoNCE preserves the highest latent structure when compared to all training objectives for both metrics. The non-contrastive CE preserves the least latent structure.

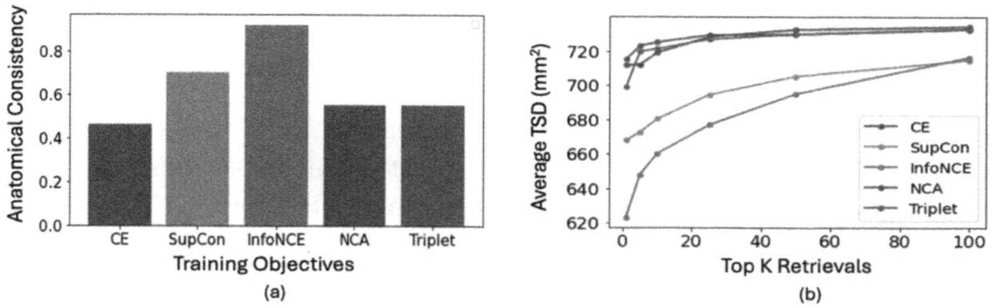

Fig. 3. Latent structure preservation results. (a): Anatomical consistency for top 5 retrievals. (b) Average tumor size difference (TSD) for top K retrievals

8.2 Classification

In Table 1, we demonstrate that NIR and I-NIR achieve comparable classification performance to KNN, and slightly exceed it for SupCon, triplet, and CE. While all training objectives are comparable for AUROC, SupCon shows a superior Kappa of 0.91. This is important because Kappa offers greater robustness to class imbalances present in the data. InfoNCE, despite preserving the highest latent structure, exhibits the lowest AUROC and Kappa, as anticipated due to its unsupervised formulation. SupCon inherits the latent structure preservation from InfoNCE while maintaining strong classification performance. SupCon matches the performance of black box algorithms, as evidenced by comparable AUROC (0.980 vs. 0.960) and Kappa scores (0.910 vs. 0.868).

Table 1. Classification Performance using EN-B2 Image Encoder and the best performing K (K = 25). BB = Black box algorithm.

Loss Functions	Ensemble AUROC				Ensemble Kappa			
	K = 25	NIR	I-NIR	BB	K = 25	NIR	I-NIR	BB
CE	0.978	**0.980**	**0.980**	0.960	0.880	0.870	0.880	0.868
InfoNCE	0.939	0.933	0.936	–	0.710	0.730	0.710	–
SupCon	0.969	**0.980**	**0.980**	–	0.910	**0.900**	**0.910**	–
Triplet	0.965	0.968	0.968	–	0.860	0.860	0.860	–
NCA	0.972	0.968	0.969	–	0.880	0.880	0.880	–

8.3 LLM Results

We perform the following experiments using only models trained with SupCon, due to the success of that training objective on retrieval and classification.

Summarization. While GPT-4 outperforms GPT-3.5 head-to-head in every respect, both LLMs produce more accurate, complete and relevant summaries for plain and normal reports (Fig. 4a). For sets of alternative reports, both often fail to perform correct mathematical operations (mean, min, and max) for tumor sizes expressed in different units, and to group together synonyms for the same anatomical plane.

Q&A. We note that GPT-4 significantly excels in the Q&A task (Fig. 4b). Q1 and Q3 involve mathematical analysis, while Q2 and Q4 focus on semantic understanding. GPT-4 is adept at simple math tasks like finding maximum values (Q3), but it struggles with more complex tasks like identifying similar tumor sizes (Q1). Both LLMs perform poorly on alternative reports, suggesting that variations in structure, units, and language challenge the LLM's ability to provide accurate answers.

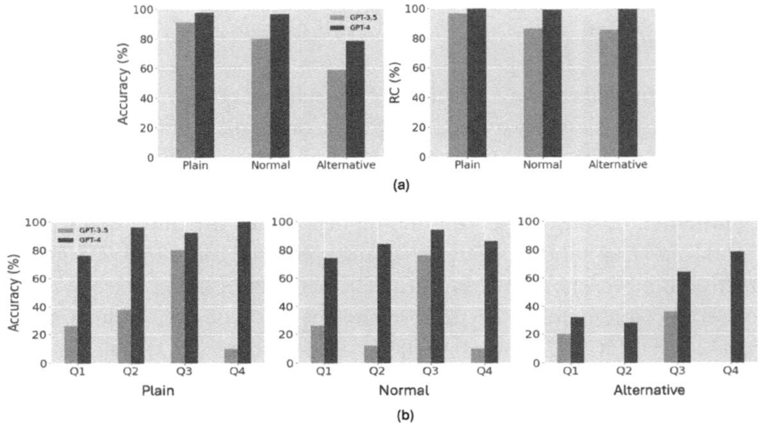

Fig. 4. LLM results. GPT-3.5 (Yellow) and GPT-4 (Green). (a) Summarization. The two subplots indicate the summarization performance using accuracy (left) and RC (right) (b) Q&A System. Each subplot in (b) indicates a type of report.

9 Discussion

In this paper, we catalog a search for the CBIR framework that maximizes interpretability while maintaining performance. For the models and objectives tests, we find that our best performing framework is an EN-B2 image encoder with interactivity enabled through GPT-4 which has SupCon training that leads to i) a structured retrieval space, and ii) maintains strong classification using NIR.

While the performance of SupCon is comparable to the black box model, its preservation of latent structure is ideal for facilitating usability and trust from clinicians. A key benefit of latent structure is that it develops automatically during training, eliminating the need for post-hoc labeling and finetuning. Future enhancements could involve clinicians ranking these retrievals to gauge their clinical relevance.

We feel that especially in the medical domain, it is important to build clinician-ML systems where models augment and support the decisions made by the doctor but are not the sole decision maker. Therefore, showing the predicted query class and confidence, the top retrievals, and summary of the retrievals on a platform where clinicians can interactively ask questions, could help those clinicians to better understand the model's reasoning (Appendix 7).

In our study, given that each patient has multiple scans, it is plausible to observe that smaller K retrievals may originate from a single patient. This situation introduces a potential bias, as the model's performance could be artificially inflated by the similarity within a single patient's scans rather than a genuine ability to generalize across diverse cases. To mitigate the influence of this bias, we extended our analysis by examining the classification performance at increased K values, ensuring our results reflect a more robust evaluation of the model's capabilities.

In our LLM experiments, we generated pseudo-text descriptions of scans due to dataset constraints, rather than using actual radiology reports. Despite this, our findings offer an initial insight into the promising integration of off-the-shelf LLMs with medical image analysis in radiology and anticipate that these trends will hold true with authentic radiology reports.

10 Conclusion and Future Work

We showcase a novel interactive CBIR system that accurately and interpretably classifies the type of brain tumor present in an MRI scan. Our usage of an off-the-shelf LLM broadens the potential for adoption of our system. While we document the clear limitations of these LLMs at present, the trend of improvement for these models inspires confidence in their future usability. Instead of using existing reports, some proposed systems [23] use multimodal LLMs to generate reports directly from medical images, but with this high level of complexity comes the risk of hallucination. We intentionally focus on leveraging an LLM strength reasoning over simple relationships–to augment interpretation of human-created content. Future work consists of applying this system to real-world users, full scan volumes, actual radiology reports, and to different modalities. This approach could be further extended in determining patient outcomes and treatment plans based on similar patients from the past, resulting in an improvement in patient care. Given that the enhanced CBIR system improves accessibility, this could open the door towards patients maintaining a more interactive role in their own care.

Appendix

1. Propensity of KNN and NIR to Overfit

Table 2 emphasizes KNN's increased propensity for overfitting the validation dataset when compared to NIR. In this experiment, a segment of the test dataset was allocated as a validation set. The KNN algorithm was applied to the validation set to determine the optimal value of K, which was subsequently utilized to compute Kappa for the test dataset. This procedure was replicated across five distinct, randomly selected validation subsets and the mean Kappa are reported. We see that the difference between Kappa for validation and testing datasets is notably greater for KNN as compared to NIR, supporting the evidence of KNN's greater tendency towards overfitting. Furthermore, NIR outperforms KNN in terms of Kappa on the test dataset.

Table 2. Assessing the propensity of KNN to overfit. This table shows the Cohen Kappa on the validation and testing.

KNN		NIR	
Validation	Testing	Validation	Testing
0.883	0.856	0.862	0.860

2. Textual Description Examples

Plain text: *The magnetic resonance imaging (MRI) T1-weighted scan in the axial. plane displays evidence of a meningioma with a size of 1097.02 mm*^{2}.

Normal report: *On axial T1 weighted MRI images, there is evidence of a meningioma with an area of approximately 1097.02 mm*^{2}. *The tumor characteristics are consistent with the typical appearance of a meningioma in this imaging modality.*

Alternate report: *On transverse T1 weighted MRI images, there is evidence of a meningeal tumor with an area of approximately 10.97 cm*^{2}. *The tumor characteristics are consistent with the typical appearance of a meningeal tumor in this imaging modality.*

3. LLM Prompts

3.1 Summarization

```
1)..Report 1 Content..
2)..Report 2 Content..
3)..Report 3 Content..
4)..Report 4 Content..
5)..Report 5 Content..
In these five reports, please identify the important at-
tributes that they have in common. Please write one sen-
tence about each of these attributes that summarizes/ag-
gregates the content in all reports. If any numbers are
involved then show the minimum, maximum and the average.
Imagine that someone doesn't have time to read all five
reports but wants to get a complete summarized version of
it. The important attributes should be, imaging modality,
imaging technique, tumor, tumor size, and plane.
```

3.2 Q&A

```
Query Report: ..Query Report..
Other Reports:
1)..Report 1 Content..
2)..Report 2 Content..
3)..Report 3 Content..
4)..Report 4 Content..
5)..Report 5 Content..
Question to ask
```

4. Metrics

4.1 Classification

Our classification objective is to assign a brain tumor type to each query image. We quantify performance using the area under the receiver operating characteristic (AUROC) and Cohen's Kappa (Kappa), a point accuracy metric adjusted for differences in class prevalence. Here, we calculate Kappa based on a soft ensemble for classification.

4.2 LLM Response Evaluation

In evaluating the tasks involving medical texts, we employ two metrics: Accuracy and relevance/completeness (RC). Accuracy assesses the correctness of the provided answers or information. It is essential in ensuring that the information conveyed is reliable and accurate, a non-negotiable requirement in medical contexts. RC evaluates whether all necessary attributes are appropriately mentioned and if the generated text pertinent to these attributes is relevant. This dual aspect of the metric not only considers the presence of key information but also scrutinizes its pertinence to the given context, thereby gauging whether a response is comprehensive and contextually appropriate. We evaluate the radiology report generation and summarization task using FA and RC and the Q&A system is using only FA. We score the Q&A system on 1 or 0—whether the LLM gets the answer right or wrong. Since the summarization consists of five attributes, we score each attribute 0 or 1. The one exception is for tumor size as we ask the LLM to provide the min, max, and the average tumor size across the five radiology reports. We weight the scores for them as 0.25, 0.25, and 0.5 respectively (total for tumor size adds up to 1). Each summary is graded on a total score of 5 for each metric.

5. Experiment Design

We perform two types of data augmentation during training, namely random Gaussian blur and random rotate, effectively increasing the size of the training data by three. The augmented images are not used for inference. As a data preprocessing step, we normalized each image between -1 and 1, resampled to a size of 1 x 224 x 224, and duplicated the first channel three times in order to support the use of pre-training weights on models originally trained on 3-channel images.

Code was implemented in Pytorch and used a NVIDIA RTX A6000 GPU for training and inference. We performed hyperparameter searches over learning rate, temperature, margin, and epoch and report the final performance of each model as the mean across 5 trials (i.e., training runs) to support reproducibility. The batch size set for training dataset was 30. For each dataset, 3-fold cross validation was done to obtain the optimal hyperparameters. Performance is reported on withheld test images not used for training or hyperparameter optimization.

6. Detailed Classification Results

We experimented with two convolutional neural network-based model architectures: Resnet50 and EF-B2 (Table 3 and Table 4). To train the image encoder, we explored classification loss (cross entropy (CE)) and metric learning losses such as triplet loss, neighborhood component analysis (NCA), InfoNCE (INCE) and supervised contrastive loss (SupCon).

Table 3. Cohen Kappa scores across model architectures. A-CK is the average Kappa while E-CK is the ensemble kappa stat across 5 trials.

		EF-B2		ResNet50	
		A-CK	E-CK	A-CK	E-CK
K = 5	CE	0.825	0.863	0.761	0.818
	NCA	0.808	0.854	0.758	0.822
	Triplet	0.808	0.868	0.792	0.837
	SupCon	0.893	**0.906**	0.799	**0.843**
	INCE	0.693	0.725	0.543	0.616
K = 25	CE	0.834	0.878	0.768	0.816
	NCA	0.818	0.858	0.765	0.832
	Triplet	0.815	0.868	0.792	0.838
	SupCon	0.894	**0.907**	0.798	**0.838**
	INCE	0.702	0.735	0.509	0.559
MLP		0.781	0.868	0.729	0.794

7. Output Example

Table 4. AUROC scores for EF-B2 model architecture across 5 different training loss functions.

Loss Function	AUROC					
	K = 5	K = 7	K = 25	NIR	I-NIR	MLP
CE	0.971	0.975	0.978	**0.98**	**0.98**	0.96
Triplet	0.964	0.964	0.965	0.968	0.968	–
NCA	0.967	0.968	0.972	0.968	0.969	–
InfoNCE	0.932	0.934	0.939	0.933	0.936	–
SupCon	0.969	0.969	0.969	**0.98**	**0.98**	–

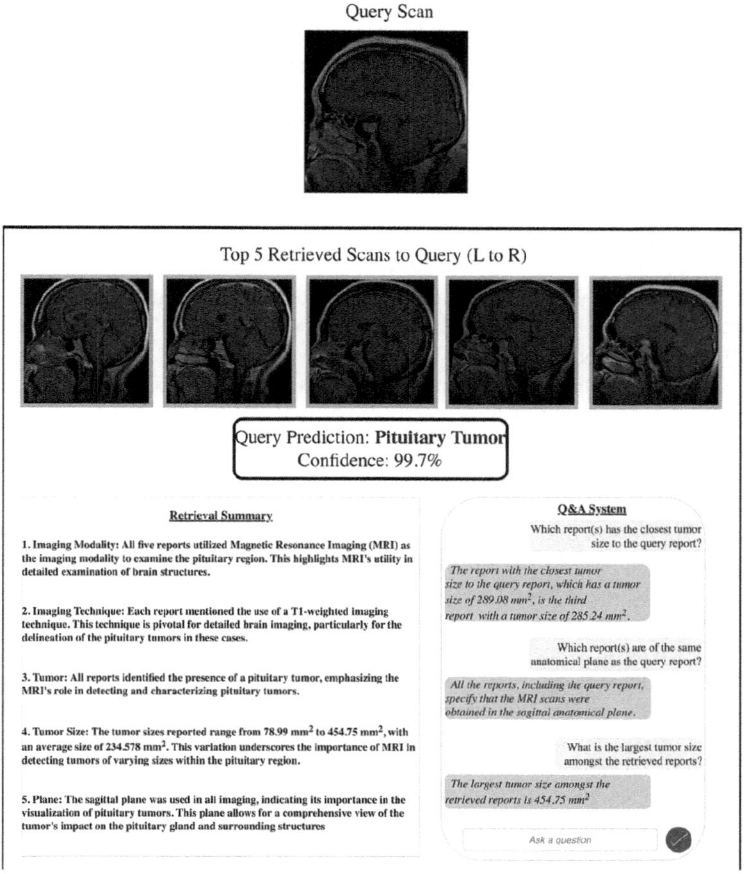

Fig. 5. Overview of the entire system: Retrieval, Classification, and LLM based Summary and Q&A

References

1. DeAngelis, L.M.: Brain tumors. N. Engl. J. Med. **344**, 114–123 (2001). https://doi.org/10.1056/NEJM200101113440207
2. Bauer, S., Wiest, R., Nolte, L.-P., Reyes, M.: A survey of MRI-based medical image analysis for brain tumor studies. Phys. Med. Biol. **58**, R97–R129 (2013). https://doi.org/10.1088/0031-9155/58/13/R97
3. Goldfarb, A., Teodoridis, F.: Why is AI adoption in health care lagging? Brookings Institution (2022)
4. Hosny, A., Parmar, C., Coroller, T.P., et al.: Deep learning for lung cancer prognostication: a retrospective multi-cohort radiomics study. PLOS Med **15**, e1002711 (2018). https://doi.org/10.1371/journal.pmed.1002711
5. Lanjewar, M.G., Parab, J.S., Shaikh, A.Y.: Development of framework by combining CNN with KNN to detect Alzheimer's disease using MRI images. Multimed Tools Appl **82**, 12699–12717 (2023). https://doi.org/10.1007/s11042-022-13935-4
6. Esteva, A., Kuprel, B., Novoa, R.A., et al.: Dermatologist-level classification of skin cancer with deep neural networks. Nature **542**, 115–118 (2017). https://doi.org/10.1038/nature21056
7. Sendak, M.P., Ratliff, W., Sarro, D., et al.: Real-world integration of a sepsis deep learning technology into routine clinical care: implementation study. JMIR Med. Inform. **8**, e15182 (2020). https://doi.org/10.2196/15182
8. Lu, S.-C., Swisher, C.L., Chung, C., et al.: On the importance of interpretable machine learning predictions to inform clinical decision making in oncology. Front. Oncol. **13**, 1129380 (2023). https://doi.org/10.3389/fonc.2023.1129380
9. Choe, J., Choi, H.Y., Lee, S.M., et al.: Evaluation of retrieval accuracy and visual similarity in content-based image retrieval of chest CT for obstructive lung disease. Sci. Rep. **14**, 4587 (2024). https://doi.org/10.1038/s41598-024-54954-5
10. Herrmann, A.E., Estrela, V.V.: Content based image retrieval (CBIR) in remote clinical diagnosis and healthcare (2016)
11. Hu, B., Vasu, B., Hoogs, A.: X-MIR: EXplainable Medical Image Retrieval. In: 2022 IEEE/CVF Winter Conference on Applications of Computer Vision (WACV), pp. 1544–1554. IEEE, Waikoloa, HI, USA (2022)
12. Qayyum, A., Anwar, S.M., Awais, M., Majid, M.: Medical image retrieval using deep convolutional neural network. Neurocomputing **266**, 8–20 (2017). https://doi.org/10.1016/j.neucom.2017.05.025
13. Owais, M., Arsalan, M., Choi, J., Park, K.R.: Effective diagnosis and treatment through content-based medical image retrieval (CBMIR) by using artificial intelligence. J. Clin. Med. **8**, 462 (2019). https://doi.org/10.3390/jcm8040462
14. Codella, N.C.F., et al.: Collaborative human-AI (CHAI): evidence-based interpretable melanoma classification in dermoscopic images (2018) https://doi.org/10.1007/978-3-030-02628-8_11
15. Arun, N., Gaw, N., Singh, P., et al.: Assessing the trustworthiness of saliency maps for localizing abnormalities in medical imaging. Radiol Artif Intell **3**, e200267 (2021). https://doi.org/10.1148/ryai.2021200267
16. Cheng, J., Huang, W., Cao, S., et al.: Enhanced performance of brain tumor classification via tumor region augmentation and partition. PLoS ONE **10**, e0140381 (2015). https://doi.org/10.1371/journal.pone.0140381
17. Schroff, F., Kalenichenko, D., Philbin, J.: FaceNet: a unified embedding for face recognition and clustering. In: 2015 IEEE Conference on Computer Vision and Pattern Recognition (CVPR), pp. 815–823 (2015)

18. Goldberger, J., Hinton, G.E., Roweis, S., Salakhutdinov, R.R.: Neighbourhood Components Analysis. In: Advances in Neural Information Processing Systems. MIT Press (2004)
19. Oord, A.V.D., Li, Y., Vinyals, O.: Representation learning with contrastive predictive coding (2019)
20. Khosla, P., et al.: Supervised contrastive learning. In: Advances in Neural Information Processing Systems. Curran Associates, Inc., pp 18661–18673 (2020)
21. Zulfiqar, F., Ijaz Bajwa, U., Mehmood, Y.: Multi-class classification of brain tumor types from MR images using EfficientNets. Biomed. Signal Process. Control **84**, 104777 (2023). https://doi.org/10.1016/j.bspc.2023.104777
22. Ramaswamy, V.V., Kim, S.S.Y., Fong, R., Russakovsky, O.: Overlooked Factors in Concept-Based Explanations: Dataset Choice, Concept Learnability, and Human Capability. In: 2023 IEEE/CVF Conference on Computer Vision and Pattern Recognition (CVPR), pp. 10932–10941. IEEE, Vancouver, BC, Canada (2023)
23. Lu, Y., Hong, S., Shah, Y., Xu, P.: Effectively fine-tune to improve large multimodal models for radiology report generation (2023)

Towards Trustworthy AI in Cardiology: A Comparative Analysis of Explainable AI Methods for Electrocardiogram Interpretation

Nils Gumpfer[1(✉)], Borislav Dinov[2], Samuel Sossalla[2], Michael Guckert[1,3(✉)], and Jennifer Hannig[1(✉)]

[1] Cognitive Information Systems, Kompetenzzentrum für Informationstechnologie, Technische Hochschule Mittelhessen, 61169 Friedberg, Germany
{nils.gumpfer,michael.guckert,jennifer.hannig}@kite.thm.de

[2] Department of Internal Medicine I, Cardiology, Justus-Liebig-University Giessen, 35392 Giessen, Germany

[3] Department of MND - Mathematik, Naturwissenschaften und Datenverarbeitung, Technische Hochschule Mittelhessen, 61169 Friedberg, Germany

Abstract. Deep learning models have successfully been applied for medical diagnoses. However, such models are perceived as opaque black boxes and their decisions are not comprehensible for human users. This intransparency and the potential resulting lack of trust is still an obstacle for a wider use of deep learning models in medicine. Explainable Artificial Intelligence (XAI) is expected to become an important factor for trust, transparency, and accountability of AI models. We present a comparison of state-of-the-art XAI methods in terms of their ability to recognize medically relevant patterns in electrocardiogram (ECG) data according to criteria defined in established medical guidelines. We trained separate convolutional neural networks for four common cardiac pathologies and applied state-of-the-art XAI methods to compare the input features identified as important by these methods. Further analyses assessed the patterns identified as relevant in terms of their distribution across leads and their overall overlap with guideline criteria. Across most XAI methods, we observed high conformance between the relevance assignment to the patterns and ECG leads listed in the referenced guidelines. However, it turned out that the explanations varied significantly depending on the choice of the XAI method. This study points out strengths and weaknesses of the XAI methods under examination. Overall, our experiments show that XAI can increase the transparency of deep learning models in showing how patterns relevant for the algorithm relate to human decision making. This may serve as an essential aspect for the implementation of trustworthy AI in cardiology.

Supported by the Hessian Ministry of Digital Strategy and Development (21_0066_4A and 21_0093_2A) and the German Federal Ministry of Education and Research via de.NBI Cloud within the German Network for Bioinformatics Infrastructure (031A532B, 031A533A, 031A533B, 031A534A, 031A535A, 031A537A, 031A537B, 031A537C, 031A537D, 031A538A).

J. Finkelstein et al. (Eds.): AIME 2024, LNAI 14845, pp. 350–361, 2024.
https://doi.org/10.1007/978-3-031-66535-6_36

Keywords: AI · XAI · Electrocardiogram · Cardiology

1 Introduction

The electrocardiogram (ECG) is a widely used tool with ubiquitous availability in hospitals as well as in ambulatory medical service. ECGs are comprised of leads (combinations of electrodes), which allow for different perspectives on the electrical activity of the heart. The correct analysis of the dimensions (time, amplitude, lead) constitutes a complex task requiring skill and experience [4]. Several studies have shown that ECG-based AI models can compete with the performance of medical experts [11]. However, such systems are usually still perceived as black boxes and are thus not adequately interpretable for human users. This lack of transparency can limit trustworthiness and thus become an obstacle for the adoption of such systems in practice [8]. AI systems can play an important role in supporting human decision-making, especially in the analysis of complex data but only if they are perceived as trustworthy. Therefore, trust is an essential aspect of the human-machine relationship and crucial for the acceptance of intelligent devices [5,8]. Explainable AI (XAI) plays an important role for increasing transparency and can be one of the key factors to facilitate trust in AI systems [1], especially when highlighting decision-relevant patterns that correspond to the pattern catalog of the domain expert [5].

The majority of studies addressing AI-based ECG analysis does not include XAI to explain its predictions or applies only a single method [12]. As the computational differences between XAI methods are substantial, they potentially provide different explanations for the same prediction [1,7]. One of the most important open questions in XAI research is therefore how well explanations meet the requirements of domain experts such as healthcare professionals [1,8].

In this study, we compared frequently used state-of-the-art XAI methods [16] based on computation of gradients (including Gradient [20], Gradient×Input [14], LRP (Layer-wise Relevance Propagation) [2], Deep- and GradSHAP (SHapley Additive exPlanations) [9], and Grad-CAM (Gradient-weighted Class Activation Mapping) [13] along with our recently proposed SIGN (Sign-based Improvement of Gradient-based explaNations) [7] methods, namely Gradient×SIGN and LRP SIGN. The comparison assesses the functionality of different XAI methods to highlight medically relevant patterns in ECG time series data relative to established medical guidelines [3,15,17]. The XAI methods were applied to four Convolutional Neural Networks (CNN) trained to classify ECGs with the following cardiac pathologies: Myocardial ischemia (ISCH), atrioventricular block (AVB), right bundle branch block (RBBB), and left bundle branch block (LBBB).

2 Data

Data from PTB-XL was used to train and evaluate the neural networks [18]. The PTB-XL dataset contains 21,799 ECG recordings (10 s, 12 leads, 500 Hz) and covers 71 diagnostic statements which are provided with a confidence score

(0–100%). We excluded 426 records of patients with an age below 18 years or unknown age. We labeled the records according to the diagnostic statements provided in the dataset and applied a confidence threshold of 80%. We excluded 6 records labeled below this threshold. 30% of the selected PTB-XL records were used as holdout test dataset and an additional 20% from the remaining 70% were used for validation. The splits were stratified by sex and pathology label. Table 1 in the appendix outlines the number of positives and negatives per pathology for each split. We filtered the ECGs using a 40 Hz Butterworth low-pass filter, 50 Hz and 0.05 Hz second-order Infinite Impulse Response (IIR) notch filters, and a moving-average baseline-correction. We used 4 s subsamples as model input, which were drawn from each sample using overlapping extraction windows and uniform strides.

3 Cardiac Pathologies

Four cardiac pathologies that show characteristic patterns in the ECG were chosen for this study. The usual abbreviations for the phases of heart activity are used in the following: P denotes atrial depolarization, Q the first negative wave after P, R any positive wave between P and T, S negative waves following R, and T denotes ventricular repolarization. Depending on their relative amplitudes (high/low), waves are either denoted with upper or lower case letters, e.g., R/r, S/s, respectively [4].

Atrioventricular Block. An atrioventricular block (AVB) is a cardiac conduction disorder characterized by impaired signal transmission between atria and ventricles [3]. This can result in delayed or complete blockage of signals, leading to a reduction of heart rate. A prolongation of the PR interval (time between atrial (P) and ventricular (R) contraction) >200 ms in lead I or other leads where the p-wave can be identified (e.g., II, V1, V2) is an important indicator for diagnosis of AVB (see Fig. 1). The severity of AVB can vary from first degree (minimal delay) to third degree (complete block) [3].

Myocardial Ischemia. Myocardial ischemia (ISCH) is a cardiac condition where blood supply of the myocardium (heart muscle) is reduced, typically due to narrowed or blocked coronary arteries [17]. Affected ischemic areas manifest in the ECG in anatomically related leads. The most important ECG criterion for myocardial ischemia is the presence of ST-segment changes (elevations and depressions) in one or multiple anatomically related leads. These changes are primarily assessed in the precordial leads (V1-V6) in relation to anterior, lateral, and septal ischemia, as they provide positional information and allow precise localization of the affected arteries [17].

Right Bundle Branch Block. A right bundle branch block (RBBB) is a cardiac conduction disorder characterized by delayed signals in the right bundle branch of the conduction system, causing an abnormal pattern of intraventricular excitation [15]. Similar to AVB, RBBB can occur as complete or incomplete blockage.

The most important ECG criteria for RBBB are a prolonged QRS duration (>120 ms), a so-called 'M'-shaped intraventricular excitation pattern in V1-V2, and a wide S-wave in V5-6 (see Fig. 1) [15].

Left Bundle Branch Block. Similarly, a left bundle branch block (LBBB) is a conduction disorder affecting the *left* bundle branch [15]. LBBBs are also differentiated in complete and incomplete types. ECG patterns in LBBB differ significantly from those in RBBB, as a different area of the conduction system is affected, however, they show a similar morphology. Important leads for identification of LBBB are V1-2 and V5-6. Besides a prolonged QRS duration (>120 ms), a deep S-wave in V1-2 and a wide R-wave in V5-6 are essential diagnostic criteria for LBBB (see Fig. 1).

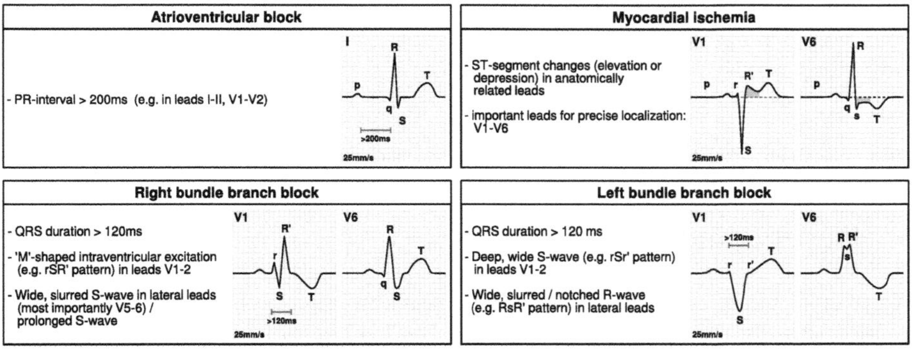

Fig. 1. ECG criteria and exemplary patterns for the examined pathologies according to established medical guidelines [3,15,17]. (Color figure online)

4 Evaluation

The experiments described in this work focus on the analysis of positive relevance values for correctly classified pathological examples (true positives). Based on the explanations generated by each XAI method for each pathology, we performed a detailed, quantitative evaluation on lead level for one exemplary pattern. Further, we selected examples showing the visual differences in pattern highlighting between the different XAI methods.

The below definitions are used in the subsequent descriptions: An XAI method assigns a relevance value $R_i \in \mathbb{R}$ to each $x_i \in x$ with respect to a specific target class c for a prediction function $x \mapsto f_c(x)$. The relevance map R contains $R_i \in R$ with the same dimensionality as the input. For ECG time series data, $x \in \mathbb{R}^{W \times D}$ is an input with the width W and the number of leads D. R^+ is a filtered representation of R containing only relevance values in the range $[0, \infty^+]$. For visualization purposes, we normalized R^+ to $[0, 1]$ and filtered out $R_i^+ \leq 0.2$ to yield $\widetilde{R}$. In order to visualize these values with respect to the input, we used dots, proportionally sized and colored in relation to $\widetilde{R}_i \in \widetilde{R}$. The distribution of relevance over leads was evaluated by the share of summed relevance attributed to each lead j over all explained examples n:

$$L_j = \frac{\sum_{k=1}^{n} \sum_{i=1}^{W} \widetilde{R}_{kij}}{\sum_{o=1}^{n} \sum_{p=1}^{W} \sum_{q=1}^{D} \widetilde{R}_{opq}} \cdot 100. \tag{1}$$

In lack of a human-defined ground-truth, we used *NeuroKit2* [10] for ECG delineation to approximate feature bounds. We focused on ST-segments, as the *NeuroKit2* delineation algorithm extracts them with high reliability. The ST-boundaries were used to filter the respective relevance map R to yield R^{ST}. Based on this filtered representation, we analyzed the share of relevance attributed to ST-segments. We both used the sum of relevance and the count of positive relevance to calculate the shares, given by

$$R^{\mathrm{ST}}_{\Sigma} = \frac{\sum_i \widetilde{R}_i^{\mathrm{ST}}}{\sum_i \widetilde{R}_i} \cdot 100 \qquad \text{and} \qquad R^{\mathrm{ST}}_{\#} = \frac{|\widetilde{R}_i^{\mathrm{ST}}|}{|\widetilde{R}_i|} \cdot 100, \tag{2}$$

where $|\cdot|$ denotes the cardinality of the respective value.

5 Results and Discussion

We trained binary classification models based on our previously published CNN architecture [6] on the splits listed in Table 1 in the appendix. To compensate unbalanced class-distributions, we trained the models using class-weight vectors [19]. Models were selected based on the highest area under the receiver operating characteristic curve (AUROC) having sensitivity and specificity above 0.7 on the validation set. For these models, the respective performance on the test set was calculated. All classifiers reached high AUROC values (0.93–1.0) and adequate sensitivity (0.89–1.0) and specificity (0.82–0.98), see Table 2 in the appendix.

For XAI evaluation, we applied every method to all true positive subsamples for each pathology. We used the following method variants and parameterizations: Gradient [20], Gradient×Input [14], Gradient×SIGN [7], LRP-$\alpha_1\beta_0$ [2], LRP-ϵ with $\epsilon = 0.5 \cdot \sigma(x)$ [2], LRP-ϵ SIGN with $\epsilon = 0.5 \cdot \sigma(x)$ and SIGN as input layer rule [7], DeepSHAP [9], GradSHAP [9], and Grad-CAM [13]. For training of the SHAP explainers, we drew a representative, stratified background ($n = 500$) from the respective training dataset of each pathology. For more information about the methods, we refer to the descriptions in [16] and [7].

Distribution of Relevance over Leads. Based on Eq. 1, we calculated the distribution of relevance over leads for all explained samples taken from the test datasets for each method separately. These distributions are depicted as radar plots in Fig. 3 in the appendix. For each pathology, the relevant leads according to guideline criteria [3,15,17] are marked red.

For AVB, the prolongation of the PR-interval can best be observed in leads I-II or V1-V2 [3]. We observed a strong focus on lead I for all methods except LRP-$\alpha_1\beta_0$ and Grad-CAM. Lead V1 was among the three most relevant leads across all methods, same for lead II, except for Gradient. Computationally, Grad-CAM does not differentiate between leads, as this information is not considered in the computation of the partial derivatives, what explains the uniform distribution of relevance over leads throughout all four pathologies. For AVB, the focus on leads across most XAI methods appears reasonable with respect to guideline recommendations [3].

The precordial leads V1-V6 provide important localization information for the diagnostic interpretation of ST-segment changes and their relationship to myocardial ischemia [17]. PTB-XL contains a high amount of anterolateral ischemia, which relates to leads I, aVL, V3-V4 (anterior), and V5-V6 (lateral) [4]. Lead V5 was among the three most relevant leads across all methods. For Gradient, Gradient×Input, Gradient×SIGN, and LRP-ϵ SIGN, lead V6 was largely relevant as well. We generally observed a focus on precordial leads throughout all methods. Interestingly, the SIGN-based variants involved a strong focus on lead I besides precordial leads as well, what is in accordance with changes caused by anterolateral ischemia [4].

The leads V1-V2 and V5-V6 are important for the identification of RBBB, as the relevant patterns manifest most clearly in these leads [15]. Throughout all methods, we observed a strong focus on lead V1, while other leads received a much lower share of relevance. Whether this focus is related to the XAI methods or the model itself cannot be answered based on this analysis, but the results suggest that a sole focus on lead V1 deemed sufficient for correct classification, as there was a strong agreement on V1 across all methods.

Similarly, leads V1-V2 and V5-V6 are important for diagnosis of LBBB as well [15]. Except for LRP-$\alpha\beta$, we observed a strong focus on lead V6 throughout all methods. Further, a large amount of relevance was assigned to leads V1-V2 by Gradient×Input, LRP-$\alpha\beta$, LRP-ϵ, DeepSHAP, and GradSHAP. For Gradient, Gradient×SIGN, and LRP-ϵ SIGN, we further observed a substantial share of relevance on lead I, what can be related to the substantial share of lateral vectors measurable by leads I and V6 [4].

The distribution of relevance over leads generally aligned with guideline criteria [3,15,17]. However, significant differences were observed among the XAI methods regarding overall lead relevance. In AVB, leads I-II and V1 were prominently relevant across most methods, consistent with diagnostic usage. Grad-CAM always exhibited uniform lead relevance, limiting its capability for lead differentiation. In ISCH, precordial leads, particularly V5 and V6, received the majority of relevance across all methods, reflecting their role in detecting and localizing myocardial ischemia. SIGN-based variants further emphasized lead I, complementing lateral involvement in diagnosing anterolateral ischemia. For RBBB detection, V1 was predominantly relevant across all methods, indicating its crucial role in identifying RBBB-related features. However, the specific reasons for this emphasis require further investigation. In LBBB detection, V6

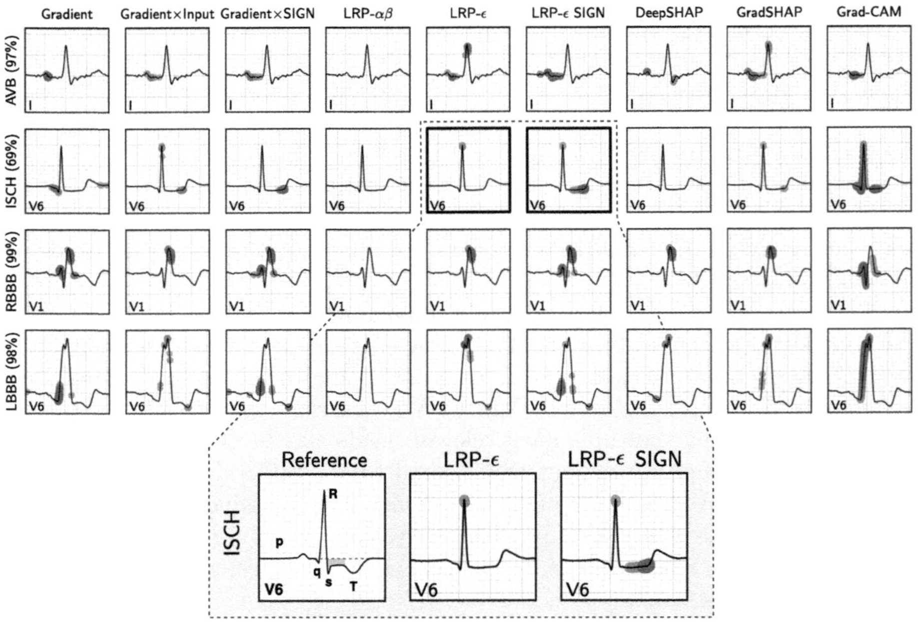

Fig. 2. Exemplary explanations for each pathology and method of selected examples from the PTB-XL test dataset. One exemplary comparison of LRP-ϵ, LRP-ϵ SIGN, and the corresponding reference pattern from Fig. 1 are highlighted. Scale: 25 mm/s, 10 mm/mV. Prediction value for pathology class given in parentheses. (Abbr.: AVB = AV-Block, ISCH = Myocardial ischemia, RBBB = Right bundle branch block, LBBB = Left bundle branch block). (Color figure online)

consistently emerged as relevant, with certain methods attributing additional relevance to leads V1-V2, which is reasonable according to guidelines [15]. Generally, we did not observe full compliance of lead usage and guideline-based lead requirements in all cases. We assume this to be reasonable, as not all leads have to be used by the model, as long as patterns are sufficiently present in some of the leads.

Patterns of Relevance. To illustrate heatmaps for all methods, we extracted exemplary QRS complexes for each pathology (see Fig. 2). The excerpts based on LRP-$\alpha\beta$ do not contain highlighted points, as the relevance values were sparsely distributed for this method. The complete ECGs for all excerpts can be found in the GitHub repository accompanying this paper[1].

In the AVB example in Fig. 2, the prolonged PR-interval is clearly visible. The majority of heatmaps provides relevance attribution for the PR-segment, the P-wave, or the R-peak as well. The features involved are therefore interpretable according to the relevant pattern according to guideline criteria [3].

[1] https://github.com/nilsgumpfer/AIME2024.

The ISCH excerpt shows a prominent ST-segment depression in lead V6. Gradient, LRP-$\alpha\beta$, and DeepSHAP did not attribute significant relevance to this segment, whereas Gradient×Input, Gradient×SIGN, LRP-ϵ SIGN, GradSHAP, and Grad-CAM partially highlighted the ST-segment. An exemplary comparison of LRP-ϵ and LRP-ϵ SIGN (see Fig. 2) shows potential influence of input value scales on explanations (for more information on correlation of input and relevance see [7]). Relevance attribution to the low-valued ST-segment might have been suppressed by higher absolute relevance attributed to the R-peak. Analyzing the share of relevance attributed to ST-segments in comparison to overall relevance for all true positive ISCH cases (see Table 3 in the appendix) revealed that the shares of $R^{\mathrm{ST}}_{\sum}$ and $R^{\mathrm{ST}}_{\#}$ were highest for LRP-ϵ SIGN, but lowest for LRP-ϵ. SIGNed variants and Grad-CAM showed a high share of ST-based relevance (>17%), while methods based on multiplication by input values (Gradient×Input, LRP-$\alpha\beta$, LRP-ϵ, DeepSHAP, GradSHAP) reached low (< 7 %) shares of relevance on ST-segments. These findings support the assumption of a bias induced by multiplication with input values, leading to proportionally low relevance for low input values such as the ST-segment.

The alteration of the QRS complex related to RBBB can distinctively be seen in Fig. 2. Besides a prolonged QRS-duration, the 'M'-shaped QRS-complex is significantly present in lead V1. The methods Gradient, Gradient×SIGN, LRP-ϵ SIGN, and Grad-CAM assigned substantial relevance to the corresponding rSR'-pattern. However, Gradient×Input, LRP-ϵ, DeepSHAP, and GradSHAP attribute high relevance to the R-peak. Similarly, these methods also show high relevance for high absolute values in the LBBB example, concisely the notched R-peak in V6. The flanks of the broadened QRS complex, which represent low absolute values, are highlighted by Gradient, Gradient×SIGN, LRP-ϵ SIGN, and Grad-CAM.

The highlighted areas often showed interpretable patterns in regard of guideline criteria for respective pathologies. However, we observed a suppression of medically relevant patterns for methods based on input multiplication. Contrasting, SIGN-based variants provided interpretable explanations with reasonable patterns throughout all cases. Explanations by Grad-CAM partly appeared diffuse and contained less detail than others. Overall, our results exemplify the ability of XAI methods to highlight medically relevant patterns in ECG time series data. These patterns can help to establish a trust relationship between users and intelligent systems by enabling an exchange of knowledge and a progression of beliefs [5]. In medical practice, this can help to use AI-based systems for access to cardiologic expertise and for educational aspects.

6 Conclusions and Future Work

In this study, we conducted a comparative analysis of nine XAI methods for ECG analysis, regarding their ability to highlight medically relevant patterns associated with four cardiac pathologies. Our results suggest that the distribution of relevance over ECG leads observed in our experiments generally aligned with guideline criteria. However, we found notable differences among XAI methods in lead relevance across pathologies. Grad-CAM exhibited uniform lead relevance, hence limiting its capabilities for usage with multi-channel time series data. We further observed significant differences between SIGN-based methods and methods based on a multiplication by input values. We noticed that methods based on input multiplication suppress low-valued input features such as ST-segments. In contrast, methods incorporating SIGN-based adaptions showed more consistent and interpretable explanations across all pathologies.

In general, our results show that state-of-the-art XAI methods can be used to highlight medically interpretable patterns for predictions based on ECG time series data. As PTB-XL represents a mid-european cohort [18], results may have limited transferability to differing ethnic groups, however enable to showcase first insights. Overall, our study contributes to advancing the understanding of XAI methods for ECG interpretation and emphasizes the importance of selecting appropriate methods that can provide accurate and clinically relevant explanations, but also highlights the pitfalls of poor method selection. We showed that the patterns used by the AI models were close to guideline criteria. Whether the patterns highlighted can intuitively be interpreted by domain experts and whether they correspond to their respective personal pattern catalog [5] needs to be investigated further in complementary user studies. Future research should therefore validate XAI methods on larger, diverse data sets that cover different cardiac pathologies and clinical scenarios and include user-centered explanatory requirements and assessments.

Overall, the results of our experiments can help to promote the use of XAI for AI-based ECG analysis and thereby enable a closer exchange between AI researchers and medical practitioners on the patterns used for accurate diagnosis. This pattern-centered assessment suggests a possible path towards the practical adoption of AI and the creation of trustworthy AI in practice.

A Appendix

Table 1. Overview of the data used for the experiments, differentiated by pathology, class, and split.

(Abbr.: AVB = AV-Block, ISCH = Myocardial ischemia, RBBB = Right bundle branch block, LBBB = Left bundle branch block)

Pathology	Training		Validation		Test	
	Positives	Negatives	Positives	Negatives	Positives	Negatives
AVB	443	11,521	111	2,881	237	6,174
ISCH	1,345	10,619	336	2,656	721	5,690
RBBB	901	11,063	225	2,767	482	5,929
LBBB	326	11,638	81	2,911	175	6,236

Table 2. Overview of model performance metrics evaluated for each pathology on different splits.

(Abbr.: AVB = AV-Block, ISCH = Myocardial ischemia, RBBB = Right bundle branch block, LBBB = Left bundle branch block, AUROC = Area under the receiver operating characteristic curve)

Pathology	Validation			Test		
	AUROC	Sensitivity	Specificity	AUROC	Sensitivity	Specificity
AVB	0.94	0.92	0.86	0.94	0.93	0.85
ISCH	0.96	0.98	0.75	0.93	0.89	0.82
RBBB	0.99	0.97	0.94	0.99	0.98	0.94
LBBB	1.00	1.00	0.98	0.99	0.96	0.98

Table 3. Results of ST relevance analysis regarding true positive examples of myocardial ischemia (ISCH) from the test dataset ($n = 710$) based on Eq. 2.

(Abbr.: SD = Standard deviation)

Method	$\frac{\sum_i \widetilde{R}_i^{ST}}{\sum_i \widetilde{R}_i} \cdot 100$		$\frac{\lvert\widetilde{R}_i^{ST}\rvert}{\lvert\widetilde{R}_i\rvert} \cdot 100$		$\sum_i \widetilde{R}_i^{ST}$		$\sum_i \widetilde{R}_i$		$\lvert\widetilde{R}^{ST}\rvert$		$\lvert\widetilde{R}\rvert$	
	Mean	SD	Mean	SD	Mean	SD	Mean	SD	Mean	SD	Mean	SD
Gradient	8.15	9.66	8.20	9.35	63.81	77.26	769.81	355.21	105.09	122.86	1253.06	567.15
Gradient×Input	5.69	9.99	6.20	10.61	10.35	22.40	115.49	77.16	17.56	37.29	177.11	122.35
Gradient×SIGN	17.57	18.36	17.55	17.97	120.63	138.73	657.76	299.04	190.59	213.32	1048.94	473.91
LRP-$\alpha\beta$	4.14	15.06	4.12	14.69	1.64	11.73	49.33	99.80	2.64	19.67	74.99	154.52
LRP-ϵ	0.24	2.52	0.25	2.55	0.20	1.95	78.95	34.45	0.30	2.80	113.46	50.07
LRP-ϵ SIGN	22.71	17.33	23.38	17.43	152.21	152.95	600.73	299.91	249.07	241.06	964.32	465.39
DeepSHAP	1.08	4.33	1.12	4.35	1.19	4.92	88.44	44.83	1.92	7.69	129.25	68.33
GradSHAP	1.25	3.92	1.45	4.35	1.93	7.10	98.91	59.27	3.34	12.02	146.07	91.84
Grad-CAM	21.26	16.05	21.99	15.94	1183.01	991.76	5201.66	1733.97	1536.39	1169.29	6691.99	2096.21

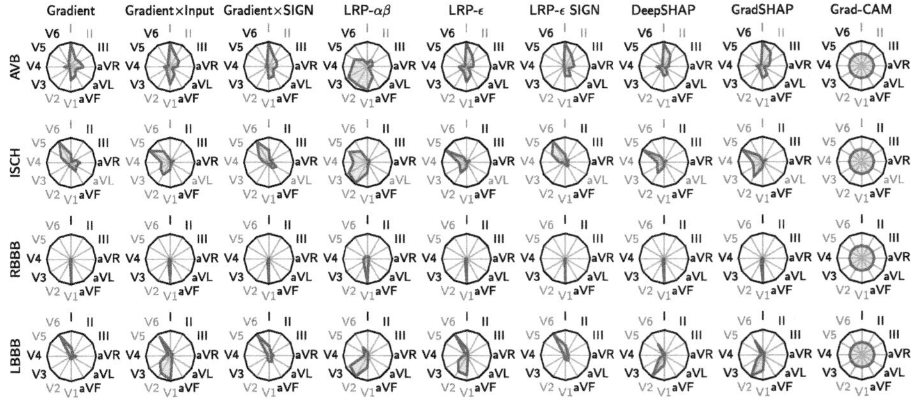

Fig. 3. Distribution of relevance over ECG leads (see Eq. 1) for each pathology and XAI method evaluated on the PTB-XL test dataset. Expected leads according to [3,4,15,17] are marked red.

(Abbr.: AVB = AV-Block, ISCH = Myocardial ischemia, RBBB = Right bundle branch block, LBBB = Left bundle branch block). (Color figure online)

References

1. Ali, S., et al.: Explainable Artificial Intelligence (XAI): what we know and what is left to attain trustworthy artificial intelligence. Inf. Fusion **99**, 101805 (2023)
2. Bach, S., Binder, A., Montavon, G., Klauschen, F., Müller, K.R., Samek, W.: On pixel-wise explanations for non-linear classifier decisions by layer-wise relevance propagation. PLoS ONE **10**(7), 1–46 (2015)
3. Epstein, A.E., et al.: ACC/AHA/HRS 2008 guidelines for device-based therapy of cardiac rhythm abnormalities. Circulation **117**(21), e350–e408 (2008)
4. Gorgels, A.P.M.: Electrocardiography. In: Willerson, J.T., Wellens, H.J.J., Cohn, J.N., Holmes, D.R. (eds.) Cardiovascular Medicine, pp. 43–77. Springer, London (2007). https://doi.org/10.1007/978-1-84628-715-2_3
5. Guckert, M., Gumpfer, N., Hannig, J., Keller, T., Urquhart, N.: A conceptual framework for establishing trust in real world intelligent systems. Cogn. Syst. Res. **68**, 143–155 (2021)
6. Gumpfer, N., Grün, D., Hannig, J., Keller, T., Guckert, M.: Detecting myocardial scar using electrocardiogram data and deep neural networks. Biol. Chem. **402**(8), 911–923 (2020)
7. Gumpfer, N., Prim, J., Keller, T., Seeger, B., Guckert, M., Hannig, J.: SIGNed explanations: unveiling relevant features by reducing bias. Inf. Fusion **99**, 101883 (2023)
8. Loh, H.W., Ooi, C.P., Seoni, S., Barua, P.D., Molinari, F., Acharya, U.R.: Application of explainable artificial intelligence for healthcare: a systematic review of the last decade (2011–2022). Comput. Methods Programs Biomed. **226**, 107161 (2022)
9. Lundberg, S.M., Lee, S.: A unified approach to interpreting model predictions. In: Guyon, I., et al. (eds.) Advances in Neural Information Processing Systems: Annual Conference on Neural Information Processing Systems 2017, 4–9 December 2017, Long Beach, CA, USA, vol. 30, pp. 4765–4774 (2017)
10. Makowski, D., et al.: NeuroKit2: a Python toolbox for neurophysiological signal processing. Behav. Res. Methods **53**(4), 1689–1696 (2021)
11. Muzammil, M.A., et al.: Artificial intelligence-enhanced electrocardiography for accurate diagnosis and management of cardiovascular diseases. J. Electrocardiol. **83**, 30–40 (2024)
12. Salih, A., et al.: A review of evaluation approaches for explainable AI with applications in cardiology (2023). techRxiv preprint, 24573304
13. Selvaraju, R.R., Cogswell, M., Das, A., Vedantam, R., Parikh, D., Batra, D.: Grad-CAM: visual explanations from deep networks via gradient-based localization. Int. J. Comput. Vision **128**(2), 336–359 (2020)
14. Shrikumar, A., Greenside, P., Kundaje, A.: Learning important features through propagating activation differences. In: Precup, D., Teh, Y.W. (eds.) Proceedings of the 34th International Conference on Machine Learning, ICML 2017, Sydney, NSW, Australia, 6–11 August 2017. Proceedings of Machine Learning Research, vol. 70, pp. 3145–3153. PMLR (2017)
15. Surawicz, B., Childers, R., Deal, B.J., Gettes, L.S.: AHA/ACCF/HRS recommendations for the standardization and interpretation of the electrocardiogram - Part III: intraventricular conduction disturbances. Circulation **119**(10), e235–e240 (2009)
16. Theissler, A., Spinnato, F., Schlegel, U., Guidotti, R.: Explainable AI for time series classification: a review, taxonomy and research directions. IEEE Access **10**, 100700–100724 (2022)

17. Wagner, G.S., et al.: AHA/ACCF/HRS recommendations for the standardization and interpretation of the electrocardiogram - Part VI: acute ischemia/infarction. Circulation **119**(10), e262–e270 (2009)
18. Wagner, P., et al.: PTB-XL, a large publicly available electrocardiography dataset. Sci. Data **7**(1), 154 (2020)
19. Wang, S., Liu, W., Wu, J., Cao, L., Meng, Q., Kennedy, P.J.: Training deep neural networks on imbalanced data sets. In: 2016 International Joint Conference on Neural Networks (IJCNN), pp. 4368–4374 (2016)
20. Zurada, J., Malinowski, A., Cloete, I.: Sensitivity analysis for minimization of input data dimension for feedforward neural network. In: Proceedings of IEEE International Symposium on Circuits and Systems - ISCAS 1994, vol. 6, pp. 447–450 (1994)

Author Index

J. Finkelstein et al. (Eds.): AIME 2024, LNAI 14845, pp. 363–366, 2024.
https://doi.org/10.1007/978-3-031-66535-6

Printed by Printforce, the Netherlands